VOLUME NINETY EIGHT

ADVANCES IN PROTEIN CHEMISTRY AND STRUCTURAL BIOLOGY

Protein and Peptide Nanoparticles for Drug Delivery

VOLUME NINETY EIGHT

ADVANCES IN PROTEIN CHEMISTRY AND STRUCTURAL BIOLOGY

Protein and Peptide Nanoparticles for Drug Delivery

Edited by

ROSSEN DONEV

Biomed Consult Ltd
Swansea, United Kingdom

ELSEVIER

AMSTERDAM • BOSTON • HEIDELBERG • LONDON
NEW YORK • OXFORD • PARIS • SAN DIEGO
SAN FRANCISCO • SINGAPORE • SYDNEY • TOKYO
Academic Press is an imprint of Elsevier

Academic Press is an imprint of Elsevier
225 Wyman Street, Waltham, MA 02451, USA
525 B Street, Suite 1800, San Diego, CA 92101-4495, USA
125 London Wall, London, EC2Y 5AS, UK
The Boulevard, Langford Lane, Kidlington, Oxford OX5 1GB, UK

First edition 2015

Notices

Knowledge and best practice in this field are constantly changing. As new research and experience broaden our understanding, changes in research methods, professional practices, or medical treatment may become necessary.

Practitioners and researchers must always rely on their own experience and knowledge in evaluating and using any information, methods, compounds, or experiments described herein. In using such information or methods they should be mindful of their own safety and the safety of others, including parties for whom they have a professional responsibility.

To the fullest extent of the law, neither the Publisher nor the authors, contributors, or editors, assume any liability for any injury and/or damage to persons or property as a matter of products liability, negligence or otherwise, or from any use or operation of any methods, products, instructions, or ideas contained in the material herein.

ISBN: 978-0-12-802828-5
ISSN: 1876-1623

CONTENTS

CONTRIBUTORS

Priyanka Bhatt
Pharmacy Department, Faculty of Technology and Engineering, The Maharaja Sayajirao University of Baroda, Vadodara, Gujarat, India

Gabriel Davidov-Pardo
Department of Food Science, University of Massachusetts, Amherst, Massachusetts, USA

Marlise A. dos Santos
Joan-Vernikos Aerospace Pharmacy Laboratory, School of Pharmacy, Microgravity Centre, Pontifical Catholic University of Rio Grande do Sul PUCRS, Porto Alegre, Brazil

Mayada M. Elgohary
Cancer Nanotechnology Research Laboratory (CNRL), and Department of Industrial Pharmacy, Faculty of Pharmacy, Alexandria University, Alexandria, Egypt

Ahmed O. Elzoghby
Cancer Nanotechnology Research Laboratory (CNRL), and Department of Industrial Pharmacy, Faculty of Pharmacy, Alexandria University, Alexandria, Egypt

Ana Grenha
CBME—Centre for Molecular and Structural Biomedicine/IBB—Institute for Biotechnology and Bioengineering, Faculty of Sciences and Technology, University of Algarve, Campus de Gambelas, Faro, Portugal

Andrew P. Jallouk
Consortium for Translational Research in Advanced Imaging and Nanomedicine, Department of Medicine, Division of Cardiology, Washington University in St. Louis, St. Louis, Missouri, USA

Iris J. Joye
Department of Food Science, University of Massachusetts, Amherst, Massachusetts, USA, and Department of Microbial and Molecular Systems, Faculty of Bioscience Engineering, KU Leuven, Leuven, Belgium

Nayra M. Kamel
Cancer Nanotechnology Research Laboratory (CNRL), and Department of Industrial Pharmacy, Faculty of Pharmacy, Alexandria University, Alexandria, Egypt

Rebecca Klippstein
CABIMER-Andalusian Center for Molecular Biology and Regenerative Medicine (CSIC-University of Seville-UPO), and Department of Medical Biochemistry, Molecular Biology and Immunology, The University of Seville Medical School, Seville, Spain

Neli S. Koseva
Institute of Polymers, Bulgarian Academy of Sciences, Sofia, Bulgaria

Hasan Kouchakzadeh
Biotechnology Group, Chemical Engineering Faculty, Tarbiat Modares University, Tehran, Iran

David Julian McClements
Department of Food Science, University of Massachusetts, Amherst, Massachusetts, USA, and Department of Biochemistry, Faculty of Science, King Abdulaziz University, Jeddah, Saudi Arabia

Ambikanandan Misra
Pharmacy Department, Faculty of Technology and Engineering, The Maharaja Sayajirao University of Baroda, Vadodara, Gujarat, India

Violeta A. Mitova
Institute of Polymers, Bulgarian Academy of Sciences, Sofia, Bulgaria

Rohun U. Palekar
Consortium for Translational Research in Advanced Imaging and Nanomedicine, Department of Medicine, Division of Cardiology, Washington University in St. Louis, St. Louis, Missouri, USA

Hua Pan
Consortium for Translational Research in Advanced Imaging and Nanomedicine, Department of Medicine, Division of Cardiology, Washington University in St. Louis, St. Louis, Missouri, USA

Sushilkumar Patil
Pharmacy Department, Faculty of Technology and Engineering, The Maharaja Sayajirao University of Baroda, Vadodara, Gujarat, India

David Pozo
CABIMER-Andalusian Center for Molecular Biology and Regenerative Medicine (CSIC-University of Seville-UPO), and Department of Medical Biochemistry, Molecular Biology and Immunology, The University of Seville Medical School, Seville, Spain

Joanna Rydz
Institute of Polymers, Bulgarian Academy of Sciences, Sofia, Bulgaria, and Centre of Polymer and Carbon Materials, Polish Academy of Sciences, Zabrze, Poland

Maryam Sadat Safavi
Biotechnology Group, Chemical Engineering Faculty, Tarbiat Modares University, Tehran, Iran

Paul H. Schlesinger
Department of Cell Biology and Physiology, Washington University in St. Louis, St. Louis, Missouri, USA

Seyed Abbas Shojaosadati
Biotechnology Group, Chemical Engineering Faculty, Tarbiat Modares University, Tehran, Iran

Ekaterina V. Stoyanova
Institute of Polymers, Bulgarian Academy of Sciences, Sofia, Bulgaria

Imran Vhora
Pharmacy Department, Faculty of Technology and Engineering, The Maharaja Sayajirao University of Baroda, Vadodara, Gujarat, India

Fei Wang
Silk Biotechnology Laboratory, School of Biological and Basic Medical Sciences, Soochow University, Suzhou, PR China

Samuel A. Wickline
Consortium for Translational Research in Advanced Imaging and Nanomedicine, Department of Medicine, Division of Cardiology, and Department of Cell Biology and Physiology, Washington University in St. Louis, St. Louis, Missouri, USA

Yu-Qing Zhang
Silk Biotechnology Laboratory, School of Biological and Basic Medical Sciences, Soochow University, Suzhou, PR China

PREFACE

Nanotechnology includes the engineering of different functional systems at molecular scale. Such systems are characterized by their unique attributes that can be attractive for a number of disciplines ranging from materials science to nanomedicine. Drug delivery via nanoparticle systems has already become a major pillar of nanomedicine. In that field, the technological innovations in the diagnostics and therapy refer to the development of useful nanovehicles or nanodevices carrying specific load (drug, imaging agent, biomolecule, etc.) to be delivered to the target site in the organism at relevant times and doses. Various nanoscale architectures have been designed—solid nanospheres or porous nanoparticles, nanotubes, nanocapsules, nanoassemblies (liposomes, micelles, and aggregates), branched macromolecules, etc. The engineered nanoparticles are smaller in size than human cells. Having dimensions less than 100 nm, they can circulate through the body, enter the cells, and interact with the therapeutic targets. The strategic rationale for application of drug delivery nanoconstructs arises from the findings made in genomics and proteomics providing the society with powerful tools in the combat against cancer, cardiovascular and neurodegenerative diseases, viral infectious diseases, etc. (Donev, Koseva, Petrov, Kowalczuk, & Thome, 2011; Koseva, Rydz, Stoyanova, & Mitova, 2015). Considering the importance of the above research field and the great interest in its practical application in biomedicine, we have focused this volume of the *Advances in Protein Chemistry and Structural Biology* on Protein and Peptide Nanoparticles for Drug Delivery.

The review in Chapter 1 "Protein– and Peptide–Drug Conjugates: An Emerging Drug Delivery Technology" by Vhora et al. focuses on various protein– and peptide–drug conjugates along with their physicochemical and pharmacological properties. Additionally, emphasis has been given to the conjugation techniques along with different linkers available for construction of conjugation and challenges involved. In Chapter 2 "Modifications of Natural Peptides for Nanoparticle and Drug Design" by Jallouk et al., authors thoroughly review the role of natural peptides in the design and creation of nanomedicines, with a particular focus on cell-penetrating peptides, antimicrobial peptides, and peptide toxins. They discuss in details the great promise of the application of natural peptides in conjunction with nanoparticle delivery systems for the development of new therapeutic formulations as well as novel platforms for the delivery of various cargoes.

The combination of polymers of biological and synthetic origin is an appealing modern strategy for the production of novel nanocarriers with unprecedented properties. Proteins' interface can play an important role in determining bioactivity and toxicity and gives perspective for future development of the polymer-based nanoparticles. The design of hybrid constructs composed of synthetic polymer and biological molecules such as proteins can be considered as a straightforward tool to integrate a broad spectrum of properties and biofunctions into a single device. All this is presented in Chapter 3 "Hybrid Protein–Synthetic Polymer Nanoparticles for Drug Delivery" by Koseva et al. which provides a detailed discussion on hybrid protein–synthetic polymer nanoparticles with different structure and level in complexity and functionality, in view of their applications as drug delivery systems.

Chapter 4 "Efficient Delivery of Therapeutic Agents by Using Targeted Albumin Nanoparticles" by Kouchakzadeh et al. explores thoroughly the current targeted and nontargeted albumin-based nanoparticles that are in various stages of development for the delivery of therapeutic agents in order to enhance the efficacy of cancer treatment.

Chapter 5 "Vasoactive Intestinal Peptide (VIP) Nanoparticles for Diagnostics and for Controlled and Targeted Drug Delivery" by Klippstein and Pozo focuses on recent developments in the use of nanotechnology-based approaches as a tool for VIP-controlled drug delivery, an excellent example on how nano-enabled methods can fill the gap between current knowledge on the biological role of neuropeptides and their translational potential to health-care solutions. Authors present recent studies aimed at developing nanostructured organic and inorganic systems either for the appropriate delivery of VIP or for VIP targeting. These technologies stand as an alternative starting point for chemical manipulations of the neuropeptides in order to improve potency, selectivity, or pharmacokinetic parameters.

Chapter 6 "Implications of Protein- and Peptide-Based Nanoparticles as Potential Vehicles for Anticancer Drugs" by Elzoghby et al. highlights the applications of protein nanoparticles for delivery of anticancer drugs. Various techniques that have been utilized for preparation of anticancer drug-loaded protein nanoparticles are examined. The major outcomes of some *in vitro* and *in vivo* investigations of surface-modified tumor-targeted protein nanoparticles are discussed here.

Last three chapters are focusing on specific subgroups of materials for preparation of nanoparticles used for drug delivery. In Chapter 7 "Polysaccharide Nanoparticles for Protein and Peptide Delivery: Exploring Less-Known Materials" by dos Santos et al., polysaccharide nanoparticles developed for protein and peptide delivery are described. The emphasis is placed on materials that in general are less explored or that are not so well-known regarding the development of nanoparticulate systems, thus excluding polymers like chitosan, alginate, cyclodextrins, and hyaluronic acid. A detailed report on the main properties and their applications in the production of nanoparticles aimed at protein delivery is presented. Chapter 8 "Bioconjugation of Silk Fibroin Nanoparticles with Enzyme and Peptide and Their Characterization" by Wang et al. explores in great details the bioconjugation of silk fibroin nanoparticles with glucose oxidase, superoxidase, β-glucosidase, L-asparaginase, neutral protease, and insulin, as well as their characterization. Finally, Chapter 9 "Food-Grade Protein-Based Nanoparticles and Microparticles for Bioactive Delivery: Fabrication, Characterization, and Utilization" by Davidov-Pardo et al. provides an overview of the functional properties of food proteins that can be used to assemble nanoparticles and microparticles, the fabrication techniques available to create those particles, the factors that influence their stability, and their potential applications within the food industry.

The aim of this volume is to promote further basic studies and clinical trials in diverse groups of protein and peptide nanoparticles as a promising tool for targeted delivery of therapeutics for treatment of various diseases and disorders.

Rossen Donev
Biomed Consult Ltd
Swansea, United Kingdom

REFERENCES

Donev, R., Koseva, N., Petrov, P., Kowalczuk, A., & Thome, J. (2011). Characterisation of different nanoparticles with a potential use for drug delivery in neuropsychiatric disorders. *The World Journal of Biological Psychiatry*, *12*(Suppl. 1), 44–51.

Koseva, N. S., Rydz, J., Stoyanova, E. V., & Mitova, V. A. (2015). Hybrid protein-synthetic polymer nanoparticles for drug delivery. *Advances in Protein Chemistry and Structural Biology*, *98*, 93–119.

CHAPTER ONE

Protein- and Peptide–Drug Conjugates: An Emerging Drug Delivery Technology

Imran Vhora, Sushilkumar Patil, Priyanka Bhatt, Ambikanandan Misra[1]

Pharmacy Department, Faculty of Technology and Engineering, The Maharaja Sayajirao University of Baroda, Vadodara, Gujarat, India

[1]Corresponding author: e-mail addresses: misraan@hotmail.com; a.n.r.misra-pharmacy@msubaroda.ac.in

Contents

Abstract

Protein- and peptide–drug conjugates hold a promising stance in the delivery of therapeutic agents by providing distinct advantage of improving therapeutic potential of drugs. Recent advancements in the proteomics and recombinant DNA technology, by enabling identification of distinct structural features of proteins and making it feasible to introduce specific functionalities in protein/peptide structure, has made it possible to synthesize high quality protein- and peptide–drug conjugates though a wide variety of coupling techniques. Additionally, use of specialized linkers makes them unique in their *in vivo* therapeutic application by providing target tissue-specific release of drug. Several protein- and peptide–drug conjugates are currently under clinical trials

Advances in Protein Chemistry and Structural Biology, Volume 98
ISSN 1876-1623
http://dx.doi.org/10.1016/bs.apcsb.2014.11.001

warranting their huge market potential in near future. Increased understanding in this field will surely enable us to produce high quality protein- and peptide–drug conjugates which will serve therapeutic needs demanded from drug delivery systems in clinical settings.

1. INTRODUCTION

Targeted therapeutics has recently gained extensive attention of drug delivery scientists. Various drug delivery systems like liposomes, micelles, dendrimers, solid lipid nanoparticles, polyester nanoparticles and others offer high drug loading and improved therapeutic potential of drugs through passive targeting in cancer and other inflammatory diseases. Further, such delivery systems can also be conjugated to proteins and peptides for actively targeted therapy. However, challenges related to their surface conjugation with targeting ligands and constraints related to their formulation aspects like uniformity of production, large-scale preparation, heterogeneity and regulatory requirements, have hindered their development. This has been reflected by a low number of such carrier systems in clinical trials.

Unlike conjugating drug-loaded nanocarrier systems to proteins and peptides, which require surface localized conjugation of ligands for better outcomes, drugs can directly be conjugated with proteins and peptides usually with a suitable linker (Fig. 1). This usually accompanies the problem of low drug loading, i.e., lower number of drug molecules conjugated to one molecule of protein. However, many-a-times this can be surpassed to an extent through use of special techniques like Universal Linkage System™, use of dendritic linkers and conjugation on several sites of protein molecules. Additionally, such conjugates retain the therapeutic potential through passive targeting due to nanoscale size (<20 nm) and improved circulation in blood due to increased molecular weight of drugs and targeting through passive mechanism or through protein/peptide targeted against specific transmembrane protein. Additionally, such conjugates can be devised using specific linker molecules which offer manifold advantages like target organ-specific cleavage of conjugate, improved flexibility of protein/peptide which allow its unhindered binding to receptors and improvement of drug hydrophilicity and water solubility. As of now, more than 25 antibody–drug conjugates and more than 5 peptide–drug conjugates are under different phases of clinical trials on cancer, pain management, heart diseases, etc.

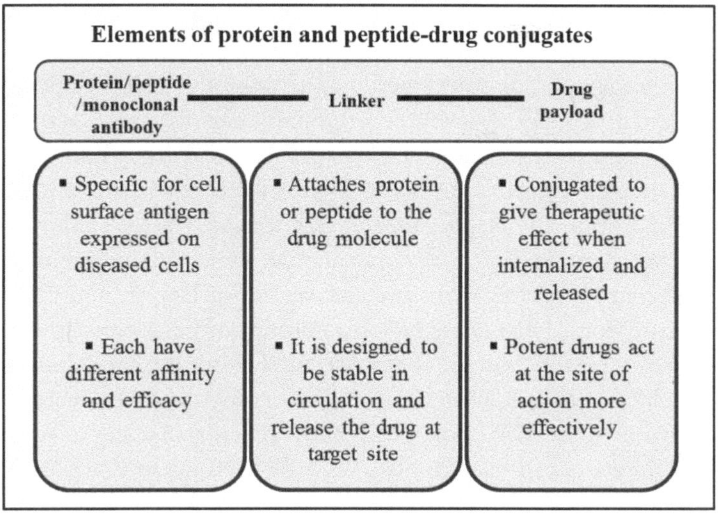

Figure 1 Elements of protein and peptide–drug conjugates.

Recent advancements in genomics and proteomics have enabled the deeper understanding of target sites which can be exploited for drug delivery and structural as well as physicochemical properties of proteins and peptides for development of conjugates. This has led to development of techniques and strategies that allow site-selective attachment of drugs to proteins/peptides. Some of such techniques are albumin-binding prodrug technology which exploits cysteine-34 of human serum albumin for *in situ* conjugation of cytotoxic drugs to albumin, albumin fusion technology which allow biosynthesis of albumin-fused therapeutic proteins/peptides, THIOMAB technology which allow site-specific drug loading on antibodies. Seeing the current advancements in the protein and peptide–drug conjugate field and their clinical scenario, the field seems to holds a promise in future drug delivery.

The review focuses on various protein– and peptide–drug conjugates along with their physicochemical and pharmacological properties. Additionally, emphasis has been given to the conjugation techniques along with different linkers available for construction of conjugation and challenges involved.

2. PROTEIN–DRUG CONJUGATES

Plasma protein binding was known to alter distribution and therapeutic characteristics of the drugs. This phenomenon was then successfully exploited by protein–drug conjugates to improve pharmacokinetics behavior of drugs for imparting longer blood residence and reduce unwanted toxicity through selective distribution to target tissue only. The strategy was very much effective in cancer therapy due to the very nature of cancer cells of uptaking and utilizing plasma proteins as a source of amino acids for their rapid growth. After conjugation to high molecular weight proteins, drugs inherit the long circulation characteristics of carrier protein which results in longer tumor exposure (Matsumura & Maeda, 1986). Additionally, Maeda et al. who demonstrated that tumors have leaky vasculature which allow permeation of higher molecular weight proteins and particles inside tumor, while the defective lymphatic drainage system lead to retention of permeated proteins and particles inside tumor, formed basis for passive tumor targeting (Maeda, Wu, Sawa, Matsumura, & Hori, 2000). Tumor vasculature has pores ranging in the size from 100 to 1200 nm, which allow macromolecules to accumulate inside the tumor interstitium (Hobbs et al., 1998; Noguchi et al., 1998; Yuan et al., 1995). In contrast, pore size of other nonsinusoidal vascular bed ranges between 5 and 12 nm and could be as low as 1 nm which excludes the entry for larger-sized macromolecules inside the normal tissues' interstitia and corresponding lymphatic counterparts (Sarin, 2010).

Additionally, threshold level of excretion of polymers from kidney is $\geq$40–45 kDa as has been established for globular proteins ($\geq$45 kDa), *N*-(2-hydroxypropyl)methacrylamide-HPMA polymer ($\geq$40 kDa) (Seymour, Duncan, Strohalm, & Kopecek, 1987), and PEG ($\geq$40 kDa) (Singh et al., 2012). Similarly, it has been shown that clearance from tumor tissues for different polymers follow same kind of trend. Globular proteins with molecular weight (MW) higher than $\geq$40 kDa, PEG carriers with MW $\geq$ 40 kDa and HPMA with MW $\geq$ 78 kDa exhibited significantly higher enhanced permeation and retention (EPR) effect (Singh et al., 2012). In terms of particle size, particles with hydrodynamic diameter $\geq$8 nm have shown improved tumor distribution through EPR effect (Singh et al., 2012). These physicochemical criteria can be used to identify the lower limits for the proteins that can be used to achieve long blood residence as well as enhanced tumor retention. Table 1 shows the list of the proteins that fit in the criteria and hence, land themselves into the category of proteins for development of protein–drug conjugates. Some preclinically and clinically evaluated protein–drug conjugates (except antibody–drug conjugates (ADCs)) are described in Table 2.

2.1. Albumin–drug conjugates

As of now, albumin is the most extensively explored protein for preparing drug–protein conjugates especially in cancer. Albumin conjugates can

Table 1 Physicochemical properties and blood half-lives of some proteins used for protein conjugates

Protein	Molecular weight (kDa)	Hydrodynamic diameter (nm)	Half-life
Albumin	66.5	7.2	15–20 days
Immunoglobulins (IgG)	150	10.58	7 to >20 days
Immunoglobulins (IgM)	950	25.3	5–8 days
Transferrin	80	7.44	7–10 days
Gelatin	~300	25–200	>5 h for 90 kDa
Human haptoglobin	100–900	9.4–50	4.5 days
Hemoglobin A	65	6.36	120 days
Fibrinogen	34	21.9	39.5 h

further be modified by PEGylation to impart hydrophilicity to improve the blood residence of conjugates (Temming, Lacombe, Schaapveld, et al., 2006; Temming, Lacombe, van der Hoeven, et al., 2006). Additionally, one can also go for coupling of targeting ligands such as cyclic or acyclic RGD, lactosamine and folate to the albumin to impart targeting potential to the conjugates (Dosio et al., 2009; Fiume et al., 2005; Temming, Lacombe, Schaapveld, et al., 2006; Temming, Lacombe, van der Hoeven, et al., 2006; Temming, Meyer, Zabinski, et al., 2006). The high endogenous availability of albumin in blood has been conveniently exploited to prepare *in situ* albumin–drug conjugates using chemically modified drugs with maleimide group which binds to serum albumin when injected in blood stream. This strategy of albumin-binding prodrugs has been specifically devised and explored by Kratz et al., which is discussed under Section 4.

2.2. Transferrin–drug conjugates

Transferrin is an endogenous iron-transporting protein which functions through transferrin receptor (TfR). Differential levels and heterogeneity of expression have been reported for TfR in cancerous as well as normal cells (Gatter, Brown, Trowbridge, Woolston, & Mason, 1983). Increased iron requirement in tumor tissues causes 10× higher expression of TfR in cancer cells as compared to normal cells (Inoue, Cavanaugh, Steck, Brunner, & Nicolson, 1993). Various cancers showing higher expression of TfR include breast cancer, leukemia, colon cancer, skin cancer, glioblasoma, and

Table 2 Protein–drug conjugates (except antibody–drug conjugates)

Conjugate type	Drug	Characteristics	Refs.
Albumin	Methotrexate	• No stoichiometric loading • Drug molecules/albumin molecule affected the tumor response in cancer rat model	Stehle et al. (1997)
	Methotrexate	• High response as compared to drug in soft-tissue, prostate, and bone cancer model • Drug and conjugate not effective in other cancer	Burger, Hartung, Stehle, Sinn, and Fiebig (2001)
	Methotrexate	• Phase I study in renal cell carcinoma showed response to therapy • Phase II studies failed to show response • Stomatitis above 50 mg/m^2 dose	Hartung et al. (1999)
	Methotrexate	• Phase II study in bladder carcinoma in combination with cisplatin showed complete and partial remission	Bolling et al. (2006)
	Paclitaxel	• Non-PEGylated and PEGylated albumin (albumin-PEG_{2000} and albumin-PEG_{5000}) was conjugated • Reduced toxicity and higher cytotoxicity in colorectal and epithelial adenocarcinoma cells and melanoma cells • PEGylation reduced clearance rate with PEG_{5000} providing highest blood retention	Dosio, Brusa, Crosasso, Arpicco, and Cattel (1997)
	Paclitaxel	• Non-PEGylated and PEGylated folate conjugated albumin (albumin-PEG_{3400}-folate) was conjugated • Higher activity of targeted conjugates in Folate receptor positive cells (KB Fr-α +ve cell) than in folate negative cells (HT-29 Fr-α −ve cells)	Dosio, Arpicco, Stella, Brusa, and Cattel (2009)

	Doxorubicin	• Activity in parental and resistant hepatoma cell lines by avoiding multidrug resistance	Hatano, Ohkawa, and Matsuda (1993)
	Doxorubicin	• Activity in parental and resistant myelogenous leukemia cell lines by avoiding multidrug resistance	Hatano et al. (1993) and Takahashi, Asakura, and Ohkawa (1996)
	Doxorubicin	• Gly-Pro-Leu-Gly–Ile-Ala-Gly-Gln and Gly-Pro-Gln-Arg-Ile-Ala-Gly-Gln linkers showed MMP-2- and MMP-9-specific cleavage in cancer cells • Ile-Ala-Gly-Gln-doxorubicin liberated after cleavage is less active than free drug (IC_{50} of conjugates with two linkers is $>20\times$ and $>300\times$ higher than free drug, respectively)	Kratz et al. (2001)
	Cisplatin	• Cisplatin loading of 7 mol/mol of albumin • Extreme reduction in activity of free cisplatin as compared to transferrin conjugate in epithelial adenocarcinoma cells	Hoshino et al. (1995a)
	VEGF-R kinase inhibitor PTK787	• PTK787 conjugated using *Universal Linkage System-ULS*™ ([Pt(*cis*-ethylenediamine)nitrate chloride]) at 1:1 molar ratio • Conjugates with *RGD grafting* and with/without PEG_{5000} linker were prepared • PEG linker improved water solubility of conjugate • *Integrin-mediated cellular uptake* in umbilical vein endothelial cells • Conjugate without PEG conjugation showed highest cytotoxic activity	Temming, Lacombe, Schaapveld, et al. (2006)
	p38 MAP kinase inhibitor SB202190	• SB202190 conjugated using *Universal Linkage System*™ • ULS yielded average of nine SB202190 and *six RGD-PEG_{5000} loading* per albumin • Glutathione (sulfur donor) mediated drug release • Integrin-mediated targeting of HUVEC cells but no toxicity to cells • Conjugates provide a novel mean to treat *inflammation* disorders such as rheumatoid arthritis	Temming, Lacombe, van der Hoeven, et al. (2006)

Continued

Table 2 Protein–drug conjugates (except antibody–drug conjugates)—cont'd

Conjugate type	Drug	Characteristics	Refs.
	Auristatin E	• Conjugates with *c(RGDfK) grafting* and with/without PEG linker • High drug loading (average 4.1 mol/mol of albumin) and high loading of RGD peptide (5–7 mol/mol) • *Integrin*-mediated uptake in HUVEC cells *in vitro* with very high cytotoxicity than parent drug • Prolonged circulation and delayed degradation in plasma • High tumor tissue accumulation in colon carcinoma xenograft mice model • Antivascular and antitumor activity	Temming, Meyer, Zabinski, et al. (2006b)
Transferrin	Doxorubicin	• Glutaraldehyde or dextran as linker • Activity in parental and resistant myelogenous leukemia cell lines	Hatano et al. (1993)
	Doxorubicin	• Tumor-specific release of drug through *acid-labile* benzoyl hydrazone or phenyl acetyl hydrazone linkers • Benzoyl hydrazone containing conjugate showed highest activity	Kratz et al. (1998)
	Doxorubicin	• Higher cytotoxicity with increasing *conjugation extent* and time of exposure	Yeh and Faulk (1984)
	Doxorubicin	• Higher cytotoxicity than doxorubicin in myelogenous leukemia cells as well as in normal blood mononuclear cells • Cytotoxicity without intercalation in DNA (hence, can act in MDR)	Sizensky, Barabas, and Faulk (1992)
	Doxorubicin	• Conjugate showed more potency against resistant cell lines than free drug in human colon adenocarcinoma, promyelocytic leukemia, lung mesothelioma, and epithelioma type-2 cells	Singh, Atwal, and Micetich (1998)
	Artemisinin	• Selective uptake through *transferrin receptors* in breast cancer cells • Release of iron and artemisinin from conjugate causes formation of free radicals and causes high toxicity in leukemic and breast cancer cells	Lai, Nakase, Lacoste, Singh, and Sasaki (2009) and Lai, Sasaki, Singh, and Messay (2005)

	N-alkylisatin derivatives	• Selective uptake through *transferrin receptors* in breast cancer cells • *Acid catalyzed* hydrolytic release of isatin derivative through *para*-phenylpropionic acid inside tumor • 10 × more effective than free drug *in vivo*	Indira Chandran et al. (2012)
	Tumor necrosis factor-α	• Dual advantage of PEG linker (improved pharmacokinetics) and targeting through *transferrin receptors* • Effective in myelogenous leukemia, HeLa derivative cells, and in sarcoma	Jiang, Liu, Hong, Zhu, and Pei (2007)
	HSV-TK gene	• Higher transfection efficiency than *to lipofectin* and retroviral vectors in myelogenous leukemia, human *colonic and gastric cancer cells* • *Biotin–streptavidin linker* prevent aggregation of conjugate and permits release of DNA from conjugate	Sato et al. (2000)
	Mitomycin-C	• Selective uptake in mouse sarcoma cell line • Increase in drug molecules loaded per transferrin molecule decreased binding efficiency targeting • Retained binding efficiency half that of transferrin below 10 mol drug/1 mol of transferrin	Tanaka, Kaneo, and Miyashita (1996)
	Cisplatin	• Reduced rate of lymphoma cell growth by conjugate • Conjugate showed cytotoxicity to HeLa cells in 7 days • Complete remission in breast cancer rat model • Notable activity in human breast cancer patients	Elliott, Stjernholm, and Elliott (1988)
	Cisplatin	• Increasing drug loading per molecule of transferrin reduced cytotoxicity in epithelial adenocarcinoma cells while increasing their plasma circulation time in melanoma bearing mice	Hoshino et al. (1995a, 1995b)

Continued

Table 2 Protein–drug conjugates (except antibody–drug conjugates)—cont'd

Conjugate type	Drug	Characteristics	Refs.
	Chlorambucil	• Acetaldehyde carboxylic hydrazine showed lysozymal cleavage • 3–18 × lower cytotoxicity than chlorambucil in breast adenocarcinoma and leukemia cells	Beyer et al. (1998)
	CRM107 (a variant of diphtheria toxin)	• In Phase I trial 60% patients showed reduction in tumor volume to half • Two complete remissions and activity in refractory tumor • No systemic toxicity on regional perfusion	Laske, Youle, and Oldfield (1997)
	Insulin	• Permeation through transcytosis in Caco-2 cells *in vitro* • Slow but long-lasting hypoglycemic effect *in vivo* in streptozotocin-induced diabetic rats • Oral absorption of conjugate in rats	Xia, Wang, and Shen (2000)
Transferrin and oligomeric transferrin	Insulin	• Oligomeric transferrin showed extensive plasma retention and higher hypoglycemic activity as compared to monomeric transferrin in streptozotocin-induced diabetic rats	Lim and Shen (2005)
	Granulocyte colony-stimulating factor G-CSF	• Leu-Glu was used as a linker • Transferrin receptor-mediated transfer *through Caco-2 cells in vitro* • Oral administration provided prolonged activity for 3 days on oral administration and increased neutrophil count comparable to commercial G-CSF through SC route	Bai, Ann, and Shen (2005)
Gelatin	Doxorubicin	• MMP-2- and MMP-9-mediated mPEG-doxorubicin release *in vitro* in lung carcinoma cell line	Cho et al. (2002)

	Doxorubicin	• PEGylated and non-PEGylated gelatin were used for conjugation • Non-PEGylated conjugates cleared from blood rapidly in murine squamous cell carcinoma bearing mice • Conjugates showed poor tumor selectivity	Cho et al. (2007)
	Gallic acid	• Conjugate showed higher cytotoxicity as compared to gallic acid in prostate and *renal cell cancer* cells • Conjugate also inhibited the activity of α-amylase and acetylcholine esterase *in vitro* evidencing the activity of conjugate for treatment of diabetes and Alzheimer's disease, respectively	Cirillo et al. (2010)
Plasminogen activator inhibitor-2 (PAI-2)	*N*-alkylisatin derivatives	• Selective cytotoxicity to breast cancer cell lines • *In vivo* efficacy at (1/20)th dose of free drug	Indira Chandran (2012)
	2′-deoxy-5-fluorouridine (5-FUdr)	• Urokinase targeted conjugate with preferential cytotoxicity to *urokinase expressing cancer* cells	Vine, Locke, Bremner, Pyne, and Ranson (2010)
Fibrinogen	Methotrexate	• Loading of 5–8 mol of methotrexate/mol of fibrinogen • Low *in vitro* cytotoxicity of conjugate in rectum adenocarcinoma, colon adenocarcinoma, bladder cancer, and nonsmall cell lung cancer cells • Higher antitumor activity *in vivo* in leukemia cells bearing mice • Narrow therapeutic index *in vivo*	Boratynski, Opolski, Wietrzyk, Gorski, and Radzikowski (2000)
Hydrolyzed fibrinogen	Methotrexate	• Loading of 159 mol methotrexate/1 mol of proteins • Low *in vitro* cytotoxicity but higher *in vivo* antitumor activity	Goszczynski, Nevozhay, Wietrzyk, Omar, and Boratynski (2013)

Continued

Table 2 Protein–drug conjugates (except antibody–drug conjugates)—cont'd

Conjugate type	Drug	Characteristics	Refs.
Insulin	Mitoxantrone	• Higher *blood residence time*, higher tumor targeting and low distribution in normal tissue cells in hepatoma bearing mice	Liu, Huang, and Zhang (2003)
	5-Fluorouracil	• *Dendritic linker* was used to attach several drug molecules to insulin • Stable in acidic media while drug release at *alkaline pH* and in presence of plasma ensuring controlled release of drug in blood	Huang, Wang, Gong, and Zhang (2007)
Hemoglobin	Ribavirin	• 6–8 ribavirin mol/mol of hemoglobin • Selective uptake in haptoglobin–haptoglobin receptor expressing cells • *Enzymatically cleaved* ribavirin showed equipotent *antiviral activity in vitro*	Brookes et al. (2006)
	Ribavirin	• Loading up to 15 mol of ribavirin/mol of hemoglobin retained binding capacity to haptoglobin and *haptoglobin receptors* • Improved survival, behavior, and reduced histological signs of disease in *hepatitis* mouse model • Inhibition of viral replication and inflammatory cytokine release *in vitro*	Levy et al. (2006)
	5-Floxuridine	• Anticancer activity in cytotoxicity against *lymphoma* cells expressing haptoglobin receptors	Same as above
	Dichloroacetic acid (DCA)	• 12 DCA molecules per hemoglobin tetramer • *Haptoglobin receptor*-mediated uptake by *monocytic cancer* cells • Activity retained as that of free DCA	Zhang and Palmer (2011)

lymphoma (Calzolari et al., 2007). Apart from cancer, use of transferrin has also been exploited for oral delivery of drugs (Lim & Shen, 2005; Xia et al., 2000) due to expression of TfRs in GI tract. It has been reported that TfR are expressed in duodenal epithelial cells (Pietrangelo et al., 1992) and are responsible for receptor-mediated endocytosis for absorption of TfR-targeted drug delivery systems (Du et al., 2013). This warrants the potential of transferrin as a targeting ligand for devising drug–transferrin conjugates.

2.3. Gelatin–drug conjugates

Gelatin is another protein that has been explored for development of drug delivery systems ranging from microspheres to nanoparticles. Gelatin is a substrate for matrix metalloproteinase enzymes, MMP-2 (gelatinase A) and MMP-9 (gelatinase B) (Nagase, 2001). MMP-2 and/or MMP-9 is/are overexpressed in several cancer tissues (Roomi et al., 2009). This makes the basis that if gelatin is conjugated to drugs, it would be preferentially cleaved inside tumors and release active drug avoiding toxicity to normal cells where expression of aforementioned enzymes is low/absent. This can also be viewed as there is no requirement of tumor-specific linkers as gelatin serves the function of both a carrier as well as a substrate releasing drug specifically in tumors.

2.4. Antibody–drug conjugates

The face of targeted drug delivery was drastically changed by development of antibodies against any identified target antigen, both as therapeutic agents as well as targeting ligands for drugs and drug delivery systems. In drug–protein conjugate field, ADCs are the most explored as evident from the large number of ADCs currently entering/under clinical trials. Antibody serves a dual function providing cell specific endocytic uptake in targeted tissue as well as carrier function by offering improved pharmacokinetics (Beck et al., 2010). Additionally, antibodies may further augment cell killing through complement-dependent cytotoxicity; however, such activity may also get diminished or abolished after conjugation to other therapeutic agents.

Initial development in the field was hindered by several drawbacks, i.e., immunogenicity of murine antibodies, lack of potency, sufficient selectivity of ADCs, etc., and led to limited success in clinical trials. However, replacement of murine antibodies by humanized antibodies was a landmark in the field of antibody therapeutics. Humanization of antibodies led to reduced immunogenicity and thereby better half-life as compared to murine antibodies, i.e., approved murine IgGs, muromononab-CD3, ibritumomab, etc. show half-lives of ~1 day in patients, while approved human IgGs demonstrated half-lives of ~25 days (Wang, Wang, & Balthasar, 2008). Among several classes and subclasses of human immunoglobulins also, only IgG1, IgG2, and IgG4 are unique for ADCs due to longer half-lives (>20 days against 5–8 days of IgA and IgM and ~7 days of IgG3). Additionally, serum half-lives of antibodies can be increased by PEGylation of amino groups of the antibodies which reduce the immunogenicity as well (Chapman, 2002). However, such modification may affect the complement-dependent cytotoxicity (Anderson & Tomasi, 1988) as well as its antigen-binding properties (Suzuki, Ikeda, & Tomono, 1989; Suzuki, Kanbara, Tomono, Hayashi, & Shinohara, 1984) and hence, needs a careful consideration. Additionally, the challenge of low potency of antibodies in cancer treatment was overcome by conjugating highly cytotoxic drug classes like auristatin, calicheamycin, maytansine, etc. (Lash, 2011).

The uniqueness of ADCs lies in the fact that almost any antigen target can be utilized for drug delivery using antibodies (Fig. 2). Today there are nearly 300 mAbs under clinical development and 40 mAbs have been approved for various indications ranging from cancer, rheumatoid arthritis, osteoporosis, asthma, etc.; however, ADCs are currently being developed for cancer only. ADCs that are currently approved/under clinical development are given in Table 3.

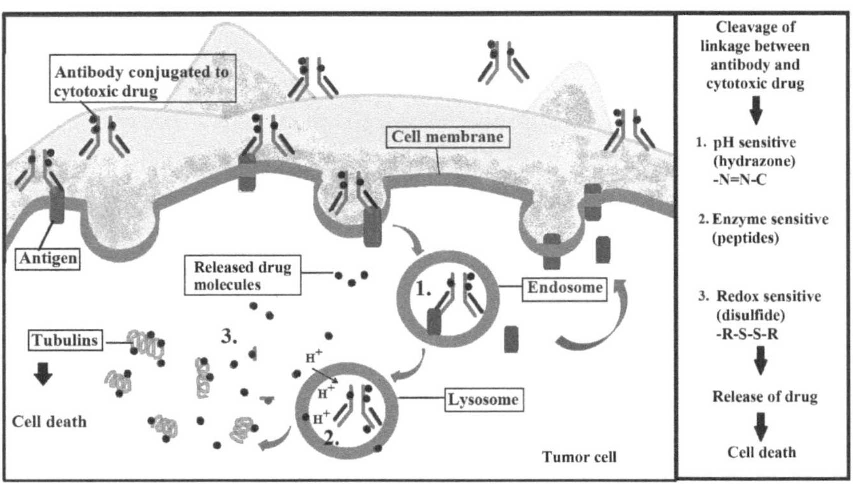

Figure 2 Internalization and release of drug from ADCs for cancer.

Table 3 Antibody–drug conjugates and their clinical status

Conjugate	Antibody class	Drug (drug class)	Disease	Clinical phase	Company
Adcetris	Anti-CD30 mAb (cIgG1)	Auristatin E (auristatin)	Non-Hodgkin's lymphoma (NHL)	Approved in Aug 2011	Seattle Generics
AGS-16M8F	Anti-AGS16 mAb (hIgG1)	Auristatin F (auristatin)	Advanced renal cell carcinoma	I	Seattle Generics/ Agensys
AGS-5ME	Anti-SLC44A4 mAb (hIgG2)	Auristatin E (auristatin)	Gastric and pancreatic neoplasm	I	Seattle Generics
SGN-75	Anti-CD70 mAb (hIgG1)	Auristatin F (auristatin)	Renal cell carcinoma, NHL	I	Seattle Generics
SGN-LIV1A	Anti-LIV1 mAb (hIgG1)	Auristatin E (auristatin)	Metastatic breast cancer	I	Seattle Generics
SGN-CD19A	Anti-CD19 mAb (hIgG1)	Auristatin E (auristatin)	Acute lymphoblastic leukemia and B-cell NHL	I	Seattle Generics
SGN-22ME	Anti-Nectin-4 mAb (hIgG1)	Auristatin E (auristatin)	Solid tumors	I	Seattle Generics/ Agensys
SGN-15ME	Anti-SLTRK6 mAb (hIgG1)	Auristatin E (auristatin)	Bladder cancer	I	Seattle Generics/ Agensys

Continued

Table 3 Antibody–drug conjugates and their clinical status—cont'd

Conjugate	Antibody class	Drug (drug class)	Disease	Clinical phase	Company
BAY 79-4620	Anti-CA9 mAb (hIgG1)	Auristatin E (auristatin)	Solid tumors	I	Bayer Healthcare/ Seattle Generics
CDX-011	Anti-GPNMB mAb (hIgG2)	Auristatin E (auristatin)	Breast cancer	II	Celldex Therapeutics/ Seattle Generics
PSMA-ADC	Anti-PSMA mAb (hIgG1)	Auristatin E (auristatin)	Prostate cancer	I	Progenics/ Seattle Generics
RG7593	Anti-CD22 mAb (hIgG1)	Auristatin E (auristatin)	B-cell NHL	I/II	Roche
RG7596	Anti-CD79b mAb (hIgG1)	Auristatin E (auristatin E)	B-cell NHL	I/II	Roche
BAY 94-9343	Anti-mesothelin mAb (hIgG1)	DM4-ravtansine (maytansine)	Solid tumors	I	Immunogen
BIIB-015	Anti-cripto mAb (hIgG1)	DM4 (maytansine)	Breast cancer	I	Imunogen/ Biogen
BT-062	Anti-CD138 mAb (cIgG4)	DM4 (maytansine)	Multiple myeloma	I/II	Biotest/ Immunogen
IMGN-901	Anti-CD56 mAb (hIgG1)	DM1 (maytansine)	Multiple myeloma, nonsmall cell lung cancer	I	Immunogen

IMGN-388	Anti-α-integrin mAb (hIgG1)	DM4 (maytansine)	Solid tumors	I	Immunogen/ Centocor
Kadcyla	Anti-HER2 mAb (hIgG1)	Emtansine-DM1 (maytansine)	Metastatic breast cancer	Approved in Feb 2013	Genentech
SAR3419	Anti-CD19 mAb (hIgG1)	DM4 (maytansine)	Diffuse large B-cell lymphoma, B-cell acute lymphoblastic lymphoma	II	Sanofi
			B-cell NHL	I	Sanofi
SAR-566658	Anti-DS6 mAb (hIgG1)	DM4 (maytansine)	CA6 positive and refractory solid tumors	I	Sanofi
IMGN242	Anti-CanAg mAb (hIgG1)	DM4 (maytansine)	Noncolorectal cancer and pancreatic cancer	I	Immunogen
Inotuzumab ozogamicin (CMC-544)	Anti-CD22 mAb (hIgG4)	Ozogamicin (calichieamicin)	NHL	III	Pfizer
Gemtuzumab ozogamicin	Anti-CD22 mAb (hIgG4)	Ozogamicin (calichieamicin)	Acute myelogenous leukemia	Approved in 2000, withdrawn in 2010	Pfizer
SGN-CD33A	Anti-CD33 mAb (hIgG1)	Pyrrolobenzodiazepine (DNA cross-linking agents)	Acute myeloid leukemia	I	Seattle Generics

Continued

Table 3 Antibody–drug conjugates and their clinical status—cont'd

Conjugate	Antibody class	Drug (drug class)	Disease	Clinical phase	Company
SGN-CD70A	Anti-CD70 mAb (hIgG1)	Pyrrolobenzodiazepine (DNA cross-linking agents)	Renal cell carcinoma or NHL	I	Seattle Generics
MDX-1203	Anti-CD70 mAb (hIgG1)	Duocarmycin	Renal cell carcinoma and NHL	I	Bristol-Myers Squibb
hLL1-DOX	Anti-CD74 mAb (hIgG1)	Doxorubicin (DNA intercalating and topoisomerase I inhibitor)	Relapsed multiple myeloma	I/II	Immunomedics
IMMU-130	Anti-CEACAM5 mAb (hIgG1)	SN-38 (irinotecan metabolite)	Colorectal cancer	I/II	Immunomedics
IMMU-132	Anti-TROP-2 mAb (mIgG1)	SN-38 (irinotecan metabolite)	Triple negative breast cancer	II	Immunomedics
			Several types of cancer	I/II	Immunomedics

CEACAM5 = carcinoembryonic cell adhesion molecule 5.
Source: Clinicaltrials.gov and company reports.

2.5. Other protein–drug conjugates

Hemoglobin (Hb), fibrinogen, insulin, etc. are other proteins that are used for formulating protein–drug conjugates. Hb–drug conjugates utilize the natural Hb scavenging activity of serum haptoglobin directed through bHb scavenger receptor CD163 located on monocyte and macrophages. Thus, Hb acts as a natural ligand for targeting macrophages and monocytes (Zhang & Palmer, 2012) involved in moonocytic leukemias. Additionally, liver macrophages and Kupffer cells are activated in viral hepatitis (Hiraoka et al., 2005) and are implicated in liver cirrhosis and in hepatocellular carcinoma (Kuang & Zheng, 2012) and hence the application of Hb–drug conjugates can be extended to the treatment of hepatitis and hepatocellular carcinoma as well.

Fibrinogen is another endogenous protein that can be used for drug delivery to cancer tissues. Tumor cells secrete several proteinases and under the influence of these proteinases cell surfaces show enhanced cell adhesion to fibrinogen and vironectin though integrin binding (Fujii & Imamura, 1995; Zhang, Ozdemir, Chung, Robertson, & Dong, 2011). This makes it suitable for targeting tumors which show overexpression of integrins.

Insulin is a natural hormone which shows its activity through insulin receptors and is useful in cancer targeting. Higher expression of insulin receptors has been noted in breast cancer, prostate cancer, cervical cancer, and hepatocellular carcinoma (Cox et al., 2009; Kurtaran et al., 1995; Papa et al., 1990; Wang et al., 2013). Additionally, Ou and Kuang, through their experimental evaluation of insulin covalently conjugated with some compounds, have reported that insulin retains its activity to bind with insulin receptors (Ou & Kuang, 2000).

3. PEPTIDE–DRUG CONJUGATES

Peptide–drug conjugates are on the other domain of drug delivery systems that offer improved therapeutic outcome from existing drugs. These conjugates share the similar attributes with protein–drug conjugates with respect to their chemistry. However, they differ from protein–drug conjugates in their pharmacological outcomes.

Peptide–drug conjugates are prepared based on the rationale of improving therapeutic efficacy of conjugated drugs and reducing side effects thereof through targeting of specific cells. However, peptides, being small amino acid sequences, do not allow attaching high payloads of drugs on the peptide molecule. Additionally, their low molecular weights make them vulnerable to clearance and hence, showing low half-lives in blood. However, the relatively rapid tissue distribution may compensate for the short half-life in blood (Burkhart, Kalet, Coleman, Post, & Koch, 2004).

Apart from targeted therapy, some peptide–drug conjugates are also designed as prodrugs which show therapeutic action after cleavage inside the tumor conferring broadened therapeutic window to cytotoxic drugs. There may also happen complete change in therapeutic outcome of drugs (i.e., release of drugs with linked amino acid residues may show different activity than free drug) or the conjugate itself may become highly active (Yang et al., 2014). Peptide–drug conjugates, with PEG chains as a linker have flexibility to allow interaction of targeting peptide with its receptors and also improve the hydrophilicity and provide longer circulation time to conjugates (Ravel et al., 2008; Tai, Shukla, Qin, Li, & Cheng, 2011). Few thoroughly researched peptide–drug conjugates are discussed below and physicochemical properties and pharmacological outcomes of some are depicted in Table 4.

3.1. RGD peptide–drug conjugates

Peptide sequences with arginine-glycine-aspartic acid (RGD) motif show a strong affinity for integrins particularly $\alpha_v\beta_3$. Integrins are transmembrane glycoproteins involved in interaction of cells with other cells or with extracellular matrix. They are actively expressed on surfaces of vascular endothelial cells and play important role in angiogenesis, leucocyte migration, and tumor metastasis. This makes integrins a suitable target for treatment of various inflammatory diseases and cancer. Several RGD motif peptides are available which can be used for synthesizing peptide–drug conjugates which include c(N-Me-VRGDf)—cyclo(MeVal-Arg-Gly-Asp-D-Phe), RGD-4C—Cys-Phe-Cys-Asp-Gly-Arg-Cys-Asp-Cys, c(RGDyK)—cyclo(Arg-Gly-Asp-D-Tyr-Lys), c(RGDfC)—cyclo(Arg-Gly-Asp-D-Phe-Cys), c(RGDfK)—cyclo(Arg-Gly-Asp-D-Phe-Lys), etc. Apart from the RGD sequence peptides, NGR sequence peptides have also shown targeted delivery to integrin expressing vasculature and integrin positive tumor cells which can be used as alternatives to RGD peptides.

3.2. Cell penetrating peptide–drug conjugates

Identification of 16 amino acid peptide sequence, penetratin, with exceptional cell penetrating capability led to development of so-called "cell

Table 4 Peptide–drug conjugates and their physicochemical and biological properties

Peptide–drug conjugate	Physicochemical and biological properties	Refs.
RGD peptide and related conjugates		
Acyclic RGD4C-doxorubicin	• Necropsy in *breast cancer* xenograft mice model showed higher antitumor activity with significantly reduced lymph node and lung metastasis as compared to free doxorubicin • Antitumor activity through RGD-mediated tumor vasculature targeting as well as uptake in *integrin-mediated uptake* in integrin expressing cancer cells	Arap, Pasqualini, and Ruoslahti (1998)
Acyclic RGD4C-5″-formyldoxsaliform and cyclic-(N-Me-VRGDf)-5″-formyldoxsaliform	• $\alpha_v\beta_3$-mediated uptake through tumor vasculature followed by hydrolysis of the *N*-Mannich base releasing the more active doxsaliform extracellularly • Instability of acyclic RGD-4C due to four sulfhydryl groups • Cyclic RGD showed 0.9-fold higher cytotoxicity than acyclic RGD conjugate having same cytotoxicity as free doxsaliform (IC_{50} 40–60 nM) in breast cancer cell line	Burkhart et al. (2004)
Acyclic RGD4C-doxorubicin	• Low cytotoxicity *in vitro* in hepatoma cells as compared to free doxorubicin • $\sim 3\times$ tumor volume reduction in hepatoma xenograft mice model	Kim and Lee (2004)
RGD4C-Aca-Ala-Phy-Lys-PABC-doxorubicin	• Tumor-associated plasmin cleavable Ala-Phy-Sys linker and self-immolative PABC provided release of free doxorubicin • $\alpha_v\beta_3$- and $\alpha_v\beta_5$-mediated uptake in human umbilical vein endothelial cells • Cytotoxicity equivalent to free doxorubicin in presence of plasmin to HUVECs and fibrosarcoma cells	de Groot et al. (2002)
CNGRC peptide-GG-KLAKLAKKLAKLAK and RGD4C-GG-KLAKLAKKLAKLAK	• KLAKLAKKLAKLAK was used as a pro-apoptotic mitochondrial membrane disrupting domain • GG (glycinylglycine) bridge was used to impart flexibility and minimize steric effect on activity of CNGRC peptide and membrane disrupting peptide • Specific killing of angiogenic cells but not of angiostatic cells • Significant reduction in tumor and metastatic burden in breast cancer mice model • *In vivo* studies in mice showed conjugates' nontoxicity and nonimmunogenicity	Ellerby et al. (1999)

Continued

Table 4 Peptide–drug conjugates and their physicochemical and biological properties—cont'd

Peptide–drug conjugate	Physicochemical and biological properties	Refs.
E[c(RGDyK)]$_2$-paclitaxel	• Conjugation of paclitaxel slightly reduced the binding affinity of peptide • Higher cytotoxicity of conjugate as compared to paclitaxel • Higher tumor distribution in breast cancer xenograft mice model as compared to other organs	Chen, Plasencia, Hou, and Neamati (2005)
E[c(RGDyK)]$_2$-paclitaxel	• Higher tumor uptake and retention for long time in breast cancer xenograft mice model as compared to free drug • Conjugate and paclitaxel both reduced tumor growth but tumor still showed growth after multiple dosing	Cao et al. (2008)
c(RGDyk)-PEG-PD0325901	• Conjugates of alkoxamine analogues of MEK1/2 kinase inhibitor—PD0325901 with mono-RGD or dimer-RGD and with/without PEG linker were synthesized • PEGylated conjugate showed highest kinase inhibition as compared to equipotent conjugates with mono-RGD and dimer-RGD in skin melanoma cell line • RGD-conjugated drugs were less cytotoxic than parent drug but RGD-mediated cellular uptake and PEG-mediated higher kinase inhibition improved activity	Li et al. (2013)
CNGRC peptide-doxorubicin	• Antitumor activity through NGR-mediated tumor vasculature targeting • 4–5 × reduced tumor size in breast cancer xenograft mice model	Arap et al. (1998)
Cell penetrating peptide conjugates		
Octaarginine-doxorubicin (VelPPP)$_3$-doxorubicin	• Cationic octaargninine CPP and amphipathic (VelPPP)$_3$ CPP conjugated with doxorubicin though oxime linkage • Glutathione-mediated drug release of free doxorubicin and thiol containing doxorubicin (active metabolite) • Higher cytotoxicity as compared to doxorubicin in breast cancer and colon cancer cells	Lelle, Frick, Steinbrink, and Peneva (2014)

Heptaarginine-cyclosporin A	• Conjugation through alkaline pH sensitive linker • Conjugate was water soluble and stable in physiological condition • Penetration through human skin and mouse skin after dermal application • Conjugate showed decreased inflammation in contact dermatitis mouse model • Phase I and II trials showed that slow release, though higher and rapid uptake, could not compete with clearance	Rothbard et al. (2000)
KAI-9803	• Conjugate of δPKC inhibitor with 11 amino acid arginine rich TAT_{47-57} peptide via disulfide bond feasible for intravenous administration in acute myocardial infarction • TAT-mediated uptake in cardiomyocytes • In Phase I and II studies, patients experienced less damage to cardiac tissue • Conjugate has the Fast Track designation from FDA for this indication • In Phase IIb trials for ST-segment elevation myocardial infarction, • Conjugate showed acceptable safety and tolerability for intracoronary injection during percutaneous coronary intervention and have shown cardioprotective effect	Bates et al. (2008), Miyaji et al. (2011), and Zijlstra (2008)
KAI-1678	• Conjugate of ϵPKC inhibitor with 11 amino acid arginine rich TAT_{47-57} peptide via disulfide bond feasible for subcutaneous administration • Conjugate was highly effective in preclinical models of neuropathic and inflammatory pain • In Phase I/IIa studies it showed that it is safe and tolerable but does not improve pain due to postherpetic neuralgia and postoperative orthopedic pain	Cousins, Pickthorn, Huang, Critchley, and Bell (2013) and Moodie, Bisley, Huang, Pickthorn, and Bell (2013)
KAI-1455	• Conjugate of ϵPKC inhibitor with 11 amino acid arginine rich TAT_{47-57} peptide via disulfide bond feasible for intravenous administration to reduce ischemic injury during coronary artery bypass grafting, vascular surgery, and pediatric cardiac surgery • Preclinical models have shown reduced injury to cardiac tissue • Currently it is to enter Phase I clinical trials	BusinessWire Report (2007)

Continued

Table 4 Peptide–drug conjugates and their physicochemical and biological properties—cont'd

Peptide–drug conjugate	Physicochemical and biological properties	Refs.
$(DGG)_3DG$-PLGLAG-R_9C-doxorubicin	• $(DGG)_3DG$ acts as an attenuating anionic peptide which binds with cationic R_9 though *PLGLAG linker* which is *MMP-2/9 cleavable* • Breast cancer cell culture pretreated with MMP-2 showed three-fold higher uptake of doxorubicin due to enzyme cleaved activated R_9C-doxorubicin conjugate • Equitoxic as compared to doxorubicin in fibrosarcoma and breast carcinoma cell lines	Shi, Gao, Xiang, and Qi (2012)
DPV1047-SN48 (DTS-108)	• Active metabolite of irinotecan (SN48) conjugated to a 20 amino acid Vectocell peptide via *esterase cleavable* linker • Provides high levels of SN48 as compared to free irinotecan without causing gastrointestinal disturbances • Synergistic antitumor activity with 5-FU in colon carcinoma xenograft mice model • Increased therapeutic window in preclinical models and in Phase I trials with no dose limiting diarrhea	Meyer-Losic et al. (2008) and Romain Coriat et al. (2012)
DPV15-paclitaxel DPV3-paclitaxel	• Water-soluble conjugate of paclitaxel and DPV via an ester linker • More activity than paclitaxel in paclitaxel resistant uterine sarcoma cells but was less active than paclitaxel in paclitaxel sensitive ovarian carcinoma and nonsmall cell lung cell carcinoma cells *in vitro* • Possibility to overcome drug resistance using conjugates	Avrameas (2005)
Other peptide–drug conjugates		
HyM-10-chlorambucil and HQ-10-chlorambucil	• Peptide driven highly efficient nucleic acid degradation • Peptide-mediated binding to DNA places chlorambucil close to the cleavage site causing sequence-specific degradation of DNA	Yang et al. (2014)
DTS-201	• Endopepeptidases in vicinity of tumor release of Ala-Leu-doxorubicin and Leu-doxorubicin which converts to free doxorubicin after cell uptake • Improved efficacy in prostate, breast, and lung tumor xenograft models • Phase I studies have shown its safety and efficacy	Drugs.com Report (2007)

N-glutaryl-(4-hydroxyprolyl)-Ala-Ser-cyclohexaglycyl-Gln-Ser-Leu-doxorubicin (L-377,202)	• Release of cytotoxic Leu-doxorubicin by PSA secreting prostate cancer cell lines • *In vivo* activity in xenograft mice model with PSA secreting tumor only • Conjugate was nontoxic and with improved therapeutic index as well as low cardiac accumulation (low cardiotoxicity)	DeFeo-Jones et al. (2000)
KCCYSL-GGG-SSKYQSL-TGXD1	• The conjugate was HER2 targeting peptide-GGG peptide spacer-PSA cleavable linker–self-cyclizing linker-phosphoinositide 3 kinases β inhibitor (TGX-D1) • PSA-mediated release of SL-TGXD1 which undergoes self-cyclization reaction to release free drug • HER2-mediated targeting and cytotoxicity to prostate cancer cells SL-TGXD1 exhibited a higher cellular uptake than SSKYQSL-TGXD1 and TGXD1 *in vitro*	Tai et al. (2011)
cIBR peptide-doxorubicin, cIBR peptide-FITC, and cIBR-PEG-doxorubicin	• Passive diffusion of doxorubicin conjugate through leukemia and HUVEC cells due to high lipophilicity • Uptake was not mediated by leukocyte function antigen-1 but only by passive diffusion	Majumdar, Kobayashi, Krise, and Siahaan (2007) and Majumdar et al. (2009)
D-(Cys-Ser-Lys-Cys)-IRS1 siRNA	• siRNA was found to be conjugated to peptide mimetic of insulin like growth factor-1 • *In vitro* transfection was achieved without use of any cationic phospholipids or physical techniques like electroporation	Cesarone, Edupuganti, Chen, and Wickstrom (2007)

RGD4C = Cys-Phe-Cys-Asp-Gly-Arg-Cys-Asp-Cys, CNRGC peptide = CNGRCVSGCAGRC, C(RGDyK) = cyclo(Arg-Gly-Asp-D-Tyr-Lys), c(RGDfC) = cyclo(Arg-Gly-Asp-D-Phe-Cys), c(RGDfK) = cyclo(Arg-Gly-Asp-D-Phe-Lys), cIBR = cyclo(1,12)PenPRGGSVLVTGC, octaarginine peptide—Cys-Arg-Arg-Arg-Arg-Arg-Arg-Arg-Arg, (VelPPP)3 = CGGWVELPPPVELPPPVEPPP, HyM-10 = Met-Hyp-Arg-Lys-(Py)4-Lys-Arg, HQ-10 = Hyp-Arg-Lys-(Py)4-Lys-Arg, DTS-201 = *N*-succinyl-β-alanyl-L-leucyl-L-alanyl-L-leucyl-doxorubicin, DPV-1047 = VKRGLKLRHVRPRVTRMDV, DPV-15 = LRRERQSRLRRERQSR, DPV-3 = RKKRRRESRKKRRRES-Cys.

penetrating peptides". These peptides share common attributes of hydrophobicity, cationic nature, and amphipathicity and provide the means for transporting drugs across the cellular membrane. Diverse class of drugs that can be attached to CPPs include pDNAs, siRNAs, proteins, peptides, low molecular weight drugs and antibiotics to larger carrier systems. Among several available CPPs, few of prime importance are R_9, TAT, pVEC, penetratin, transportan, Vectocell® (Diatos Peptide Vectors), etc. (Madani, Lindberg, Langel, Futaki, & Graslund, 2011). These arginine-based peptide sequences show cellular uptake and cytosolic release predominantly by direct transmembrane travel or by endocytosis (Guterstam et al., 2009; Madani et al., 2011). Some novel human derived cell penetrating peptide sequences like Vectocell showed potential for nuclear delivery (De Coupade et al., 2005).

The only disadvantage of the CPP-based conjugates is their nonspecific cell uptake due to their cationic nature, by almost all cells in the body. Use of activable CPPs consisting of cationic portion counteracted with anionic peptide sequence can reduce its uptake by normal cells and provide better penetration to tumor cells (Shi et al., 2012). Such strategy can be further augmented by placing a tumor-specific enzyme cleavable linker between cationic CPP and anionic peptide which would trigger the release of cationic CPP conjugated drug inside the cancer cells only. Additionally, an arginine oligopeptide-based conjugate has shown enhanced penetration through skin barriers on dermal application making them feasible ligands for enhancing topical absorption of poorly absorbable drugs (Rothbard et al., 2000).

4. STRATEGIES FOR CHEMICAL CONJUGATION

Although replete with variety of amino acids, from the perspective of conjugation, the chemical groups on proteins available are limited, unless supported by artificial protein engineering methods (discussed later). Further, the primary sequence, steric hindrance, and its three-dimensional structure influences the conjugation procedures adopted. The target amino acids should be reactive enough such that the procedure would not require to expose protein to harsh conditions and affect its quaternary structure and receptor affinity (Pisal, Kosloski, & Balu-Iyer, 2010). Therefore, nonpolar amino acids are of no interest for conjugation. The aliphatic amino acids along with alanine, methionine, and phenyl alanine do not bear any reactive functional groups. Serine and threonine, although having hydroxyl group, are also less important due to lower nucleophilicity. The asparagine and arginine also play limited role in conjugation. Cysteines are preferred due to requirements of mild condition for reduction and subsequent conjugation (Jung & Theato, 2013).

In addition, every protein/peptide has a preferred location within them and desired amino acid for conjugation should be available on the surface so as to be accessible to reagents (Stevens & Arkin, 1999). Therefore, amino acids with higher average surface accessibility like, lysine, aspartic acids, glutamine, glutamic acid, etc., along with N- and C-terminal amino acids are preferred for conjugation. Lysine side chain flaunts a reactive primary amine group for conjugation with added advantage of not being, generally, a part of recognition regions (Jung & Theato, 2013). However, the number of lysine sites may frequently exceed the desired protein:drug ratio leading to a heterogenous product due to lack of site-specific control (An, 2009). Therefore, it could be advantageous to target scarce sites such as cysteine to afford a homogenous product. Moreover, some structural features specific to certain proteins, i.e., cysteine-34 of albumin and sulfhydryl groups generated from disulfide bonds of antibodies are common targets for conjugation of drugs.

Conjugation to selected site can be carried out by direct and indirect methods (attachment of drug to protein/peptide with/without a linker, respectively), e.g., carbodiimide coupling, using activated esters, reductive amination, azide coupling, conjugations via isothiocyanate, maleimide or acryloyl moiety. Methotrexate–albumin conjugate (MTX-HSA) was developed by directly coupling MTX to lysine residues of HSA for collagen-induced arthritis (Wunder et al., 2003). Additionally, direct coupling technique using heterobifunctional linkers (heterobifunctional PEG linkers like Maleimide-PEG_{5000}-NHS, etc.) which have two reacting end-groups to react directly with the drug after activation with DDC/EDC and protein through maleimide-mediated thiol conjugation can also be employed (Kim et al., 2010). Such conjugation techniques can be used for conjugation of transferrin, albumin, antibodies, gelatin as well as other proteins and peptides.

Conjugates of transferrin have been synthesized using glutaraldehyde-mediated conjugation and subsequent termination through ethanolamine addition (Singh et al., 1998; Sizensky et al., 1992; Yeh & Faulk, 1984). Technique of conjugation that involves reaction of thiolated protein with maleimide derivative of drug has been exploited extensively. This strategy involves thiolation of the carrier protein to generate the protein with free sulfhydryl group using a thiolating agent followed by reaction of thiolated protein with drug which has been modified with maleimide group to form disulfide bond. This strategy has been used for transferrin, albumin, antibodies as well as other peptides. In case of transferrin, initial thiolation is required (Kratz et al., 1998); however, in case of albumin, structural elucidation has showed availability of inherently present cysteine-34 which allows direct conjugation with maleimide modified drugs and hence this strategy was further developed as albumin-binding prodrug technique (Table 5) which allowed intravenous administration of maleimide modified drugs, which on reaction with human serum albumin produces conjugate *in situ* within minutes (Kratz, 2008). Additionally, this technique developed by Kratz et al. allows formation of homogeneous stoichiometrically defined conjugates without affecting the three-dimensional protein structure. In case of antibodies, nonselective reduction of disulfides and subsequent conjugation produce nondefined antibody–drug conjugates which by their nature show different therapeutic potential, pharmacokinetic behavior, safety profile (Hamblett et al., 2004). In earlier efforts, one or more interchain cysteines were replaced with serine, thereby limiting available conjugation sites. These were used to create homogeneous ADC with two or four drug molecules per antibody (McDonagh et al., 2006). This led to development of newer novel technology called THIOMAB platform by Genentech's Timothy Fong which involves selective inclusion of two free cysteines in antibodies allowing controlled formation of thiol groups which form stoichiometrically defined conjugates (Junutula et al., 2008). Currently, there are three strategies (Fig. 3) that will add on to the next generation of ADCs' synthesis. These are insertion of cysteine residues in the antibody sequence by mutation or insertion, insertion of an unnatural amino acid with a bioorthogonal reactive handle and enzymatic conjugation to produce homogeneous ADCs (Perez et al., 2014).

Some drugs also couple directly with the proteins *in vitro*. Example includes cisplatin–transferrin conjugate which is prepared by direct reaction of cisplatin in presence of bicarbonate with iron-free transferrin (which has ability to bind with metal ions) (Stjernholm, 1986). This requires presence of externally added excess of cysteine which protects transferrin's sulfhydryl groups. This technique yields drug loading of 2 mol of cisplatin/mol of transferrin.

Another strategy that is specifically directed to coupling of therapeutically active peptides/proteins is use of recombinant DNA technology which is termed protein fusion technology. Through DNA recombination, one

Table 5 Albumin-binding prodrugs

Prodrug (conjugating moiety-linker drug)	Drug release mechanism	Physicochemical and biological properties	Refs.
Maleimidoalkanoyl-hydrazone-doxorubicin	Acid catalyzed hydrolysis	Alkanoyl group with $(CH_2)_2$, $(CH_2)_3$, $(CH_2)_5$, and $(CH_2)_7$ aliphatic spacer Prodrug with $(CH_2)_5$ (maleimidocaproyl-hydrazone-doxorubicin (EMC-hydrazone-doxorubicin)) was rapidly and selectively bound to cysteine-34 of albumin Controlled release of free doxorubicin Higher IC_{50} as compared to free drug due to controlled release Higher efficacy in breast cancer and renal cancer mouse xenograft models	Kratz et al. (2002)
Maleimide-oligopolyehtyleneglycol-camptothecin	Hydrolysis/ enzymatic cleavage	3–6 (OCH_2CH_2) units were introduced in the prodrug Introduction of ethyloxy groups increased water solubility of prodrugs Major metabolic products of conjugate are camtothecin and camptothecin-20-*O*-glycinate Moderate activity in human colon cancer xenograft model in mice as compared to	Warnecke and Kratz (2003)
Maleimido-oxacaproyl-cyclobutanedicarboxylate-carboplatin analogues	—	Low cytotoxicity as compared to carboplatin *in vitro* 75 mg/kg carboplatin equivalent dose showed antitumor activity equipotent to carboplatin in breast cancer xenograft mice model 100 mg/kg carboplatin equivalent dose showed 2.5- to 4-fold antitumor activity as compared to carboplatin	Warnecke, Fichtner, Garmann, Jaehde, and Kratz (2004)

Continued

Table 5 Albumin-binding prodrugs—cont'd

Prodrug (conjugating moiety-linker drug)	Drug release mechanism	Physicochemical and biological properties	Refs.
MT-Ser-Ser-Tyr-Tyr-Ser-Gly-doxorubicin, MT-Asn-Ser-Ser-Tyr-Phe-Gln-doxorubicin MC-Arg-Arg-Ser-Ser-Tyr-Tyr_Ser-Gly-doxorubicin	PSA	Conjugates containing arginine residues showed better water solubility for intravenous administration Controlled release of less active Ser-Gly-doxorubicin or Phe-Gln-doxorubicin by PSA but no release of free doxorubicin Conjugate was stable in human plasma	Kratz et al. (2005)
EMC-Gly-Gly-Gly-Arg-Arg-doxorubicin	Urokinase-type plasminogen activator (uPA)	Conjugate was stable in human plasma Controlled release of Arg-doxorubicin by uPA *in vitro* Major metabolic product of Arg-doxorubicin in breast cancer carcinoma xenograft tumor tissue homogenates at pH 7.4 LD_{50} of 72 mg/kg of doxorubicin equivalents in mice ($4.5\times$ higher than standard doxorubicin dose of used in tumor xenograft model)	Chung and Kratz (2006)
EMC-Arg-Arg-Ala-Phe-Met-Ala-camptothecin, EMC-Arg-Arg-Phe-Tyr-Met-Ala-camptothecin	Nonspecific proteases and esterases	Incorporation of two arginine residues improved water solubility Use of alanine residue improved release of drug Not selectively cleaved in colon tumor but showed enhanced antitumor activity in colon cancer xenograft model Major metabolic product of hydroxycarboxylate form of camptothecin and Ala-CPT in colon cancer xenograft tumor tissue homogenates Overall degradation of conjugates was higher at pH 5 (>70%) than at pH 7.4 (>50%)	Schmid, Warnecke, Fichtner, Jung, and Kratz (2007)

EMC-Arg-Arg-Ala-Leu-Ala-Leu-doxorubicin	Cathepsin B	Cleaved to H-Leu-Ala-Leu-doxorubicin, H-Leu-doxorubicin, and doxorubicin Improved therapeutic index H-Leu-doxorubicin active against breast cancer cells, fibrosarcoma cells, and prostate cancer cells H-Leu-doxorubicin less cardiotoxic than doxorubicin in animal models (de Jong, Geijssen, Munniksma, Vermorken, & van der Vijgh, 1992; Deprez-DeCampeneere, Jaenke, & Trouet, 1982) Equipotent to free doxorubicin in breast cancer xenograft model	Schmid, Chung, Warnecke, Fichtner, and Kratz (2007)
EMC-Arg-Arg-Ala-Leu-Ala-Leu-Ala-camptothecin	Cathepsin B	Cleaved to camptothecin-peptides and camptothecin Improved therapeutic index Superior to free camptothecin in breast cancer xenograft model	Schmid, Chung, et al. (2007)
EMC-hydrazone-doxorubicin	Acid catalyzed hydrolysis	$>5\times$ higher LD_{50} in mice, >2.2–$4.4\times$ higher LD_{50} in rats, of conjugated doxorubicin as compared to free doxorubicin Delayed peripheral neuropathy NOAEL of 10 mg/kg of doxorubicin equivalent conjugate in rats (which is equivalent to MTD of doxorubicin) NOAEL of 3 mg/kg of doxorubicin equivalent conjugate in Beagle dogs (which is equivalent to twice the MTD of doxorubicin)	Kratz, Ehling, Kauffmann, and Unger (2007)

Continued

Table 5 Albumin-binding prodrugs—cont'd

Prodrug (conjugating moiety-linker drug)	Drug release mechanism	Physicochemical and biological properties	Refs.
EMC-hydrazone-doxorubicin	Acid catalyzed hydrolysis	Phase I study in cancer patients showed that: it was well tolerated up to 200 mg/m^2 doxorubicin equivalents without drug related side effects No cardiotoxicity was observed Conjugated showed low clearance, a small volume of distribution, and high plasma AUC Antitumor activity spectrum similar to doxorubicin	Unger et al. (2007)
EMC-Ala-Phe-Lys-Lys-methotexate	Cathepsin B and plasmin	Cleaved in cathepsin B and plasmin and in ovarian carcinoma xenograft tissue homogenate in Lys-methotrexate Lys-methotrexate was more active than methotrexate in xenograft model 60 mg/kg methotrexate equivalents of conjugate showed higher antitumor efficacy than methotrexate at a dose of 400 mg/kg	Warnecke, Fichtner, Sass, and Kratz (2007)
EMC-Arg-Ser-Ser-Tyr-Tyr-Ser-Arg-doxorubicin	Prostate-specific antigen (PSA)	Albumin conjugate released H-Ser-Arg-doxorubicin in PSA positive cancer cells Conjugate was 100-fold less active against LNCaP cells than the free drug *in vitro* High efficacy in prostate cancer xenograft model with significant reduction in metastasis in lung	Graeser et al. (2008)
EMC-Phe-Lys-PABC-doxorubicin	Cathepsin B	PABC-doxorubicin acts as a self-immolative spacer Breast cancer and pancreatic cancers were $\sim 6\times$ less sensitive to conjugate than free doxorubicin MTD of conjugate was 72 mg/kg of doxorubicin equivalent conjugate in nude mice breast cancer model (which is $4.5\times$ the MTD of doxorubicin)	Abu Ajaj, Graeser, Fichtner, and Kratz (2009)

EMC-Arg-Ser-Ser-Tyr-Tyr-Ser-Leu-PABC-paclitaxel	PSA	PSA-mediated release to Ser-Leu-PABC-paclitaxel and then to PABC-paclitaxel which released free paclitaxel Albumin bound drug was 3 × more active in killing PSA positive prostate cancer cells *in vitro* MTD twice that of paclitaxel Antitumor activity at 72 mg paclitaxel equivalent/kg in mice prostate cancer model equipotent to MTD of paclitaxel (36 mg/kg) reduced lung metastasis but no effect on bone metastasis	Elsadek et al. (2010)
EMC-Arg-Ser-Ser-Tyr-Tyr-Ser-Arg-PABC-doxorubicin	PSA	PSA released Ser-Leu-PABC-doxorubicin which was further degraded to PABC-doxorubicin by adventitious tumor proteases. PABC-doxorubicin spontaneously releases free doxorubicin Ser-Leu-PABC-doxorubicin was more active than Ser-Arg-doxorubicin and Ser-Leu-doxorubicin without PABC spacer No activity of free doxorubicin in primary cancer while conjugate showed 30–50% reduction in primary tumor and significantly reduced lung and bone metastasis	Elsadek et al. (2011)

MT, maleimidotriethylenglycol acid; EMC, ε-maleimidocaproic acid; PABC, *para*-aminobenzyloxycarbonyl.

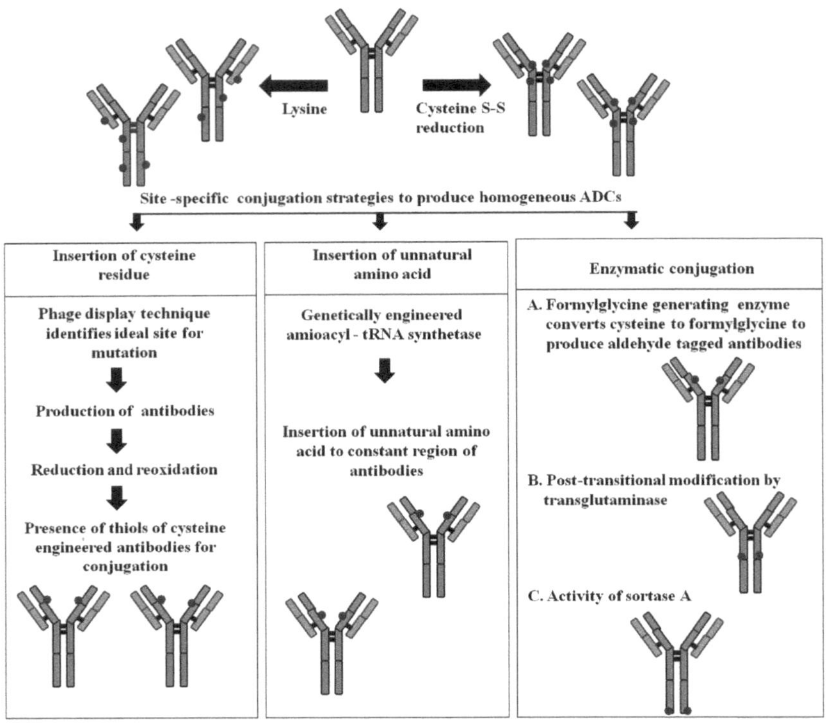

Figure 3 Site-specific conjugation strategies to produce homogeneous ADCs.

can easily synthesize bioactive cytokines (interferons, interleukins, etc.) directly fused with carrier proteins which could be either albumin or antibodies (Andrady, Sharma, & Chester, 2011). This also enables introduction of binding sites in proteins that can be utilized for conjugation of drugs (Andrady et al., 2011). Several examples of protein–drug conjugates synthesized using this technology include albumin-interferon α-2b fusion protein (Albuferon), balbumin-intrleukin-2 fusion protein (Albeleukin), and TIMP-2-albumin fusion protein (fusion protein of tissue MMP inhibitor), a albumin-antibody fusion protein targeting epidermal growth factor receptors and HER-3 (MM-111) and transferrin-IgG fusion protein. These fusion proteins inhibit plasma clearance of therapeutic proteins, protect them from peptidases and improve therapeutic potential (Lee et al., 2012). Additionally, such fusion protein, if prepared with specific targeting protein, also shows targeting potential to specific organ, i.e., transferrin-IgG fusion protein has shown improved brain targeting (Shin et al., 1995). Such technology can afford targeting of immunoglobulins in brain disorders.

As in case of proteins, similar strategies of conjugation apply for peptide–drug conjugates. However, type of conjugation may have significant effect on maintaining peptide binding affinity, drug activity and conjugate stability. Compared to proteins, peptides offer wide choice of chemistries ranging from esters, hydrazones, amides and thioethers to carbamates. Amide and thioether-linked conjugates are known for their high stability and may even imped the drug release, in contrast, ester-linked conjugates are less stable and may release drug before reaching target due to esterase activity in plasma. Carbamates have considerably more stability than ester-based conjugates and can be used for drugs with hydroxyl groups (Majumdar & Siahaan, 2012). However, key consideration has to be given to not to affect the original conformation of peptide after conjugation, therefore cyclic RGDs are preferred due to their high conformational rigidity. Regioselective oximation reaction between hydroxylamine ether coupled to the carboxyl terminus of peptide with formyl group of drugs (salicylamide groups of doxsaliform) (Burkhart et al., 2004). As in case of fusion technology used for proteins, therapeutic peptides can be coupled with carrier peptides but via synthetic routes. Such fusion peptides can be designed using computer aided designing of homing peptide (targeting peptide), a linking peptide (flexible linker), and therapeutic peptide and unlike antibodies, this avoids use of phase display and extensive immunization procedures (Ellerby et al., 1999). Additionally, it has been shown that one should consider the stereochemistry of peptides as well. Peptides with L-amino acids are involved in receptor interaction (Ellerby et al., 1999) and hence, should be used for targeting peptide synthesis and those with D-amino acids show protease resistance (Bessalle, Kapitkovsky, Gorea, Shalit, & Fridkin, 1990; Dathe et al., 1997). Additionally, activation of drugs with succinic anhydride to form active esters of drug which can further be coupled with EDC/NHS-mediated conjugation with protein have also been used (Cao et al., 2008). Heterobifunctional linkers which flaunt reactive groups at two ends can also be used for conjugation; one group reacts with the drug and other with peptide (Lelle et al., 2014).

4.1. Linker technologies

Soon after development of chemical conjugation strategies it was realized that choice of linker that covalently tethers antibody to cytotoxic drug was critical in development of successful conjugates owing to its effect on stability, modulation of release and pharmacokinetics. As with conjugation techniques, since each drug molecule and targets have different chemical constraints, the choice of linker is target dependent and influenced by the drug molecule used, site of conjugation, number of attachment sites available as well as the cleavability and the polarity of the linker. To simplify, these linkers can be broadly categories as cleavable and noncleavable.

4.1.1 Cleavable linkers

Cleavable linkers provide additional specificity to the conjugates through specific enzyme-mediated cleavage of drug/drug metabolites inside the target tissue. Most common chemically labile cleavable linkers include hydrazone and disulfide linkers which cleave in response to different physiologic conditions in plasma and site of action. The hydrazone linker responds to the low pH in endosomes to trigger the hydrolysis (van Der Velden et al., 2001). A number of early ADCs of different cytotoxic drugs (daunomycin, vinca alkaloid vinblastine DAVLB) used acid-labile linkages between the drug and carbohydrate residues of monoclonal antibodies. Pfizer's gemtuzumab ozogamicin (Mylotarg®), using the hydrazone linker technology, was the first ADC to successfully complete clinical trials (Sievers et al., 1999) and received regulatory approval by the U.S. Food and Drug Administration (FDA) in 2000 for use in relapsed acute myelocytic leukemia (AML). However, Mylotarg® showed limited success and was withdrawn from the market in 2010 due to a narrow therapeutic window and lack of target dependence.

Another chemically labile linker extensively explored in the development of protein or peptide–drug conjugates is disulfide bond. Disulfides are thermodynamically stable in nonreducing environment of plasma, however, trigger the drug release upon internalization inside cells, where the cytosol provides a significantly more reducing environment (Erickson et al., 2006; Meister & Anderson, 1983). The scission of disulfide bonds requires the presence of a cytoplasmic thiol-cofactor, such as (reduced) glutathione (GSH), which exists at a 1000-fold excess concentration in cytoplasm than in plasma (Erickson et al., 2006; Kovtun et al., 2006; Lewis Phillips et al., 2008). In addition, this is advantageous in cancer where hypoxic conditions in tumors result in enhanced activity of reductive enzymes and therefore higher glutathione concentrations (de Groot, Damen, & Scheeren, 2001; Russo, DeGraff, Friedman, & Mitchell, 1986). *N*-succinimidyl-4-(2-pyridyldithio)pentanoate (SPP), *N*-succinimidyl-4-(2-pyridyldithio)butyrate (SPDB) and 4-(4′-acetylphenoxy)butanoic acid (AcBut) linkers are disulfide containing linkers (Kitson et al., 2013). Some researchers have also developed ADCs using hydrazone linkage between the drug (doxorubicin) and thiolated lysine residue of an antibody with two cleavable sites containing a disulfide linkage in addition to the hydrazone in the linker arm (Greenfield et al., 1990). Although these conjugates were active in an antigen-specific manner, they had poor *in vivo* potency (Braslawsky, Edson, Pearce, Kaneko, & Greenfield, 1990; Trail et al., 1992). Chemically labile linkers are mostly associated with limited stability in circulation. However, their stability can be manipulated by introducing steric hindrance by using substituents near the linkage (Thorpe et al., 1987).

Enzymatically cleavable peptide linkers possess better plasma stability and can be triggered by lysosomal proteases such as cathepsin B and plasmin. These proteases have very low activities in blood due to endogenous inhibitors and unfavorably high pH of blood (Ciechanover, 2005). Early peptide linkers were tetrapeptides like, Gly-Phe-Leu-Gly (Koblinski, Ahram, & Sloane, 2000) and Ala-Leu-Ala-Leu (Studer, Kroger, DeNardo, Kukis, & Meares, 1992; Versluis, Rump, Rensen, Van Berkel, & Bijsterbosch, 1998), but, they release drug slowly and their hydrophobic nature leads to aggregation. On the other hand, dipeptide-based linkers Val-Cit and Phe-Lys are reasonably stable under physiological conditions and undergo rapid hydrolysis in the presence of lysosomal cathepsin B (Kirschke, Barrett, Rawlings, et al., 1995; Otto & Schirmeister, 1997). *In vivo* half-life of drug from Va-Cit and Phe-Lys peptide-based conjugates were higher than the hydrazone linkers (230, 80, and 2.6 days, respectively) (Castaneda et al., 2013; Senter, 2009). Seattle Genetics' Brentuximab-Vedotin (Adcetris™) which was granted accelerated regulatory approval by USFDA in 2011 contains Val-Cit peptide linker. However, direct attachment of the drug to such linkers may result in slower hydrolysis, especially with bulkier drugs and may result in release of amino acid adduct of the drug, thereby perhaps reducing the cytotoxic activity. To overcome such effect, a self-immolative spacer was designed to spatially separate the drug from the site of enzymatic cleavage (Fig. 4). The subsequent crumple of the incorporated spacer releases fully active, chemically unmodified drug. One of the most commonly used spacers is the bifunctional *p*-aminobenzyl alcohol group, which is linked to the peptide through the amino group, forming an amide bond, while amine containing cytotoxic drugs are attached through carbamate functionalities to the benzylic hydroxyl group of the linker (PABC). The resulting prodrugs are activated upon protease-mediated cleavage, leading to a 1,6-elimination reaction releasing

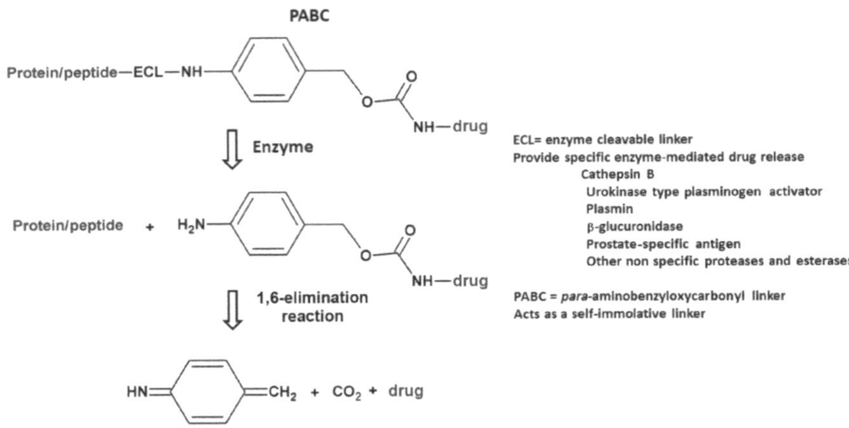

Figure 4 Release of drug from conjugate through enzyme cleavable linker attached with drug via self-immolative PABC linker.

the unmodified drug, carbon dioxide, and remnants of the linker group (Fig. 4; Toki, Cerveny, Wahl, & Senter, 2002).

β-Glucuronic acid-based linker makes use of lysosomal β-glucuronidase which is overexpressed in some tumor types (Albin et al., 1993; Jeffrey et al., 2006), while the enzyme activity outside cells is low. Furthermore, incorporating the highly hydrophilic nature of β-glucuronic acid circumvented the tendency of some ADCs to undergo aggregation and improved stability, especially for those with highly hydrophobic drugs (e.g., DNA minor-groove binders) (Jeffrey et al., 2005).

4.1.2 Noncleavable linkers

The phenomena that, upon binding to the transmembrane target some ADCs get rapidly internalized and once internalized, the antibody can be degraded, amino acid by amino acid, to the point where the active drug is exposed, forms the basis for use of noncleavable linker. Noncleavable linkers include amide moieties; SMCC (succinimidyl-4-[*N*-maleimidomethyl] cyclohexane-1-carboxylate) and maleimidocaproyl (MC) moieties (Kitson et al., 2013; Nolting, 2013). Initially, researchers were convinced that the cleaving of the linker was the most reasonable way to free the drug (Lehr, 2011). A noncleavable linker can be successfully utilized only if the released drug metabolite functions as an active cytotoxic drug component (Doronina et al., 2003; Lewis Phillips et al., 2008). Further, noncleavable linkers are limited to specific tumors only, as it requires good internalization for degradation within the cell to become active, as there is no mechanism for extracellular cleavage of the linker and subsequent permeation of the drug into the cell (passive diffusion). The amino acid–drug metabolites from conjugates with noncleavable linkers are more hydrophilic and much less membrane permeable, which leads to less bystander effects and less nonspecific toxicities compared to conjugates with a cleavable linker (Erickson et al., 2006; Okeley et al., 2010). Early examples for immunoconjugates with noncleavable linkers include immunoconjugates of methotrexate (Endo et al., 1987), daunorubicin (Pimm, Paul, Ogumuyiwa, & Baldwin, 1988), the vinca alkaloids (Spearman, Goodwin, Apelgren, & Bumol, 1987), mitomycin C (Kato, Tsukada, Hara, & Hirai, 1983), idarubicin (Rowland, Pietersz, & McKenzie, 1993), and *N*-acetyl melphalan (Smyth, Pietersz, & McKenzie, 1987) via amide or succinimide spacers to different murine monoclonal antibodies. At this time, the most commonly used noncleavable linkages in antibody–drug conjugates are succinimide–thioether bonds, which are formed by the reaction of maleimides with thiols. This methodology has been applied to both of the two currently mostly used classes of toxic moieties, maytansinoids and auristatins. Taken together, the reported findings (Doronina et al., 2006) suggest that auristatin-based antibody–drug conjugates using a noncleavable linker may have broad utility for the treatment of human carcinomas.

5. CHALLENGES

Throughout the last decade, pharmaceutical scientists are captivated by the quest for enhanced local delivery to cellular compartment of diseased cell which will significantly change the dosage regimen of modern therapeutics. The peptide–drug conjugates have the potential to address this problem driven by their specificity, robust physical nature and comparatively straightforward manufacture. However, the clinical scenario reveals that there are some hurdles which are limiting the emergence of newer peptide–drug conjugates. Major drawback of the peptide drugs is high proteolytic instability which results into poor pharmacokinetic aspects with rapid elimination due to small sizes through renal excretory system. Chemical modification of the peptide either within the sequence or through conjugation to polymer or lipid can be considered as an approach to solve the stability issues (Otvos & Wade, 2014). Serum stability assays can be used as an indicator of their stability and pharmacokinetic behavior (Powell, Grey, Gaeta, Sette, & Colon, 1992). Incorporation of peptide–drug conjugate into a biodegradable sustained release matrix like poly-D,L-lactide-*co*-glycolide (PLGA) polymer or a hydrogel can be used to provide programmed biphasic or multiphasic release in plasma. Improvements in peptide penetration through biological barriers can be achieved by adding passive or active transporter, e.g., incorporation of positively charged amino acids at terminal positions.

Additionally, for peptide–drug conjugates, an alteration of receptor binding affinity of peptide has been observed after conjugation. Conjugation of DOX to bombesin receptor type 2 antagonist-RC-3095 resulted in twice reduced affinity while conjugation to the receptor agonist-RC-3094 resulted in 1000 × increase in affinity. Therefore, peptide–drug conjugated may not continue the same binding properties of the original peptide due to inherent inability of the short peptide to retain the original structure and conformation at the receptor binding sites after conjugation (Majumdar & Siahaan, 2012). The problem can be overcome by use of flexible linker such as short PEG chain to allow receptor interaction or use of conformationally rigid cyclic RGD peptides which mostly retain original conformation after conjugation (Majumdar & Siahaan, 2012). Further, the conjugate formation can also result in altered rate of internalization of conjugate formed based on the epitope of target antigen, the affinity of interaction and intracellular trafficking pattern. For example, ligand-activated EGFR monomer internalizes at 100-fold greater rate than carcinoembryonic antigen. This results in the increased rate of catabolism of antibodies leading to increase in potency of drug than those targeting slow internalizing antigens. Therefore, attempts are being made with approaches that can bypass internalization, e.g., targeting extradomain-B (ED-B).

Any protein–drug conjugate should normally inherit the pharmacokinetic properties of protein component and therefore exhibit relatively low clearance and long half-lives. However, conjugates can exhibit higher clearance for higher drug:antibody ratio (DAR) than with lower ones (Hamblett et al., 2004). The variable DAR, due to heterogenous nature of product, leads to substantial alteration of clearance and overall pharmacokinetic properties of drug conjugates than unconjugated proteins. This can potentially affect the efficacy, safety, and tolerability (Lu, Sahasranaman, Zhang, & Girish, 2013). The clearance of anti-CD30-valine-citrulline-MMAE ADCs containing DAR of eight was found to be twice that of four-loaded counterpart and much more than unmodified ones. Additionally, since maximum tolerated dose is calculated from drug loading, twice as much four-loaded ADC can be administered as eight-loaded ADC providing a longer tumor exposure at optimum level. Thus, drug loading should be optimized as a mean to control the tolerability and therapeutic window of ADCs.

Formulation of ADCs with good physical stability is also a challenging task. This becomes predominant especially when attempts are made to achieve higher DAR with hydrophobic drugs. The high DAR makes the surface hydrophobic and induces attractive interaction resulting in antibody aggregation during preparation and storage (Wakankar et al., 2010). However, it has been shown that addition of short hydrophilic PEG chains as linker between antibody and cleavage element results in ADCs that retain the hydrophilicity and solubility of conjugate avoiding aggregation even at DAR of 16 drug molecules per antibody. Another linker, bac-PEG4-ValLys, when used instead of mcValLys linker, resulted in 95% reduced aggregation (Jones & McKnight, 2013).

Common challenge to both the protein and peptide–drug conjugates is that they are orally inactive and need invasive intravenous administration. The size, hydrophobicity and polarity of peptides hinder their passage across intestinal mucosa. In addition, the high metabolic capacity of intestinal proteases can digest protein and peptide backbones converting them into inactive small fragments (Majumdar & Siahaan, 2012). In spite of this, a few protein–drug conjugates have been developed using transferrin which show TfR-mediated uptake through GI tract cells (Bai et al., 2005; Xia et al., 2000). Although it might be difficult to develop an oral formulation due to unavoidable *in vivo* barriers, alternative delivery strategies through pulmonary, nasal, and transoral route, which can bypass the metabolic barrier, could be conceived for future developments to achieve better patient compliance.

Further, the quality of protein- and peptide–drug conjugates is controlled by the manufacturing process owing to their high molecular weight inducing physical instability and the fact that they are heterogenous mixture of compounds with variable drug:protein ratios (Ducry, 2012). Further, aqueous biological buffers used in conjugation process are microbial growth supportive and potential source of endotoxins and hence, may compromise the product quality and safety (Rohrer, 2012). Therefore manufacturing procedures should be developed to ensure low contaminations within the acceptable limits (Hussong, 2010).

The sensitivity and complexity of conjugates make reliable control on critical quality attributes of heterogeneous product difficult. In such case, the sensitivity and specificity of analytical techniques become highly important. The purity and homogeneity estimation generally rely on combination of separation and quantification methods like size-exclusion chromatography followed by light scattering, ultra-violet or refractive index detection. Physical instability of aggregation can be detected by sedimentation analysis, ultracentrifugation and increase in size by light scattering (Sahin & Roberts, 2012). Drug:protein ratio can be determined by UV spectrometry utilizing the differential absorption properties of carrier and drug. In case of ADCs, hydrophobic interaction chromatography is capable of separating each species differing in number of bound drug and can be used for determination of average drug molecules per antibody (Ouyang, 2013). The charge distribution revealed from capillary gel electrophoresis and isoelectric focusing can give indication of drug distribution (Wakankar, Chen, Gokarn, & Jacobson, 2011). The SDS-PAGE can allow estimating specific location of payloads on antibody chain through molecular weight-based separation of protein fragments. Finally, the functional capacity to bind the antigen is assessed by ELISA assay. The fraction of conjugated and unconjugated antibodies can be assessed from difference in the total antibody assay and ELISA-based assay (Kozak et al., 2013). The *in vitro* performance and safety can be studies by cell uptake and cytotoxicity studies, respectively.

6. FUTURE PERSPECTIVE

In spite of the complexities involved in design of protein and peptide–drug conjugates, the advantages of these therapeutic molecules offer a strong incentive for sustained interest in this field for decades. The outcomes of preclinical and clinical studies are encouraging for technological advancements in the field. The major hurdle of heterogeneity of protein–drug conjugates are among the primary focus in near future and more and more techniques for site-selective conjugation are being developed with the help of recombinant DNA technology (Perez et al., 2014). The advances in the phage display technique are good tool to identify sites for mutation which will not affect the physical stability of antibody. In addition, prospects of carbohydrates as a natural way to produce conjugates in reliable, controllable, and reproducible manner are promising. University of Georgia Research Foundation has recently patented carbohydrate conjugation method which involved modification of antibody with a carbohydrate at specific sites and utilize natural posttranslational modification of the antibody heavy chain for conjugation, i.e., a simple incubation with a glycosyl-transferase followed by metal-free click chemistry accomplishes the conjugation of biochemical compounds. Along with drugs, this method can also link other desirable biochemical compound like fluorophores, nucleic acids and proteins to the antibody.

The recent advances in the field of genetics have led to the growth of siRNA as a novel agent for targeting mRNA in cells and reducing gene expression. However, the field is struggling due to inability to deliver siRNA stably to target cells without carrier-associated toxicity. Peptide conjugate of 3′-modified siRNA can be prepared by stable thioether linkage to the guiding or passenger strand of siRNA to aid specific delivery into cell or animal tissues (Shankar et al., 2008). Bio-Synthesis, Inc. commercially provides conjugates of cell penetration peptides with siRNA sense strand (OMe/DNA or RNA) through disulfide linkage and subsequently hybridized to the complementary antisense RNA strand. Further, Lang et al. have reported improved siRNA stability after conjugation of peptide to 3′-terminus of sense strand (siRNA-pS), while same conjugation at 3′-terminus of antisense strand had no effects. This indicated that serum nucleases digest siRNA duplex through the 3′-end of sense strand and peptide conjugation at this end will confer stability as well as cellular delivery (Lang Zou et al., 2014).

Finally, the quest for ideal targeting agent has led to development of DARPins (designed ankyrin repeat proteins), which are genetically engineered nonantibody-based small proteins with highly specific and high affinity protein binding (Stumpp, Binz, & Amstutz, 2008). Their property of being derived from four to five repeat motifs having small size with molecular weights in the range of 14–18 kDa enables them to penetrate deep into solid tumors and other targets outside blood circulation. DARPins are therefore thought to be ideal protein carrier to deliver toxic agents to tumors. They can bind to different epitopes simultaneously unlike antibodies makes them more effective. In spite of being used as investigational agents, in 2010, the first of these molecules, MP0112 has entered into clinical studies.

7. CONCLUSION

Protein- and peptide–drug conjugates are ideal drug delivery technology tool to modify the therapeutic regimen of drugs and to overcome the limitations of toxic drugs with unfavorable pharmacokinetics and tissue distribution profile. The success of the therapeutics under this category is strongly influenced by the differential distribution of target antigen, the choice of targeting protein or peptide, serum stability of the conjugate and drug release characteristics of the linker used. The peptides, though small in size, harbor good tissue accessibility, while proteins owing to their large size, offer long blood residence. The selection of linkers offers a means of site-selective release, preservation of drug activity as well as improvement of physical stability. The *in vitro* results and clinical scenario reflect a promising trend. The advent of monoclonal antibodies has revolutionized the field due to its exceptions selectivity for a particular target antigen. Although challenged by problems like tissue penetrability, altered binding properties and product heterogeneity, etc., the continuous growth in biotechnology is constantly involved in solving them through recombination techniques for site-selective conjugation, application of DARPins, improvement in manufacturing quality and product characterization techniques opening new way through curtain. In future, these ongoing efforts in the field will lead to widening of the impact of protein- and peptide-drug conjugates as targeted therapeutics for cancer as well as other diseases.

REFERENCES

Abu Ajaj, K., Graeser, R., Fichtner, I., & Kratz, F. (2009). In vitro and in vivo study of an albumin-binding prodrug of doxorubicin that is cleaved by cathepsin B. *Cancer Chemotherapy and Pharmacology*, *64*(2), 413–418.

Albin, N., Massaad, L., Toussaint, C., Mathieu, M. C., Morizet, J., Parise, O., et al. (1993). Main drug-metabolizing enzyme systems in human breast tumors and peritumoral tissues. *Cancer Research*, *53*(15), 3541–3546.

An, Z. (2009). *Therapeutic monoclonal antibodies: From bench to clinic.* Hoboken, New Jersey: John Wiley & Sons, Inc.

Anderson, W. L., & Tomasi, T. B. (1988). Polymer modification of antibody to eliminate immune complex and Fc binding. *Journal of Immunological Methods*, *109*(1), 37–42.

Andrady, C., Sharma, S. K., & Chester, K. A. (2011). Antibody-enzyme fusion proteins for cancer therapy. *Immunotherapy*, *3*(2), 193–211.

Arap, W., Pasqualini, R., & Ruoslahti, E. (1998). Cancer treatment by targeted drug delivery to tumor vasculature in a mouse model. *Science*, *279*(5349), 377–380.

Avrameas, A. (2005). *Amino acid sequences facilitating penetration of a substance of interest into cells and/or cell nuclei: Google Patents.*

Bai, Y., Ann, D. K., & Shen, W.-C. (2005). Recombinant granulocyte colony-stimulating factor-transferrin fusion protein as an oral myelopoietic agent. *Proceedings of the National Academy of Sciences of the United States of America*, *102*(20), 7292–7296.

Bates, E., Bode, C., Costa, M., Gibson, C. M., Granger, C., Green, C., et al. (2008). Intracoronary KAI-9803 as an adjunct to primary percutaneous coronary intervention for acute ST-segment elevation myocardial infarction. *Circulation*, *117*(7), 886–896.

Beck, A., Haeuw, J. F., Wurch, T., Goetsch, L., Bailly, C., & Corvaia, N. (2010). The next generation of antibody-drug conjugates comes of age. *Discovery Medicine*, *10*(53), 329–339.

Bessalle, R., Kapitkovsky, A., Gorea, A., Shalit, I., & Fridkin, M. (1990). All-D-magainin: Chirality, antimicrobial activity and proteolytic resistance. *FEBS Letters*, *274*(1–2), 151–155.

Beyer, U., Roth, T., Schumacher, P., Maier, G., Unold, A., Frahm, A. W., et al. (1998). Synthesis and in vitro efficacy of transferrin conjugates of the anticancer drug chlorambucil. *Journal of Medicinal Chemistry, 41*(15), 2701–2708.

Bolling, C., Graefe, T., Lübbing, C., Jankevicius, F., Uktveris, S., Cesas, A., et al. (2006). Phase II study of MTX-HSA in combination with cisplatin as first line treatment in patients with advanced or metastatic transitional cell carcinoma. *Investigational New Drugs, 24*(6), 521–527.

Boratynski, J., Opolski, A., Wietrzyk, J., Gorski, A., & Radzikowski, C. (2000). Cytotoxic and antitumor effect of fibrinogen-methotrexate conjugate. *Cancer Letters, 148*(2), 189–195.

Braslawsky, G. R., Edson, M. A., Pearce, W., Kaneko, T., & Greenfield, R. S. (1990). Antitumor activity of adriamycin (hydrazone-linked) immunoconjugates compared with free adriamycin and specificity of tumor cell killing. *Cancer Research, 50*(20), 6608–6614.

Brookes, S., Biessels, P., Ng, N. F. L., Woods, C., Bell, D. N., & Adamson, G. (2006). Synthesis and characterization of a hemoglobin–ribavirin conjugate for targeted drug delivery. *Bioconjugate Chemistry, 17*(2), 530–537.

Burger, A. M., Hartung, G., Stehle, G., Sinn, H., & Fiebig, H. H. (2001). Pre-clinical evaluation of a methotrexate-albumin conjugate (MTX-HSA) in human tumor xenografts in vivo. *International Journal of Cancer, 92*(5), 718–724.

Burkhart, D. J., Kalet, B. T., Coleman, M. P., Post, G. C., & Koch, T. H. (2004). Doxorubicin-formaldehyde conjugates targeting αvβ3 integrin. *Molecular Cancer Therapeutics, 3*(12), 1593–1604.

Calzolari, A., Oliviero, I., Deaglio, S., Mariani, G., Biffoni, M., Sposi, N. M., et al. (2007). Transferrin receptor 2 is frequently expressed in human cancer cell lines. *Blood Cells, Molecules & Diseases, 39*(1), 82–91.

Cao, Q., Li, Z.-B., Chen, K., Wu, Z., He, L., Neamati, N., et al. (2008). Evaluation of biodistribution and anti-tumor effect of a dimeric RGD peptide–paclitaxel conjugate in mice with breast cancer. *European Journal of Nuclear Medicine and Molecular Imaging, 35*(8), 1489–1498.

Castaneda, L., Maruani, A., Schumacher, F. F., Miranda, E., Chudasama, V., Chester, K. A., et al. (2013). Acid-cleavable thiomaleamic acid linker for homogeneous antibody-drug conjugation. *Chemical Communications (Cambridge, England), 49*(74), 8187–8189.

Cesarone, G., Edupuganti, O. P., Chen, C.-P., & Wickstrom, E. (2007). Insulin receptor substrate 1 knockdown in human MCF7 ER+ breast cancer cells by nuclease-resistant IRS1 siRNA conjugated to a disulfide-bridged d-peptide analogue of insulin-like growth factor 1. *Bioconjugate Chemistry, 18*(6), 1831–1840.

Chapman, A. P. (2002). PEGylated antibodies and antibody fragments for improved therapy: A review. *Advanced Drug Delivery Reviews, 54*(4), 531–545.

Chen, X., Plasencia, C., Hou, Y., & Neamati, N. (2005). Synthesis and biological evaluation of dimeric rgd peptide–paclitaxel conjugate as a model for integrin-targeted drug delivery. *Journal of Medicinal Chemistry, 48*(4), 1098–1106.

Cho, Y. W., Park, S. A., Han, T. H., Son, D. H., Park, J. S., Oh, S. J., et al. (2007). In vivo tumor targeting and radionuclide imaging with self-assembled nanoparticles: Mechanisms, key factors, and their implications. *Biomaterials, 28*(6), 1236–1247.

Cho, S., Park, K., Kim, S. Y., Yang, J., Cho, K.-J., & Byun, Y. (2002). Gelatin-MPEG-doxorubicin conjugates for the targeting delivery based on angiogenesis. In *Paper presented at the 29th annual meeting of the Controlled Release Society Proceedings, Seoul, Korea.*

Chung, D.-E., & Kratz, F. (2006). Development of a novel albumin-binding prodrug that is cleaved by urokinase-type-plasminogen activator (uPA). *Bioorganic & Medicinal Chemistry Letters, 16*(19), 5157–5163.

Ciechanover, A. (2005). Intracellular protein degradation: From a vague idea, through the lysosome and the ubiquitin-proteasome system, and onto human diseases and drug targeting (Nobel lecture). *Angewandte Chemie (International Ed. in English)*, *44*(37), 5944–5967.

Cirillo, G., Kraemer, K., Fuessel, S., Puoci, F., Curcio, M., Spizzirri, U. G., et al. (2010). Biological activity of a gallic acid-gelatin conjugate. *Biomacromolecules*, *11*(12), 3309–3315.

Cousins, M. J., Pickthorn, K., Huang, S., Critchley, L., & Bell, G. (2013). The safety and efficacy of KAI-1678- an inhibitor of epsilon protein kinase C (epsilonPKC)-versus lidocaine and placebo for the treatment of postherpetic neuralgia: A crossover study design. *Pain Medicine*, *14*(4), 533–540.

Cox, M. E., Gleave, M. E., Zakikhani, M., Bell, R. H., Piura, E., Vickers, E., et al. (2009). Insulin receptor expression by human prostate cancers. *Prostate*, *69*(1), 33–40.

Dathe, M., Wieprecht, T., Nikolenko, H., Handel, L., Maloy, W. L., MacDonald, D. L., et al. (1997). Hydrophobicity, hydrophobic moment and angle subtended by charged residues modulate antibacterial and haemolytic activity of amphipathic helical peptides. *FEBS Letters*, *403*(2), 208–212.

De Coupade, C., Fittipaldi, A., Chagnas, V., Michel, M., Carlier, S., Tasciotti, E., et al. (2005). Novel human-derived cell-penetrating peptides for specific subcellular delivery of therapeutic biomolecules. *The Biochemical Journal*, *390*, 407–418.

DeFeo-Jones, D., Garsky, V. M., Wong, B. K., Feng, D.-M., Bolyar, T., Haskell, K., et al. (2000). A peptide-doxorubicin 'prodrug' activated by prostate-specific antigen selectively kills prostate tumor cells positive for prostate-specific antigen in vivo. *Nature Medicine*, *6*(11), 1248–1252.

de Groot, F. M. H., Broxterman, H. J., Adams, H. P. H. M., van Vliet, A., Tesser, G. I., Elderkamp, Y. W., et al. (2002). Design, synthesis, and biological evaluation of a dual tumor-specific motive containing integrin-targeted plasmin-cleavable doxorubicin prodrug. *Molecular Cancer Therapeutics*, *1*(11), 901–911, This work was partly supported by the Spinoza Award (to H. M. P.).

de Groot, F. M., Damen, E. W., & Scheeren, H. W. (2001). Anticancer prodrugs for application in monotherapy: Targeting hypoxia, tumor-associated enzymes, and receptors. *Current Medicinal Chemistry*, *8*(9), 1093–1122.

de Jong, J., Geijssen, G. J., Munniksma, C. N., Vermorken, J. B., & van der Vijgh, W. J. (1992). Plasma pharmacokinetics and pharmacodynamics of a new prodrug N-l-leucyldoxorubicin and its metabolites in a phase I clinical trial. *Journal of Clinical Oncology*, *10*(12), 1897–1906.

Deprez-DeCampeneere, D., Jaenke, R., & Trouet, A. (1982). Comparative cardiac and renal toxicity of daunorubicin in the rat and rabbit. *Cancer Treatment Reports*, *66*(2), 395–397.

Doronina, S. O., Mendelsohn, B. A., Bovee, T. D., Cerveny, C. G., Alley, S. C., Meyer, D. L., et al. (2006). Enhanced activity of monomethylauristatin F through monoclonal antibody delivery: Effects of linker technology on efficacy and toxicity. *Bioconjugate Chemistry*, *17*(1), 114–124.

Doronina, S. O., Toki, B. E., Torgov, M. Y., Mendelsohn, B. A., Cerveny, C. G., Chace, D. F., et al. (2003). Development of potent monoclonal antibody auristatin conjugates for cancer therapy. *Nature Biotechnology*, *21*(7), 778–784.

Dosio, F., Arpicco, S., Stella, B., Brusa, P., & Cattel, L. (2009). Folate-mediated targeting of albumin conjugates of paclitaxel obtained through a heterogeneous phase system. *International Journal of Pharmaceutics*, *382*(1–2), 117–123.

Dosio, F., Brusa, P., Crosasso, P., Arpicco, S., & Cattel, L. (1997). Preparation, characterization and properties in vitro and in vivo of a paclitaxel–albumin conjugate. *Journal of Controlled Release*, *47*(3), 293–304.

Diatos Announces Positive Clinical Results for its DTS-201 Doxorubicin Prodrug Program, Drugs.com, June 2007. Accessed at http://www.drugs.com/clinical_trials/diatos....dts-201-doxorubicin-prodrug-program-1121.html on 1st Oct 2014.

Du, W., Fan, Y., Zheng, N., He, B., Yuan, L., Zhang, H., et al. (2013). Transferrin receptor specific nanocarriers conjugated with functional 7 peptide for oral drug delivery. *Biomaterials*, *34*(3), 794–806.

Ducry, L. (2012). Challenges in the development and manufacturing of antibody-drug conjugates. *Methods in Molecular Biology*, *899*, 489–497.

Ellerby, H. M., Arap, W., Ellerby, L. M., Kain, R., Andrusiak, R., Rio, G. D., et al. (1999). Anti-cancer activity of targeted pro-apoptotic peptides. *Nature Medicine*, *5*(9), 1032–1038.

Elliott, R. L., Stjernholm, R., & Elliott, M. C. (1988). Preliminary evaluation of platinum transferrin (MPTC-63) as a potential nontoxic treatment for breast cancer. *Cancer Detection and Prevention*, *12*(1–6), 469–480.

Elsadek, B., Graeser, R., Esser, N., Schafer-Obodozie, C., Abu Ajaj, K., Unger, C., et al. (2010). Development of a novel prodrug of paclitaxel that is cleaved by prostate-specific antigen: An in vitro and in vivo evaluation study. *European Journal of Cancer*, *46*(18), 3434–3444.

Elsadek, B., Graeser, R., Esser, N., Schafer-Obodozie, C., Tsurumi, C., Abu Ajaj, K., et al. (2011). In vivo evaluation of a novel albumin-binding prodrug of doxorubicin in an orthotopic mouse model of prostate cancer (LNCaP). *Prostate Cancer and Prostatic Diseases*, *14*(1), 14–21.

Endo, N., Takeda, Y., Kishida, K., Kato, Y., Saito, M., Umemoto, N., et al. (1987). Target-selective cytotoxicity of methotrexate conjugated with monoclonal anti-MM46 antibody. *Cancer Immunology, Immunotherapy*, *25*(1), 1–6.

Erickson, H. K., Park, P. U., Widdison, W. C., Kovtun, Y. V., Garrett, L. M., Hoffman, K., et al. (2006). Antibody-maytansinoid conjugates are activated in targeted cancer cells by lysosomal degradation and linker-dependent intracellular processing. *Cancer Research*, *66*(8), 4426–4433.

Fiume, L., Bolondi, L., Busi, C., Chieco, P., Kratz, F., Lanza, M., et al. (2005). Doxorubicin coupled to lactosaminated albumin inhibits the growth of hepatocellular carcinomas induced in rats by diethylnitrosamine. *Journal of Hepatology*, *43*(4), 645–652.

Fujii, K., & Imamura, S. (1995). Cell surface proteolysis by serine proteinases enhances RGD-sensitive melanoma cell adhesion on fibrinogen and vitronectin. *Experimental Cell Research*, *220*(1), 201–211.

Gatter, K. C., Brown, G., Trowbridge, I., Woolston, R., & Mason, D. (1983). Transferrin receptors in human tissues: Their distribution and possible clinical relevance. *Journal of Clinical Pathology*, *36*(5), 539–545.

Goszczynski, T., Nevozhay, D., Wietrzyk, J., Omar, M. S., & Boratynski, J. (2013). The antileukemic activity of modified fibrinogen-methotrexate conjugate. *Biochimica et Biophysica Acta*, *1830*(3), 2526–2530.

Graeser, R., Chung, D. E., Esser, N., Moor, S., Schachtele, C., Unger, C., et al. (2008). Synthesis and biological evaluation of an albumin-binding prodrug of doxorubicin that is cleaved by prostate-specific antigen (PSA) in a PSA-positive orthotopic prostate carcinoma model (LNCaP). *International Journal of Cancer*, *122*(5), 1145–1154.

Greenfield, R. S., Kaneko, T., Daues, A., Edson, M. A., Fitzgerald, K. A., Olech, L. J., et al. (1990). Evaluation in vitro of adriamycin immunoconjugates synthesized using an acid-sensitive hydrazone linker. *Cancer Research*, *50*(20), 6600–6607.

Guterstam, P., Madani, F., Hirose, H., Takeuchi, T., Futaki, S., El Andaloussi, S., et al. (2009). Elucidating cell-penetrating peptide mechanisms of action for membrane interaction, cellular uptake, and translocation utilizing the hydrophobic counter-anion pyrenebutyrate. *Biochimica et Biophysica Acta (BBA)—Biomembranes*, *1788*(12), 2509–2517.

Hamblett, K. J., Senter, P. D., Chace, D. F., Sun, M. M., Lenox, J., Cerveny, C. G., et al. (2004). Effects of drug loading on the antitumor activity of a monoclonal antibody drug conjugate. *Clinical Cancer Research*, *10*(20), 7063–7070.

Hartung, G., Stehle, G., Sinn, H., Wunder, A., Schrenk, H. H., Heeger, S., et al. (1999). Phase I trial of methotrexate-albumin in a weekly intravenous bolus regimen in cancer patients. Phase I Study Group of the Association for Medical Oncology of the German Cancer Society. *Clinical Cancer Research*, *5*(4), 753–759.

Hatano, T., Ohkawa, K., & Matsuda, M. (1993). Cytotoxic effect of the protein-doxorubicin conjugates on the multidrug-resistant human myelogenous leukemia cell line, K562, in vitro. *Tumour Biology*, *14*(5), 288–294.

Hiraoka, A., Horiike, N., Akbar, S. M., Michitaka, K., Matsuyama, T., & Onji, M. (2005). Expression of CD163 in the liver of patients with viral hepatitis. *Pathology, Research and Practice*, *201*(5), 379–384.

Hobbs, S. K., Monsky, W. L., Yuan, F., Roberts, W. G., Griffith, L., Torchilin, V. P., et al. (1998). Regulation of transport pathways in tumor vessels: Role of tumor type and microenvironment. *Proceedings of the National Academy of Sciences*, *95*(8), 4607–4612.

Hoshino, T., Misaki, M., Yamamoto, M., Shimizu, H., Ogawa, Y., & Toguchi, H. (1995a). In vitro cytotoxicities and in vivo distribution of transferrin-platinum(II) complex. *Journal of Pharmaceutical Sciences*, *84*(2), 216–221.

Hoshino, T., Misaki, M., Yamamoto, M., Shimizu, H., Ogawa, Y., & Toguchi, H. (1995b). Receptor-binding, in vitro cytotoxicity, and in vivo distribution of transferrin-bound cis-platinum(II) of differing molar ratios. *Journal of Controlled Release*, *37*(1), 75–81.

Huang, J., Wang, J. W., Gong, T., & Zhang, Z. R. (2007). Synthesis and characterization of insulin-5-Fu conjugate, enabling insulin as multi-drug carrier via dendritic approach. *Chinese Chemical Letters*, *18*(3), 247–250.

Hussong, D. (2010). Sterile products: Advances and challenges in formulation, manufacturing and regulatory aspects—A regulatory review perspective. *AAPS PharmSciTech*, *11*(3), 1482–1484.

Indira Chandran, V. (2012). *Development of targeted anticancer agents using novel N-alkylisatin derivatives*. Doctor of Philosophy thesis. School of Biological Sciences, University of Wollongong. http://ro.uow.edu.au/theses/3654

Indira Chandran, V., Matesic, L., Locke, J. M., Skropeta, D., Ranson, M., & Vine, K. L. (2012). Anti-cancer activity of an acid-labile N-alkylisatin conjugate targeting the transferrin receptor. *Cancer Letters*, *316*(2), 151–156.

Inoue, T., Cavanaugh, P. G., Steck, P. A., Brunner, N., & Nicolson, G. L. (1993). Differences in transferrin response and numbers of transferrin receptors in rat and human mammary carcinoma lines of different metastatic potentials. *Journal of Cellular Physiology*, *156*(1), 212–217.

Jeffrey, S. C., Andreyka, J. B., Bernhardt, S. X., Kissler, K. M., Kline, T., Lenox, J. S., et al. (2006). Development and properties of beta-glucuronide linkers for monoclonal antibody-drug conjugates. *Bioconjugate Chemistry*, *17*(3), 831–840.

Jeffrey, S. C., Torgov, M. Y., Andreyka, J. B., Boddington, L., Cerveny, C. G., Denny, W. A., et al. (2005). Design, synthesis, and in vitro evaluation of dipeptide-based antibody minor groove binder conjugates. *Journal of Medicinal Chemistry*, *48*(5), 1344–1358.

Jiang, Y. Y., Liu, C., Hong, M. H., Zhu, S. J., & Pei, Y. Y. (2007). Tumor cell targeting of transferrin-PEG-TNF-alpha conjugate via a receptor-mediated delivery system: Design, synthesis, and biological evaluation. *Bioconjugate Chemistry*, *18*(1), 41–49.

Jones, L., & McKnight, A. J. (2013). *Biotherapeutics: Recent developments using chemical and molecular biology*. Piccadilly, London: Royal Society of Chemistry.

Jung, B., & Theato, P. (2013). Chemical strategies for the synthesis of protein–polymer conjugates. In S. Helmut (Ed.), *Bio-synthetic polymer conjugates* (pp. 37–70). Berlin, Germany: Springer-Verlag Berlin Heidelberg.

Junutula, J. R., Raab, H., Clark, S., Bhakta, S., Leipold, D. D., Weir, S., et al. (2008). Site-specific conjugation of a cytotoxic drug to an antibody improves the therapeutic index. *Nature Biotechnology*, *26*(8), 925–932.

KAI Pharmaceuticals Initiates Phase 1 Trial of KAI-1455 for Ischemic Injury, BusinessWire, May 2007. Accessed at http://www.businesswire.com/news/home/....Initiates-Phase-1-Trial-KAI-1455#.VDq_HvmUfYh on 1st Oct 2014.
Kato, Y., Tsukada, Y., Hara, T., & Hirai, H. (1983). Enhanced antitumor activity of mitomycin C conjugated with anti-alpha-fetoprotein antibody by a novel method of conjugation. *Journal of Applied Biochemistry*, *5*(4–5), 313–319.
Kim, I., Kim, T. H., Ma, K., Lee, E. S., Kim, D., Oh, K. T., et al. (2010). Synthesis and evaluation of human serum albumin-modified exendin-4 conjugate via heterobifunctional polyethylene glycol linkage with protracted hypoglycemic efficacy. *Bioconjugate Chemistry*, *21*(8), 1513–1519.
Kim, J.-W., & Lee, H.-S. (2004). Tumor targeting by doxorubicin-RGD-4C peptide conjugate in an orthotopic mouse hepatoma model. *International Journal of Molecular Medicine*, *14*(4), 529–564.
Kirschke, H., Barrett, A. J., & Rawlings, N. D. (1995). Cathepsin B in protein profiles proteinases 1. In P. Sheterline (Ed.), *Lysosomal cysteine proteinases* (pp. 1587–1643). London: Academic Press.
Kitson, S. L., Quinn, D. J., Moody, T. S., Speed, D., Watters, W., & Rozzell, D. (2013). Antibody-drug conjugates (ADCs)—Biotherapeutic bullets. *Monographic Supplement Series CROs/CMOs—Chimica Oggi—Chemistry Today*, *31*(4).
Koblinski, J. E., Ahram, M., & Sloane, B. F. (2000). Unraveling the role of proteases in cancer. *Clinica Chimica Acta*, *291*(2), 113–135.
Kovtun, Y. V., Audette, C. A., Ye, Y., Xie, H., Ruberti, M. F., Phinney, S. J., et al. (2006). Antibody-drug conjugates designed to eradicate tumors with homogeneous and heterogeneous expression of the target antigen. *Cancer Research*, *66*(6), 3214–3221.
Kozak, K. R., Tsai, S. P., Fourie-O'Donohue, A., dela Cruz Chuh, J., Roth, L., Cook, R., et al. (2013). Total antibody quantification for MMAE-conjugated antibody-drug conjugates: Impact of assay format and reagents. *Bioconjugate Chemistry*, *24*(5), 772–779.
Kratz, F. (2008). Albumin as a drug carrier: Design of prodrugs, drug conjugates and nanoparticles. *Journal of Controlled Release*, *132*(3), 171–183.
Kratz, F., Beyer, U., Roth, T., Tarasova, N., Collery, P., Lechenault, F., et al. (1998). Transferrin conjugates of doxorubicin: Synthesis, characterization, cellular uptake, and in vitro efficacy. *Journal of Pharmaceutical Sciences*, *87*(3), 338–346.
Kratz, F., Drevs, J., Bing, G., Stockmar, C., Scheuermann, K., Lazar, P., et al. (2001). Development and in vitro efficacy of novel MMP2 and MMP9 specific doxorubicin albumin conjugates. *Bioorganic & Medicinal Chemistry Letters*, *11*(15), 2001–2006.
Kratz, F., Ehling, G., Kauffmann, H.-M., & Unger, C. (2007). Acute and repeat-dose toxicity studies of the (6-maleimidocaproyl)hydrazone derivative of doxorubicin (DOXO-EMCH), an albumin-binding prodrug of the anticancer agent doxorubicin. *Human & Experimental Toxicology*, *26*(1), 19–35.
Kratz, F., Mansour, A., Soltau, J., Warnecke, A., Fichtner, I., Unger, C., et al. (2005). Development of albumin-binding doxorubicin prodrugs that are cleaved by prostate-specific antigen. *Archiv der Pharmazie*, *338*(10), 462–472.
Kratz, F., Warnecke, A., Scheuermann, K., Stockmar, C., Schwab, J., Lazar, P., et al. (2002). Probing the cysteine-34 position of endogenous serum albumin with thiol-binding doxorubicin derivatives. Improved efficacy of an acid-sensitive doxorubicin derivative with specific albumin-binding properties compared to that of the parent compound. *Journal of Medicinal Chemistry*, *45*(25), 5523–5533.
Kuang, D.-M., & Zheng, L. (2012). Immunobiology of monocytes/macrophages in hepatocellular carcinoma. In S. Biswas (Ed.), *Tumor microenvironment and myelomonocytic cells* (pp. 157–172). Croatia: InTech, ISBN: 978-953-51-0439-1.
Kurtaran, A., Li, S. R., Raderer, M., Leimer, M., Müller, C., Pidlich, J., et al. (1995). Technetium-99 m-galactosyl-neoglycoalbumin combined with iodine-123-Tyr-(A14)-

insulin visualizes human hepatocellular carcinomas. *Journal of Nuclear Medicine: Official Publication, Society of Nuclear Medicine, 36*(10), 1875–1881.
Lai, H., Nakase, I., Lacoste, E., Singh, N. P., & Sasaki, T. (2009). Artemisinin-transferrin conjugate retards growth of breast tumors in the rat. *Anticancer Research, 29*(10), 3807–3810.
Lai, H., Sasaki, T., Singh, N. P., & Messay, A. (2005). Effects of artemisinin-tagged holotransferrin on cancer cells. *Life Sciences, 76*(11), 1267–1279.
Lang Zou, Y. H., Wang, X., Ma, Y., Liu, Y., Guan, Z. U., Zhang, L., et al. (2014). Serum stability enhancement of siRNA caused by peptide conjugation at 3′-terminus of sense strand. *Journal of Chinese Pharmaceutical Sciences, 23*(4), 215–219.
Lash, A. (2011). Antibody-drug conjugates: The next generation of moving parts. *Start-up*, pp. 1–6.
Laske, D. W., Youle, R. J., & Oldfield, E. H. (1997). Tumor regression with regional distribution of the targeted toxin TF-CRM107 in patients with malignant brain tumors. *Nature Medicine, 3*(12), 1362–1368.
Lee, M.-S., Jung, J.-I., Kwon, S.-H., Lee, S.-M., Morita, K., & Her, S. (2012). TIMP-2 fusion protein with human serum albumin potentiates anti-angiogenesis-mediated inhibition of tumor growth by suppressing MMP-2 expression. *PLoS One*, 7(4), e35710.
Lehr, M. (2011). *Antibody-drug conjugates > linker chemistry*.
Lelle, M., Frick, S. U., Steinbrink, K., & Peneva, K. (2014). Novel cleavable cell-penetrating peptide–drug conjugates: Synthesis and characterization. *Journal of Peptide Science, 20*(5), 323–333.
Levy, G. A., Adamson, G., Phillips, M. J., Scrocchi, L. A., Fung, L., Biessels, P., et al. (2006). Targeted delivery of ribavirin improves outcome of murine viral fulminant hepatitis via enhanced anti-viral activity. *Hepatology, 43*(3), 581–591.
Lewis Phillips, G. D., Li, G., Dugger, D. L., Crocker, L. M., Parsons, K. L., Mai, E., et al. (2008). Targeting HER2-positive breast cancer with trastuzumab-DM1, an antibody-cytotoxic drug conjugate. *Cancer Research, 68*(22), 9280–9290.
Li, X., Hou, J., Wang, C., Liu, X., He, H., Xu, P., et al. (2013). Synthesis and biological evaluation of RGD-conjugated MEK1/2 kinase inhibitors for integrin-targeted cancer therapy. *Molecules, 18*(11), 13957–13978.
Lim, C.-J., & Shen, W.-C. (2005). Comparison of monomeric and oligomeric transferrin as potential carrier in oral delivery of protein drugs. *Journal of Controlled Release, 106*(3), 273–286.
Liu, W. S., Huang, Y., & Zhang, Z. R. (2003). Synthesis and characterization of the tumor targeting mitoxantrone-insulin conjugate. *Archives of Pharmacal Research, 26*(11), 892–897.
Lu, D., Sahasranaman, S., Zhang, Y., & Girish, S. (2013). Strategies to address drug interaction potential for antibody-drug conjugates in clinical development. *Bioanalysis, 5*(9), 1115–1130.
Madani, F., Lindberg, S., Langel, U., Futaki, S., & Graslund, A. (2011). Mechanisms of cellular uptake of cell-penetrating peptides. *Journal of Biophysics, 2011*, 414729.
Maeda, H., Wu, J., Sawa, T., Matsumura, Y., & Hori, K. (2000). Tumor vascular permeability and the EPR effect in macromolecular therapeutics: A review. *Journal of Controlled Release, 65*(1–2), 271–284.
Majumdar, S., Kobayashi, N., Krise, J. P., & Siahaan, T. J. (2007). Mechanism of internalization of an ICAM-1-derived peptide by human leukemic cell line HL-60: Influence of physicochemical properties on targeted drug delivery. *Molecular Pharmaceutics, 4*(5), 749–758.
Majumdar, S., & Siahaan, T. J. (2012). Peptide-mediated targeted drug delivery. *Medicinal Research Reviews, 32*(3), 637–658.
Majumdar, S., Tejo, B. A., Badawi, A. H., Moore, D., Krise, J. P., & Siahaan, T. J. (2009). Effect of modification of the physicochemical properties of ICAM-1-derived peptides on internalization and intracellular distribution in the human leukemic cell line HL-60. *Molecular Pharmaceutics, 6*(2), 396–406.

Matsumura, Y., & Maeda, H. (1986). A new concept for macromolecular therapeutics in cancer chemotherapy: Mechanism of tumoritropic accumulation of proteins and the antitumor agent smancs. *Cancer Research*, *46*(12 Pt 1), 6387–6392.

McDonagh, C. F., Turcott, E., Westendorf, L., Webster, J. B., Alley, S. C., Kim, K., et al. (2006). Engineered antibody-drug conjugates with defined sites and stoichiometries of drug attachment. *Protein Engineering, Design & Selection*, *19*(7), 299–307.

Meister, A., & Anderson, M. E. (1983). Glutathione. *Annual Review of Biochemistry*, *52*, 711–760.

Meyer-Losic, F., Nicolazzi, C., Quinonero, J., Ribes, F., Michel, M., Dubois, V., et al. (2008). DTS-108, a novel peptidic prodrug of SN38: In vivo efficacy and toxicokinetic studies. *Clinical Cancer Research*, *14*(7), 2145–2153.

Miyaji, Y., Walter, S., Chen, L., Kurihara, A., Ishizuka, T., Saito, M., et al. (2011). Distribution of KAI-9803, a novel delta-protein kinase C inhibitor, after intravenous administration to rats. *Drug Metabolism and Disposition*, *39*(10), 1946–1953.

Moodie, J. E., Bisley, E. J., Huang, S., Pickthorn, K., & Bell, G. (2013). A single-center, randomized, double-blind, active, and placebo-controlled study of KAI-1678, a novel PKC-epsilon inhibitor, in the treatment of acute postoperative orthopedic pain. *Pain Medicine*, *14*(6), 916–924.

Nagase, H. (2001). Substrate specificity of MMPs. In N. J. Clendeninn & K. Appelt (Eds.), *Cancer drug discovery and development: Matrix metalloproteinase inhibitors in cancer therapy* (pp. 39–66). Totowa, New Jersey: Humana Press Inc.

Noguchi, Y., Wu, J., Duncan, R., Strohalm, J., Ulbrich, K., Akaike, T., et al. (1998). Early phase tumor accumulation of macromolecules: A great difference in clearance rate between tumor and normal tissues. *Japanese Journal of Cancer Research*, *89*(3), 307–314.

Nolting, B. (2013). Linker technologies for antibody-drug conjugates. *Methods in Molecular Biology*, *1045*, 71–100.

Okeley, N. M., Miyamoto, J. B., Zhang, X., Sanderson, R. J., Benjamin, D. R., Sievers, E. L., et al. (2010). Intracellular activation of SGN-35, a potent anti-CD30 antibody-drug conjugate. *Clinical Cancer Research*, *16*(3), 888–897.

Otto, H.-H., & Schirmeister, T. (1997). Cysteine proteases and their inhibitors. *Chemical Reviews*, *97*, 133–172.

Otvos, L., Jr., & Wade, J. D. (2014). Current challenges in peptide-based drug discovery. *Frontiers in Chemistry*, *2*, 62.

Ou, X., & Kuang, A. (2000). [The use of insulin as carrier in targeted therapy]. *Sheng Wu Yi Xue Gong Cheng Xue Za Zhi*, *17*(1), 87–90.

Ouyang, J. (2013). Drug-to-antibody ratio (DAR) and drug load distribution by hydrophobic interaction chromatography and reversed phase high-performance liquid chromatography. *Methods in Molecular Biology*, *1045*, 275–283.

Papa, V., Pezzino, V., Costantino, A., Belfiore, A., Giuffrida, D., Frittitta, L., et al. (1990). Elevated insulin receptor content in human breast cancer. *The Journal of Clinical Investigation*, *86*(5), 1503–1510.

Perez, H. L., Cardarelli, P. M., Deshpande, S., Gangwar, S., Schroeder, G. M., Vite, G. D., et al. (2014). Antibody-drug conjugates: Current status and future directions. *Drug Discovery Today*, *19*(7), 869–881.

Pietrangelo, A., Rocchi, E., Casalgrandi, G., Rigo, G., Ferrari, A., Perini, M., et al. (1992). Regulation of transferrin, transferrin receptor, and ferritin genes in human duodenum. *Gastroenterology*, *102*(3), 802–809.

Pimm, M. V., Paul, M. A., Ogumuyiwa, Y., & Baldwin, R. W. (1988). Biodistribution and tumour localisation of a daunomycin-monoclonal antibody conjugate in nude mice with human tumour xenografts. *Cancer Immunology, Immunotherapy*, *27*(3), 267–271.

Pisal, D. S., Kosloski, M. P., & Balu-Iyer, S. V. (2010). Delivery of therapeutic proteins. *Journal of Pharmaceutical Sciences*, *99*(6), 2557–2575.

Powell, M. F., Grey, H., Gaeta, F., Sette, A., & Colon, S. (1992). Peptide stability in drug development: A comparison of peptide reactivity in different biological media. *Journal of Pharmaceutical Sciences*, *81*(8), 731–735.

Ravel, D., Dubois, V., Quinonero, J., Meyer-Losic, F., Delord, J., Rochaix, P., et al. (2008). Preclinical toxicity, toxicokinetics, and antitumoral efficacy studies of DTS-201, a tumor-selective peptidic prodrug of doxorubicin. *Clinical Cancer Research*, *14*(4), 1258–1265.

Rohrer, T. (2012). Consideration for the safe and effective manufacturing of antibody drug conjugates. *Chimica Oggi—Chemistry Today*, *30*, 76–79.

Romain Coriat, S. J. F., Dreyer, C., Mir, O., Bouattour, M., Goldwasser, F., & Raymond, E. (2012). First-in-human phase I and pharmacokinetic study of DTS-108 in patients with advanced carcinomas. In *Paper presented at the 2012 ASCO Annual Meeting, ASCO University, USA*.

Roomi, M. W., Monterrey, J. C., Kalinovsky, T., Rath, M., & Niedzwiecki, A. (2009). Patterns of MMP-2 and MMP-9 expression in human cancer cell lines. *Oncology Reports*, *21*(5), 1323–1333.

Rothbard, J. B., Garlington, S., Lin, Q., Kirschberg, T., Kreider, E., McGrane, P. L., et al. (2000). Conjugation of arginine oligomers to cyclosporin A facilitates topical delivery and inhibition of inflammation. *Nature Medicine*, *6*(11), 1253–1257.

Rowland, A. J., Pietersz, G. A., & McKenzie, I. F. C. (1993). Preclinical investigation of the antitumour effects of anti-CD19-idarubicin immunoconjugates. *Cancer Immunology, Immunotherapy*, *37*(3), 195–202.

Russo, A., DeGraff, W., Friedman, N., & Mitchell, J. B. (1986). Selective modulation of glutathione levels in human normal versus tumor cells and subsequent differential response to chemotherapy drugs. *Cancer Research*, *46*(6), 2845–2848.

Sahin, E., & Roberts, C. J. (2012). Size-exclusion chromatography with multi-angle light scattering for elucidating protein aggregation mechanisms. *Methods in Molecular Biology*, *899*, 403–423.

Sarin, H. (2010). Physiologic upper limits of pore size of different blood capillary types and another perspective on the dual pore theory of microvascular permeability. *Journal of Angiogenesis Research*, *2*(14), 10.1186.

Sato, Y., Yamauchi, N., Takahashi, M., Sasaki, K., Fukaura, J., Neda, H., et al. (2000). In vivo gene delivery to tumor cells by transferrin-streptavidin-DNA conjugate. *The FASEB Journal*, *14*(13), 2108–2118.

Schmid, B., Chung, D. E., Warnecke, A., Fichtner, I., & Kratz, F. (2007). Albumin-binding prodrugs of camptothecin and doxorubicin with an Ala-Leu-Ala-Leu-linker that are cleaved by cathepsin B: Synthesis and antitumor efficacy. *Bioconjugate Chemistry*, *18*(3), 702–716.

Schmid, B., Warnecke, A., Fichtner, I., Jung, M., & Kratz, F. (2007). Development of albumin-binding camptothecin prodrugs using a peptide positional scanning library. *Bioconjugate Chemistry*, *18*(6), 1786–1799.

Senter, P. D. (2009). Potent antibody drug conjugates for cancer therapy. *Current Opinion in Chemical Biology*, *13*(3), 235–244.

Seymour, L. W., Duncan, R., Strohalm, J., & Kopecek, J. (1987). Effect of molecular weight (Mw) of N-(2-hydroxypropyl)methacrylamide copolymers on body distribution and rate of excretion after subcutaneous, intraperitoneal, and intravenous administration to rats. *Journal of Biomedical Materials Research*, *21*(11), 1341–1358.

Shankar, P., Lee, S. K., Swamy, M., Kumar, P., Wu, H., & Ban, H. S. (2008). *Targeted delivery of siRNA: Google patents*.

Shi, N.-Q., Gao, W., Xiang, B., & Qi, X.-R. (2012). Enhancing cellular uptake of activable cell-penetrating peptide–doxorubicin conjugate by enzymatic cleavage. *International Journal of Nanomedicine*, 7, 1613.

Shin, S. U., Friden, P., Moran, M., Olson, T., Kang, Y. S., Pardridge, W. M., et al. (1995). Transferrin-antibody fusion proteins are effective in brain targeting. *Proceedings of the National Academy of Sciences*, *92*(7), 2820–2824.
Sievers, E. L., Appelbaum, F. R., Spielberger, R. T., Forman, S. J., Flowers, D., Smith, F. O., et al. (1999). Selective ablation of acute myeloid leukemia using antibody-targeted chemotherapy: A phase I study of an anti-CD33 calicheamicin immunoconjugate. *Blood*, *93*(11), 3678–3684.
Singh, M., Atwal, H., & Micetich, R. (1998). Transferrin directed delivery of adriamycin to human cells. *Anticancer Research*, *18*(3A), 1423–1427.
Singh, Y., Gao, D., Gu, Z., Li, S., Stein, S., & Sinko, P. J. (2012). Noninvasive detection of passively targeted poly(ethylene glycol) nanocarriers in tumors. *Molecular Pharmaceutics*, *9*(1), 144–155.
Sizensky, J. A., Barabas, K., & Faulk, W. P. (1992). Characterization of the anti-cancer activity of transferrin-adriamycin conjugates. *American Journal of Reproductive Immunology*, *27*(3–4), 163–166.
Smyth, M. J., Pietersz, G. A., & McKenzie, I. F. (1987). Selective enhancement of antitumor activity of N-acetyl melphalan upon conjugation to monoclonal antibodies. *Cancer Research*, *47*(1), 62–69.
Spearman, M. E., Goodwin, R. M., Apelgren, L. D., & Bumol, T. F. (1987). Disposition of the monoclonal antibody-vinca alkaloid conjugate KS1/4-DAVLB (LY256787) and free 4-desacetylvinblastine in tumor-bearing nude mice. *The Journal of Pharmacology and Experimental Therapeutics*, *241*(2), 695–703.
Stehle, G., Wunder, A., Sinn, H., Schrenk, H. H., Schütt, S., Frei, E., et al. (1997). Pharmacokinetics of methotrexate-albumin conjugates in tumor-bearing rats. *Anti-Cancer Drugs*, *8*(9), 835.
Stevens, T. J., & Arkin, I. T. (1999). Are membrane proteins "inside-out" proteins? *Proteins*, *36*(1), 135–143.
Stjernholm, R. L. (1986). United States Patent No. 4590001.
Studer, M., Kroger, L. A., DeNardo, S. J., Kukis, D. L., & Meares, C. F. (1992). Influence of a peptide linker on biodistribution and metabolism of antibody-conjugated benzyl-EDTA. Comparison of enzymatic digestion in vitro and in vivo. *Bioconjugate Chemistry*, *3*(5), 424–429.
Stumpp, M. T., Binz, H. K., & Amstutz, P. (2008). DARPins: A new generation of protein therapeutics. *Drug Discovery Today*, *13*(15–16), 695–701.
Suzuki, T., Ikeda, K., & Tomono, T. (1989). Physicochemical and biological properties of poly(ethylene glycol)-coupled immunoglobulin G. Part II. Effect of molecular weight of poly(ethylene glycol). *Journal of Biomaterials Science. Polymer Edition*, *1*(2), 71–84.
Suzuki, T., Kanbara, N., Tomono, T., Hayashi, N., & Shinohara, I. (1984). Physicochemical and biological properties of poly(ethylene glycol)-coupled immunoglobulin G. *Biochimica et Biophysica Acta*, *788*(2), 248–255.
Tai, W., Shukla, R. S., Qin, B., Li, B., & Cheng, K. (2011). Development of a peptide-drug conjugate for prostate cancer therapy. *Molecular Pharmaceutics*, *8*(3), 901–912.
Takahashi, N., Asakura, T., & Ohkawa, K. (1996). Pharmacokinetic analysis of protein-conjugated doxorubicin (DXR) and its degraded adducts in DXR-sensitive and -resistant rat hepatoma cells. *Anti-Cancer Drugs*, 7(6), 687–696.
Tanaka, T., Kaneo, Y., & Miyashita, M. (1996). Synthesis of transferrin-mitomycin C conjugate as a receptor-mediated drug targeting system. *Biological & Pharmaceutical Bulletin*, *19*(5), 774–777.
Temming, K., Lacombe, M., Schaapveld, R. Q. J., Orfi, L., Kéri, G., Poelstra, K., et al. (2006). Rational design of RGD–albumin conjugates for targeted delivery of the VEGF-R kinase inhibitor PTK787 to angiogenic endothelium. *ChemMedChem*, *1*(11), 1200–1203.

Temming, K., Lacombe, M., van der Hoeven, P., Prakash, J., Gonzalo, T., Dijkers, E. C. F., et al. (2006). Delivery of the p38 MAPkinase inhibitor SB202190 to angiogenic endothelial cells: Development of novel RGD-equipped and PEGylated drug–albumin conjugates using platinum(II)-based drug linker technology. *Bioconjugate Chemistry*, *17*(5), 1246–1255.

Temming, K., Meyer, D. L., Zabinski, R., Dijkers, E. C., Poelstra, K., Molema, G., et al. (2006). Evaluation of RGD-targeted albumin carriers for specific delivery of auristatin E to tumor blood vessels. *Bioconjugate Chemistry*, *17*(6), 1385–1394.

Thorpe, P. E., Wallace, P. M., Knowles, P. P., Relf, M. G., Brown, A. N., Watson, G. J., et al. (1987). New coupling agents for the synthesis of immunotoxins containing a hindered disulfide bond with improved stability in vivo. *Cancer Research*, *47*(22), 5924–5931.

Toki, B. E., Cerveny, C. G., Wahl, A. F., & Senter, P. D. (2002). Protease-mediated fragmentation of p-amidobenzyl ethers: A new strategy for the activation of anticancer prodrugs. *The Journal of Organic Chemistry*, *67*(6), 1866–1872.

Trail, P. A., Willner, D., Lasch, S. J., Henderson, A. J., Greenfield, R. S., King, D., et al. (1992). Antigen-specific activity of carcinoma-reactive BR64-doxorubicin conjugates evaluated in vitro and in human tumor xenograft models. *Cancer Research*, *52*(20), 5693–5700.

Unger, C., Häring, B., Medinger, M., Drevs, J., Steinbild, S., Kratz, F., et al. (2007). Phase I and pharmacokinetic study of the (6-maleimidocaproyl)hydrazone derivative of doxorubicin. *Clinical Cancer Research*, *13*(16), 4858–4866.

van Der Velden, V. H., te Marvelde, J. G., Hoogeveen, P. G., Bernstein, I. D., Houtsmuller, A. B., Berger, M. S., et al. (2001). Targeting of the CD33-calicheamicin immunoconjugate Mylotarg (CMA-676) in acute myeloid leukemia: In vivo and in vitro saturation and internalization by leukemic and normal myeloid cells. *Blood*, *97*(10), 3197–3204.

Versluis, A. J., Rump, E. T., Rensen, P. C., Van Berkel, T. J., & Bijsterbosch, M. K. (1998). Synthesis of a lipophilic daunorubicin derivative and its incorporation into lipidic carriers developed for LDL receptor-mediated tumor therapy. *Pharmaceutical Research*, *15*(4), 531–537.

Vine, K. L., Locke, J. M., Bremner, J. B., Pyne, S. G., & Ranson, M. (2010). Selective targeting of 2′-deoxy-5-fluorouridine to urokinase positive malignant cells in vitro. *Bioorganic & Medicinal Chemistry Letters*, *20*(9), 2908–2911.

Wakankar, A., Chen, Y., Gokarn, Y., & Jacobson, F. S. (2011). Analytical methods for physicochemical characterization of antibody drug conjugates. *MAbs*, *3*(2), 161–172.

Wakankar, A. A., Feeney, M. B., Rivera, J., Chen, Y., Kim, M., Sharma, V. K., et al. (2010). Physicochemical stability of the antibody-drug conjugate Trastuzumab-DM1: Changes due to modification and conjugation processes. *Bioconjugate Chemistry*, *21*(9), 1588–1595.

Wang, W., Wang, E. Q., & Balthasar, J. P. (2008). Monoclonal antibody pharmacokinetics and pharmacodynamics. *Clinical Pharmacology and Therapeutics*, *84*(5), 548–558.

Wang, C.-F., Zhang, G., Zhao, L.-J., Qi, W.-J., Li, X.-P., Wang, J.-L., et al. (2013). Overexpression of the insulin receptor isoform A promotes endometrial carcinoma cell growth. *PLoS One*, *8*(8), e69001.

Warnecke, A., Fichtner, I., Garmann, D., Jaehde, U., & Kratz, F. (2004). Synthesis and biological activity of water-soluble maleimide derivatives of the anticancer drug carboplatin designed as albumin-binding prodrugs. *Bioconjugate Chemistry*, *15*(6), 1349–1359.

Warnecke, A., Fichtner, I., Sass, G., & Kratz, F. (2007). Synthesis, cleavage profile, and antitumor efficacy of an albumin-binding prodrug of methotrexate that is cleaved by plasmin and cathepsin B. *Archiv der Pharmazie (Weinheim)*, *340*(8), 389–395.

Warnecke, A., & Kratz, F. (2003). Maleimide-oligo(ethylene glycol) derivatives of camptothecin as albumin-binding prodrugs: Synthesis and antitumor efficacy. *Bioconjugate Chemistry*, *14*(2), 377–387.

Wunder, A., Muller-Ladner, U., Stelzer, E. H., Funk, J., Neumann, E., Stehle, G., et al. (2003). Albumin-based drug delivery as novel therapeutic approach for rheumatoid arthritis. *Journal of Immunology*, *170*(9), 4793–4801.

Xia, C. Q., Wang, J., & Shen, W.-C. (2000). Hypoglycemic effect of insulin-transferrin conjugate in streptozotocin-induced diabetic rats. *Journal of Pharmacology and Experimental Therapeutics*, *295*(2), 594–600.

Yang, R. C. K., Huang, J. T. B., Chen, Y.-L., Hung, C.-C., Liao, M., Yao, W.-C., et al. (2014). Enthalpy-driven nuclease-like activity and mechanism of peptide-chlorambucil conjugates. *Organic & Biomolecular Chemistry*, *12*(27), 4890–4904.

Yeh, C. J., & Faulk, W. P. (1984). Killing of human tumor cells in culture with adriamycin conjugates of human transferrin. *Clinical Immunology and Immunopathology*, *32*(1), 1–11.

Yuan, F., Dellian, M., Fukumura, D., Leunig, M., Berk, D. A., Torchilin, V. P., et al. (1995). Vascular permeability in a human tumor xenograft: Molecular size dependence and cutoff size. *Cancer Research*, *55*(17), 3752–3756.

Zhang, P., Ozdemir, T., Chung, C. Y., Robertson, G. P., & Dong, C. (2011). Sequential binding of alphaVbeta3 and ICAM-1 determines fibrin-mediated melanoma capture and stable adhesion to CD11b/CD18 on neutrophils. *Journal of Immunology*, *186*(1), 242–254.

Zhang, N., & Palmer, A. F. (2011). Development of a dichloroacetic acid-hemoglobin conjugate as a potential targeted anti-cancer therapeutic. *Biotechnology and Bioengineering*, *108*(6), 1413–1420.

Zhang, N., & Palmer, A. F. (2012). Liposomes surface conjugated with human hemoglobin target delivery to macrophages. *Biotechnology and Bioengineering*, *109*(3), 823–829.

Zijlstra, F. (2008). A novel [delta]-protein-kinase inhibitor for adjunctive therapy during percutaneous coronary intervention. *Nature Clinical Practice. Cardiovascular Medicine*, *5*(8), 446–447.

CHAPTER TWO

Modifications of Natural Peptides for Nanoparticle and Drug Design

Andrew P. Jallouk*, Rohun U. Palekar*, Hua Pan*, Paul H. Schlesinger[†], Samuel A. Wickline*,[†,1]
*Consortium for Translational Research in Advanced Imaging and Nanomedicine, Department of Medicine, Division of Cardiology, Washington University in St. Louis, St. Louis, Missouri, USA
[†]Department of Cell Biology and Physiology, Washington University in St. Louis, St. Louis, Missouri, USA
[1]Corresponding author: e-mail address: wicklines@aol.com

Contents

Abstract

Natural products serve as an important source of novel compounds for drug development. Recently, peptides have emerged as a new class of therapeutic agents due to their versatility and specificity for biological targets. Yet, their effective application often requires use of a nanoparticle delivery system. In this chapter, we review the role of natural peptides in the design and creation of nanomedicines, with a particular focus on cell-penetrating peptides, antimicrobial peptides, and peptide toxins. The use of natural peptides in conjunction with nanoparticle delivery systems holds great promise for the development of new therapeutic formulations as well as novel platforms for the delivery of various cargoes.

Advances in Protein Chemistry and Structural Biology, Volume 98
ISSN 1876-1623
http://dx.doi.org/10.1016/bs.apcsb.2014.12.001

1. INTRODUCTION

Since the dawn of medicine, natural products have served as a crucial source of therapeutic compounds for drug development. From the identification of salicylic acid as the active component of willow bark extract to the discovery of penicillin produced by the mold *Penicillium rubens*, natural compounds have formed the basis of many commonly used drugs throughout history. The classical paradigm of drug development from natural products includes screening of biological extracts and identification of active components, followed by structure determination and modification. In the past 20 years, this method of developing small-molecule drugs has largely been supplanted by high-throughput screening of synthetic chemical libraries, lead compound identification, and structural optimization (Koehn & Carter, 2005). During this same time period, peptides have emerged as an important new class of drugs due to their versatility and specificity for individual targets. An enormous number of potentially therapeutic natural peptides have been identified in the course of biological research. Their clinical translation, however, has been limited by the cost of large-scale synthesis and their inefficient delivery to therapeutic sites *in vivo* (Vlieghe, Lisowski, Martinez, & Khrestchatisky, 2010). Along with continual improvements in the efficiency of peptide synthesis, nanoparticle platforms have been proposed as a strategy to facilitate delivery of peptide drugs. Conversely, natural peptides with unique biological properties have been studied as tools to enhance delivery of nanoparticles bearing other therapeutic compounds. In this chapter, we will review the use of natural peptides in the design and creation of nanomedicines, with a particular focus on short cationic or amphipathic peptides. The sequence and natural origin of several of these peptides are listed in Table 1. These compounds, which may be classified as cell-penetrating peptides (CPPs), antimicrobial peptides (AMPs), and peptide toxins, exhibit interesting interactions with lipid membranes which influence their use as components of nanoparticle drug delivery systems. It is important to note that there is significant overlap between members of these classes and that a particular peptide may behave as a member of more than one class. As a result, the following sections have been constructed based primarily on the applications being discussed rather than individual peptide identities. For each section, we provide a brief review of the structure and function of peptides used for these purposes followed by a detailed description of their application in nanoparticle delivery systems.

Table 1 Natural peptides used for nanoparticle and drug design

Peptide	Sequence	Origin
Tat peptide	YGRKKRRQRRR	HIV-1 *trans*-activator of transcription (Tat) protein
Penetratin	RQIKIWFQNRRMKWKK	*Drosophila* Antennapedia protein
Transportan	GWTLNSAGYLLGKINLKALAALAKKIL	Neuropeptide galanin/Wasp venom mastoparan
MPG	GALFLGFLGAAGSTMGAWSQPKKKRKV	HIV-1 glycoprotein 41/Simian virus 40 large T antigen
Pep-1	KETWWETWWTEWSQPKKKRKV	HIV-1 reverse transcriptase/Simian virus 40 large T antigen
Melittin	GIGAVLKVLTTGLPALISWIKRKRQQ	European honeybee venom
Magainin II	GIGKFLHSAKKFGKAFVGEIMNS	African clawed frog secretion
Cecropin A	KWKLFKKIEKVGQNIRDGIIKAGPAVAVVG QATQIAK	Cecropia moth hemolymph
Buforin II	TRSSRAGLQFPVGRVHRLLRK	Asian toad stomach tissue
LL-37	LLGDFFRKSKEKIGKEFKRIVQRIKDFLRNL VPRTES	Human cathelicidin
Chlorotoxin	MCMPCFTTDHQMARKCDDCCGGKGRGKCYG PQCLCR	Deathstalker scorpion venom

2. ROLE OF NANOPARTICLES IN PEPTIDE DRUG DELIVERY

Despite a great deal of enthusiasm, there remain several key pharmacological limitations to the development of peptide-based pharmaceutical agents (Craik, Fairlie, Liras, & Price, 2013). Short peptides (<50 amino acids) have poor pharmacokinetic profiles due to their degradation by serum proteases and rapid clearance by renal filtration. Additionally, most peptide drugs are limited to extracellular targets due to their inability to cross the plasma membrane and, like all drugs, may have off-target effects when administered systemically. Nanoparticle delivery platforms hold great promise as tools to overcome these limitations and enhance the efficacy and utility of peptide drugs. The term "nanoparticle" refers to an extremely broad range of constructs between 1 and 500 nm in diameter which possess unique physical, chemical, and biological properties as a result of their size (Caruthers, Wickline, & Lanza, 2007). Although most types of nanoparticles have been applied in some way for drug delivery, lipid and polymer nanoparticles are by far the most extensively studied and commonly used for this purpose. The rest of this chapter will focus on the use of these nanoparticles as a platform for the delivery of natural peptides and their derivatives.

Lipid nanoparticles as a class include liposomes and perfluorocarbon nanoparticles, as well as a number of other lipid-based constructs. Liposomes are among the most well-studied nanostructures for drug delivery and include several formulations currently approved for clinical use (Allen & Cullis, 2013). As shown in Fig. 1, liposomes are vesicles consisting of a phospholipid bilayer surrounding an aqueous core (Fig. 1A). Based on their hydrophobicity, small-molecule compounds may either be encapsulated within the aqueous region (e.g., liposomal doxorubicin, Doxil®) or incorporated into the lipid bilayer (e.g., liposomal amphotericin B, AmBisome®). Perfluorocarbon nanoparticles, in contrast, consist of a hydrophobic perfluorocarbon core stabilized by a phospholipid monolayer shell (Fig. 1B). These nanoparticles have been extensively used for targeted drug delivery and imaging via incorporation of targeting ligands into the phospholipid monolayer (Tran et al., 2007). For both liposomes and perfluorocarbon nanoparticles, peptides may be covalently attached to the phospholipid head groups or noncovalently associated with the lipid membrane. Due to their relatively large size, liposomes and perfluorocarbon nanoparticles alter the

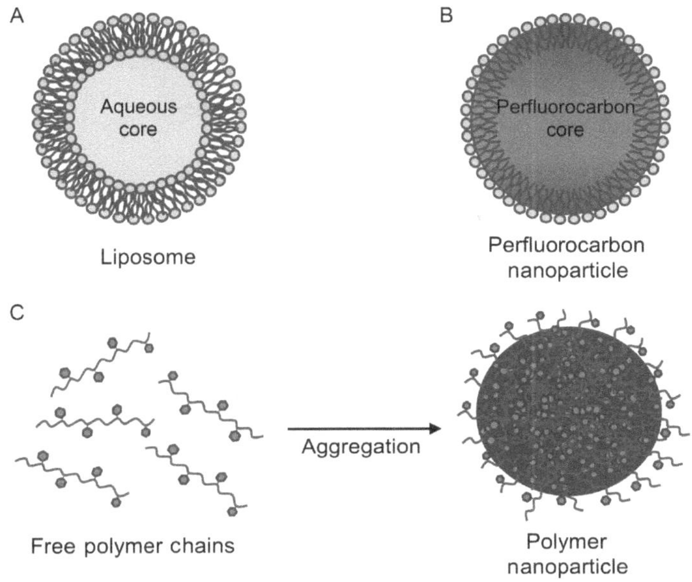

Figure 1 Structure of lipid and polymer nanoparticles. (A) Liposomes consist of a phospholipid bilayer surrounding an aqueous core. (B) Perfluorocarbon nanoparticles consist of a phospholipid monolayer surrounding a hydrophobic perfluorocarbon core. (C) Polymer nanoparticles are often formed by aggregation of hydrophobic polymer chains in an aqueous environment.

pharmacokinetic profile of attached peptides, preventing renal filtration and leading to an overall increase in the circulating peptide half-life. For instance, covalent attachment of the thrombin inhibitor D-phenylalanyl-L-prolyl-L-arginyl-chloromethylketone (PPACK) to the surface of liposomes (Palekar et al., 2013) or perfluorocarbon nanoparticles (Myerson, He, Lanza, Tollefsen, & Wickline, 2011) substantially increased its duration of antithrombotic activity relative to free PPACK. Additionally, insertion into the lipid monolayer of perfluorocarbon nanoparticles enhanced circulation time of melittin, a cytolytic peptide toxin derived from bee venom (Soman et al., 2009). Loading of melittin onto perfluorocarbon nanoparticles also protected melittin from cleavage by serum proteases and significantly reduced its hemolytic activity and toxicity when injected systemically. Interestingly, attempts to use liposomes as a melittin delivery vehicle resulted in membrane pore formation and complete disruption of liposome structure (Soman, Lanza, Heuser, Schlesinger, & Wickline, 2008). These findings illustrate the importance of selecting an appropriate nanoparticle carrier to enhance the therapeutic potential of peptide agents.

Polymer nanoparticles are extremely versatile drug carriers which may be generated using a wide variety of natural or synthetic polymers. Biodegradable synthetic polymers, such as poly(lactic-co-glycolic acid) (PLGA) and poly(ε-caprolactone), often aggregate to form nanoparticles via hydrophobic interactions (Fig. 1C). Additionally, several methods have been developed to synthesize nanoparticles composed of natural hydrophilic polymers, such as chitosan and gelatin (Soppimath, Aminabhavi, Kulkarni, & Rudzinski, 2001). One important characteristic of polymer nanoparticles is the ability to directly modify the polymer chain as needed to achieve desired functionality. For instance, PLGA nanoparticles containing polymer chains grafted with poly(ethylene glycol) (PEG) and a cyclic RGD targeting peptide demonstrated enhanced delivery of paclitaxel and doxorubicin to a variety of *in vitro* and *in vivo* tumor models (Danhier et al., 2012). Peptide cargoes may be either covalently linked to the polymer backbone or noncovalently associated with the polymer matrix via electrostatic and hydrophobic interactions. Interestingly, amphipathic peptides may self-assemble into nanoparticles, thereby facilitating their own delivery and that of any complexed cargo. CPPs, in particular, have been used in this fashion to enhance translocation of nucleotides and other peptides across the cell membrane (Heitz, Morris, & Divita, 2009).

3. CELL-PENETRATING PEPTIDES

In order for macromolecules and nanoparticles to exert therapeutic effects intracellularly, they must first cross the plasma membrane. Over the past 25 years, a large amount of work has been focused on the development of peptide agents capable of facilitating this process, leading to enhanced cellular uptake of cargoes ranging from individual protein molecules (Schwarze, Ho, Vocero-Akbani, & Dowdy, 1999) to liposomes with diameters of 200 nm (Torchilin, Rammohan, Weissig, & Levchenko, 2001). Many of these CPPs are originally derived from natural proteins which translocate across biological membranes. The first reports of this phenomenon came from studies of the HIV-1 *trans*-activator of transcription (Tat) protein, which penetrates membranes and accumulates in the nucleus by an energy-independent mechanism (Frankel & Pabo, 1988; Green & Loewenstein, 1988). Further studies identified the minimal peptide sequence required for this uptake (Vivès, Brodin, & Lebleu, 1997) and led to the discovery of other CPPs such as penetratin, derived from the homeodomain of the *Drosophila* Antennapedia protein (Derossi, Joliot,

Chassaing, & Prochiantz, 1994). Since then, a wide variety of CPPs have been identified from sources such as viral proteins (Elliott & O'Hare, 1997), animal venoms (Hou, Pan, Lanza, & Wickline, 2013), and synthetic peptide libraries (Gao, Simon, Hue, Morrison, & Banta, 2011). These diverse peptides are typically 5–30 amino acids in length and possess a variety of structural and chemical properties which impact their use as macromolecule and nanoparticle delivery agents. In this section, we will review structural characteristics of the most commonly used CPPs and examine their use in nanoparticle-based drug delivery systems. For a more comprehensive listing of CPPs, the reader is referred to the review by Milletti (2012).

3.1 Structure and properties

The number and variety of CPPs reported in the literature make precise classification difficult. CPPs may be broadly divided into groups based on multiple considerations, including origin, sequence, and proposed mechanism of internalization. In particular, classification by sequence provides a logical framework for understanding the structure and function of CPPs, as well as their role in the design of nanostructures. In this system, CPPs are classified as cationic, amphipathic, or hydrophobic.

3.1.1 Cationic CPPs

Cationic CPPs contain positively charged residues which are essential for uptake and often adopt random coil configurations both in solution and when associated with lipid membranes (Eiríksdóttir, Konate, Langel, Divita, & Deshayes, 2010). While the Tat peptide is the most commonly used member of this class, many other cationic CPPs have been identified from sources such as heparan-binding proteins (De Coupade et al., 2005), bacteriophages (Futaki et al., 2001), and transcription factors (Balayssac et al., 2006). Internalization of cationic CPPs relies heavily on the presence of arginine residues, which are thought to interact strongly with anionic cell surface proteoglycans. In fact, polyarginine sequences themselves can serve as CPPs, with several studies identifying octaarginine as the minimal sequence for cellular uptake (El-Sayed, Futaki, & Harashima, 2009; Tünnemann et al., 2008). Following interaction with the cell surface, the mechanisms of CPP uptake become controversial (Heitz et al., 2009; Richard et al., 2003). Although one or more endocytotic mechanisms are likely involved, the contribution of each mechanism may depend on the particular CPP being studied and the cargo to which it is bound (Duchardt, Fotin-Mleczek, Schwarz, Fischer, & Brock, 2007).

3.1.2 Amphipathic CPPs

Amphipathic CPPs contain both hydrophilic and hydrophobic domains which are necessary for cellular internalization. In primary amphipathic CPPs, these segments are separated based on their position along the peptide chain. This class includes several natural peptides such as pVEC, which is derived from VE-cadherin (Elmquist, Lindgren, Bartfai, & Langel, 2001). Primary amphipathic CPPs may be designed as fusion constructs containing the targeting domain of one protein and the membrane-interacting domain of another. For instance, both MPG and Pep-1 contain the cationic Simian Virus 40 (SV40) nuclear localization sequence. However, MPG contains a hydrophobic domain derived from the fusion sequence of HIV-1 glycoprotein 41, while that of Pep-1 is derived from the dimerization motif of HIV-1 reverse transcriptase (Morris, Depollier, Mery, Heitz, & Divita, 2001; Morris, Vidal, Chaloin, Heitz, & Divita, 1997). These different hydrophobic domains allow MPG to carry nucleic acid cargoes while Pep-1 preferentially complexes with other peptides and proteins (Deshayes, Morris, Divita, & Heitz, 2005).

Secondary amphipathic CPPs are another large class of peptides in which the separation between hydrophilic and hydrophobic domains occurs due to the formation of secondary structures, such as α helices or β sheets. Many of the most commonly used CPPs are members of this class. These include penetratin (Derossi et al., 1994), which is derived from the Antennapedia homeodomain, and transportan (Pooga, Hällbrink, Zorko, Uuml, & Langel, 1998), which is derived from fusion of the neuropeptide galanin with the wasp venom peptide mastoparan. Many AMPs and animal toxins also resemble amphipathic CPPs in structure. In order for these peptides to be successfully used for drug delivery, however, they must first be modified to minimize their cytotoxicity. Although most amphipathic CPPs are cationic, there are a number of neutral and anionic peptides which retain their cell-penetrating ability (Li, Nicol, & Szoka, 2004; Taylor et al., 2009). These findings indicate that cellular internalization is a consequence of these peptides' amphipathic nature, not solely their charge.

3.1.3 Hydrophobic CPPs

Hydrophobic CPPs possess a low net charge and are composed of predominantly nonpolar amino acids. Several hydrophobic CPPs are derived from natural proteins, including α1-antitrypsin (Rhee & Davis, 2006) and fibroblast growth factor 12 (Nakayama et al., 2011). Nevertheless, this class includes the fewest CPPs discovered to date and most new hydrophobic

CPPs are generated using synthetic peptide libraries (Marks, Placone, Hristova, & Wimley, 2011). This class also includes a number of chemically modified peptides, such as stapled peptides (Chu et al., 2014), prenylated peptides (Ochocki, Mullen, Wattenberg, & Distefano, 2011), and pepducins (Covic, Gresser, Talavera, Swift, & Kuliopulos, 2002). The use of these synthetic modifications to enhance peptide uptake raises exciting prospects for the development of cell-penetrating agents based on a wide variety of natural products.

3.2 Applications in nanoparticle delivery

Although nanoparticles possess several distinct advantages as drug delivery vehicles, their size generally prevents them from crossing the plasma membrane, thereby precluding their use for intracellular drug delivery. One promising approach to overcome this limitation involves incorporating CPPs directly into the nanoparticle structure. Depending on the type of nanoparticle, this conjugation may be either covalent or noncovalent.

Multiple studies have demonstrated that covalent addition of cationic or amphipathic CPPs to the surface of nanostructures is sufficient to enhance their cellular uptake. This technique has been successfully applied to a wide range of nanoparticles, including liposomes (Tseng, Liu, & Hong, 2002), iron oxide nanoparticles (Zhao, Kircher, Josephson, & Weissleder, 2002), gold nanoparticles (De la Fuente & Berry, 2005), and quantum dots (Ruan, Agrawal, Marcus, & Nie, 2007). Advantages of covalent conjugation include its simplicity, stability, and applicability to a variety of well-defined nanoparticles. Nevertheless, continued interaction between the CPP and cargo following translocation into the cell may diminish the cargo's biological activity (Juliano, Alam, Dixit, & Kang, 2008). In these cases, the use of cleavable linkers or noncovalent association of the CPP may be more appropriate.

The noncovalent approach for CPP incorporation relies on electrostatic and hydrophobic interactions between the CPP and cargo to maintain nanoparticle integrity. This technique was originally developed using the fusion peptide of influenza hemagglutinin, which condenses with DNA to form nanoscale complexes capable of escaping the endosome (Wagner, Plank, Zatloukal, Cotten, & Birnstiel, 1992). Recently, this approach has been extended to a number of other CPPs and has been successfully used for the delivery of nucleotide (Hou, Pan, Lanza, et al., 2013; Nakase et al., 2012), peptide (Grdisa, 2011), and small-molecule therapeutics (Liu et al., 2008). The wide variety of CPPs, nanoparticles, and conjugation methods available provides great freedom in the design of drug delivery systems. For the purposes of this chapter, we have classified studies based on the specific CPP used in each delivery system.

3.2.1 Tat peptide

One of the most commonly used CPPs in the field of drug delivery is the Tat peptide, an 11-mer identified as the minimal sequence necessary for membrane translocation of the HIV-1 *trans*-activator of transcription (Tat) protein (Vivès et al., 1997). The first application of this peptide to enhance nanoparticle uptake involved its covalent attachment to the surface of superparamagnetic iron oxide nanoparticles (SPIO NPs; Josephson, Tung, Moore, & Weissleder, 1999). SPIO NPs bearing the Tat peptide were internalized over 100-fold more efficiently than control nanoparticles, allowing labeled cells to be detected by magnetic resonance imaging and retained on magnetic separation columns. A similar technique was used by Torchilin et al. (2001) to enhance uptake of 200-nm liposomes by a variety of cell lines. This and other studies (Marty, Meylan, Schott, Ballmer-Hofer, & Schwendener, 2004) demonstrated that liposome uptake was dependent on the number of peptides attached to the liposome as well as their ability to interact with the plasma membrane and cell surface proteoglycans. Covalent functionalization with Tat peptide has subsequently been applied to a variety of lipid-based drug delivery systems. Liposomes are well established as carriers to increase the efficacy and reduce off-target toxicity of chemotherapeutic agents (Wang, Langer, & Farokhzad, 2012). Tseng et al. (2002) demonstrated that addition of Tat peptide to the liposomal surface increased uptake of doxorubicin by 12-fold in a number of cancer cell lines. Interestingly, however, they also found that this enhanced uptake was not accompanied by increased cytotoxicity or inhibition of tumor growth in a subcutaneous xenograft model. Other studies have reported somewhat more positive results, with Qin et al. (2011) claiming that Tat peptide-conjugated liposomes enhanced cytotoxicity and moderately improved survival in a mouse model of brain glioma. In either case, limited penetration into the tumor and slow release of encapsulated drug have been identified as key obstacles to liposomal drug delivery (Manzoor et al., 2012) and may persist even with the use of CPPs. Nanostructured lipid carriers functionalized with Tat peptide have also been used in transdermal applications to deliver the nonsteroidal anti-inflammatory drug celecoxib (Desai, Patlolla, & Singh, 2010). These agents significantly increased celecoxib accumulation in all skin layers compared to control carriers lacking the Tat peptide.

The first application of CPPs for nanoparticle-based gene delivery involved the use of lipoplexes formed by electrostatic interaction of Tat peptide-modified cationic liposomes with DNA (Torchilin et al., 2003). These nanoparticles were used to transfect fibroblasts and cardiomyocytes *in vitro*, as well as Lewis lung carcinoma cells growing subcutaneously *in vivo*. Similar lipoplexes demonstrated successful gene delivery to brain tumors (Gupta, Levchenko, & Torchilin, 2007) and ischemic myocardium (Ko, Hartner, Kale, & Torchilin, 2008) in animal models. Recently, polymer nanostructures have gained popularity as nucleotide delivery vehicles. These carriers often contain Tat peptide covalently linked to a cationic polymer which forms nanoscale complexes with nucleotides. Kleemann et al. (2005) demonstrated this concept using a complex of DNA with a Tat peptide-conjugated form of polyethyleneimine. The resulting polyplexes successfully transfected lung carcinoma cells *in vitro*, as well as bronchial and alveolar tissue following intratracheal instillation. A similar nanoparticle was later used to transfect neurotypic cells with possible implications for treatment of neurodegenerative diseases (Suk et al., 2006). Tat peptide-functionalized chitosan nanoparticles have also been used for this application and have successfully delivered siRNA to reduce *in vitro* production of ataxin, a protein overexpressed in spinocerebellar ataxia (Malhotra, Tomaro-Duchesneau, & Prakash, 2013). While Tat peptide-conjugated polymer carriers are often used for gene delivery, they have also been applied as alternatives to liposomes for the delivery of paclitaxel (Sawant & Torchilin, 2009; Zhao, Wang, et al., 2010) and doxorubicin (Lee et al., 2011) to tumors, as well as the delivery of antibiotics such as ciprofloxacin (Liu et al., 2008) across the blood–brain barrier.

One major drawback to the use of CPPs for drug delivery is their lack of specificity for a particular tissue or cell type. An early mouse study which involved intraperitoneal injection of a Tat peptide-β-galactosidase fusion protein found that this construct accumulated in all tissues analyzed, including the brain (Schwarze et al., 1999). While this property may be desirable for some applications, such as systemic gene delivery, it could result in off-target effects and increased dose requirement for many drug delivery systems. Thus, recent efforts using Tat peptide have focused on developing nanocarriers in which the internalizing effects of the CPP are activated only under specific conditions. Kale and Torchilin (2007a, 2007b) developed one such system in which liposomes were functionalized with both Tat peptide and pH-sensitive PEG chains. The PEG coating sterically hindered membrane interaction of Tat peptide, thereby preventing liposome internalization until the PEG chains were hydrolyzed in the acidic tumor microenvironment. A similar concept has been applied to liposomes bearing a monoclonal antibody to enhance interaction with tumor cells (Koren, Apte, Jani, & Torchilin, 2012). These multifunctional liposomes enhanced delivery of doxorubicin to ovarian cancer both *in vitro* and *in vivo* (Apte, Koren, Koshkaryev, & Torchilin, 2014). Drug delivery systems activated by tumor-associated proteases have also been developed. Gullotti, Park, and Yeo (2013) reported a Tat peptide-functionalized polymer nanoparticle bearing PEG chains anchored by a matrix metalloproteinase-2 (MMP-2) cleavable peptide sequence. Treatment with MMP-2 led to increased nanoparticle uptake and delivery of paclitaxel in an ovarian cancer cell line. Similarly, Liu et al. (2014) generated a Tat peptide-modified liposome activated by legumain, a protease overexpressed in a variety of tumor tissues. In this study, however, the cleavable group was covalently linked to Tat peptide, reducing its cell-penetrating activity prior to cleavage by legumain. These strategies for CPP activation may be used in conjunction with other techniques to locally enhance nanoparticle uptake. The application of ultrasound energy, for instance, improved the ability of Tat peptide-modified bubble liposomes to escape the endosome, leading to a 30-fold increase in gene delivery (Omata et al., 2011).

3.2.2 Penetratin

Penetratin, one of the first CPPs to be discovered, is a 16-mer peptide derived from the third helix of the *Drosophila* Antennapedia homeodomain (Derossi et al., 1994). It is commonly used as an alternative or complement to Tat peptide for enhancing cellular uptake of nanoparticles. Like Tat peptide, penetratin is often covalently linked to the nanoparticle surface. One of the most common applications for penetratin-modified nanoparticles is to facilitate drug delivery across the blood–brain barrier. Xia et al. (2012) found that polymer nanoparticles functionalized with penetratin demonstrated increased cellular uptake *in vitro* and improved pharmacokinetics *in vivo*. These nanoparticles accumulated to a greater extent in the brain and to a lesser extent in other tissues than unmodified nanoparticles. Similarly, addition of penetratin to the surface of liposomes improved liposomal accumulation and delivery of encapsulated doxorubicin to brain tumors (Sharma, Modgil, Zhong, Sun, & Singh, 2014). Penetratin-functionalized polymer nanoparticles have also been used to deliver peptide nucleic acids, a new class of therapeutics, to malignant lymphocytes in a mouse model of lymphoma (Babar et al., 2012). Interestingly, this study found that pre-B cells preferentially internalized nanoparticles coated with penetratin when compared to other CPPs, such as Tat peptide and polyarginine.

3.2.3 MPG/Pep-1

Covalent linkage of Tat peptide and penetratin has been shown to inhibit both cellular uptake and biological activity of certain types of cargo, particularly nucleotides (Presente & Dowdy, 2013). To avoid this limitation, several CPPs have been designed which form noncovalent nanoscale complexes with cargo via electrostatic and hydrophobic interactions. These complexes enhance cellular internalization while protecting cargo from degradation by serum nucleases and proteases. MPG is an amphipathic peptide generated by using a flexible peptide linker to join the hydrophobic fusion sequence of HIV-1 glycoprotein 41 with the cationic nuclear localization sequence of SV40 large T antigen (Morris et al., 1997). This peptide forms 200-nm complexes with nucleic acids and was initially used for the delivery of oligonucleotides and plasmid DNA into cultured cells (Morris, Chaloin, Méry, Heitz, & Divita, 1999). More recently, the sequence of this construct was optimized for siRNA delivery by introducing a point mutation to the nuclear localization sequence, thereby facilitating nanoparticle accumulation and siRNA release in the cytoplasm (Simeoni, Morris, Heitz, & Divita, 2003). Although this and other MPG derivatives have been successfully used for gene delivery to a variety of cell lines *in vitro*, it appears that endosomal entrapment remains an important limitation to transfection efficiency (Veldhoen, Laufer, Trampe, & Restle, 2006). For *in vivo* applications, biodistribution presents yet another potential obstacle to effective gene delivery. To address this issue, Crombez et al. (2009) functionalized an MPG derivative with cholesterol and demonstrated that this modification improved tissue distribution following systemic administration in mice. The use of this system to deliver cyclin B1 siRNA reduced tumor growth and increased survival following intravenous injection in a mouse xenograft cancer model.

Given the success of MPG for enhancing nucleotide delivery, Morris et al. (2001) developed a similar peptide which facilitates uptake of proteins via formation of noncovalent complexes. This peptide, known as Pep-1, contains the same hydrophilic domain and peptide linker as MPG, but a different hydrophobic domain derived from a tryptophan-rich cluster found in HIV-1 reverse transcriptase. A study of Pep-1 complexation with the cell cycle inhibitor protein $p27^{Kip1}$ found an optimal Pep-1:p27 molar mixing ratio of 20:1, which resulted in discrete nanoparticles with diameters of approximately 100 nm (Muñoz-Morris, Heitz, Divita, & Morris, 2007). Addition of these complexes to cultured fibroblasts led to cell cycle arrest in 70% of cells analyzed. Pep-1 has been used to deliver proteins to a wide variety of cell lines (Gros et al., 2006), including primary neurons (Gallo, 2003), hepatocytes (Bardag-Gorce et al., 2003), and macrophages (Garnon et al., 2005). This peptide has also been commercialized for research purposes under the trade name Chariot™ (Active Motif, Carlsbad, CA). *In vivo* applications of Pep-1 have been promising, with successful protein delivery demonstrated following intratracheal instillation in a mouse model of emphysema (Aoshiba, Yokohori, & Nagai, 2003). Recently, however, Pep-1 has primarily been used as a covalent attachment to facilitate penetration of nanoparticles into tumors (Kang, Park, Kang, Park, & Choi, 2013; Wang, Lv, et al., 2014) and generate fusion proteins with enhanced tissue accumulation (Ahn et al., 2014; He et al., 2013). These applications resemble those of other CPPs and demonstrate the versatility of Pep-1 as a tool for improving drug delivery.

3.2.4 Antimicrobial peptides

Like CPPs, AMPs often possess cationic or amphipathic segments which enhance their interaction with lipid membranes. Indeed, some authors have proposed that both CPPs and AMPs belong to a class of "membrane-active peptides," with their primary distinguishing feature being the extent of membrane translocation versus permeabilization induced at a particular peptide concentration (Henriques, Melo, & Castanho, 2006). Several AMPs have been reported to exhibit cell-penetrating properties (Splith & Neundorf, 2011). However, their practical application as drug delivery vehicles remains limited by their toxicity to mammalian cells. Judicious peptide sequence modifications may substantially reduce cytotoxicity while maintaining cell-penetrating activity. For instance, Pan et al. (2010) reported that N-terminal truncation of melittin, an AMP derived from bee venom, reduced cytotoxicity by two-orders of magnitude without significantly affecting the ability of melittin to insert into lipid membranes. Addition of arginine and histidine residues to the C-terminus of this peptide gave rise to p5RHH, a transfection agent capable of forming nanoscale complexes with siRNA (Hou, Pan, Lanza, et al., 2013). As shown in Fig. 2, these complexes are stable in serum and have been successfully used to transfect macrophages, B16 melanoma cells, and human umbilical vein endothelial cells *in vitro*. Mechanistic studies indicated that endosomal acidification led to disassembly of these complexes and lysis of the endosomal membrane,

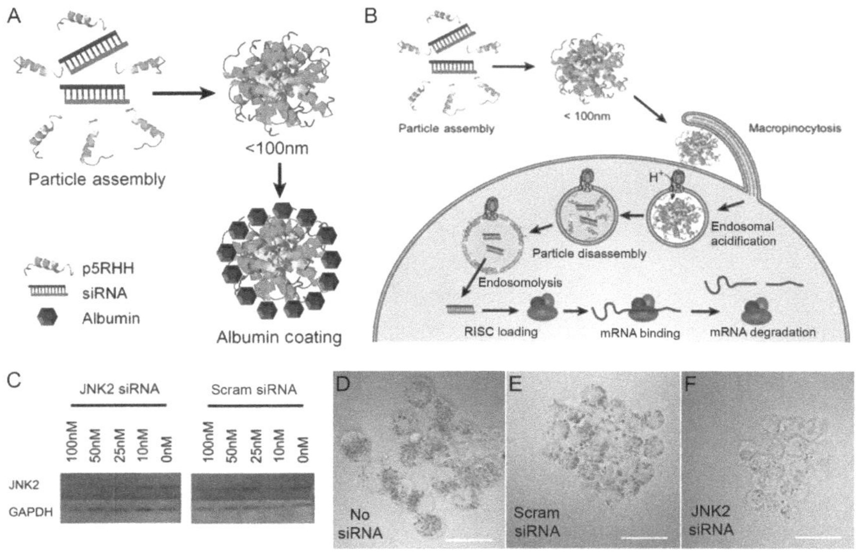

Figure 2 Self-assembly and efficacy of p5RHH–siRNA nanocomplexes. (A) Schematic diagram of p5RHH and siRNA complexation. (B) Schematic diagram of p5RHH–siRNA complex internalization and disassembly following endosomal acidification. (C) Application of p5RHH–siRNA complexes containing JNK2 siRNA lowered JNK2 expression in a macrophage cell line. (D, E, and F) Application of p5RHH–siRNA complexes containing JNK2 siRNA reduced foam cell formation following exposure to acetylated LDL. *Panels (A) and (B) adapted with permission from Hou, Pan, Ratner, et al. (2013). Copyright 2013 American Chemical Society. Panels (C–F) adapted with permission from Hou, Pan, Lanza, et al. (2013). Copyright 2013 Elsevier.*

thereby facilitating siRNA release into the cytoplasm (Hou, Pan, Ratner, Schlesinger, & Wickline, 2013). These complexes accumulated at inflammatory sites following intravenous injection in a mouse model of rheumatoid arthritis. Incorporation of siRNA targeting the p65 subunit of NF-κB led to dramatic improvements in joint inflammation without eliciting a humoral immune response or affecting p65 expression in other tissues (Zhou et al., 2014).

4. ANTIMICROBIAL PEPTIDES

With the rise of antibiotic resistance and the emergence of multidrug-resistant organisms (Andersson & Hughes, 2011; Bush et al., 2011), there is a pressing need to develop new antimicrobial agents which avoid conventional mechanisms of resistance. One promising set of compounds which could form the basis for a new class of therapeutics is AMPs. These peptides are produced by a wide variety of organisms and generally exert their antimicrobial effect by disrupting biological membranes, leading to depolarization, lysis, and cellular death. Like CPPs, AMPs are often short (10–50 amino acids), cationic (overall charge +2 to +9), and possess a number of hydrophobic residues to facilitate their interaction with lipid membranes (Hancock & Sahl, 2006). This class of compounds is also extremely diverse, with over 1200 peptides divided into five structural classes (Wang, Li, & Wang, 2009) and a variety of proposed mechanisms of action (Nguyen, Haney, & Vogel, 2011). In contrast to conventional antibiotics, AMPs do not target a specific molecule or biochemical process, giving them broad-spectrum activity and making development of resistance difficult (Zasloff, 2002). Nevertheless, there are some important limitations preventing clinical translation of these compounds. While many natural AMPs demonstrate some degree of preference for bacterial membranes, toxicity to eukaryotic cells remains a key concern (Marr, Gooderham, & Hancock, 2006). Additionally, like many short peptides, free AMPs are rapidly cleared by the kidney and possess a poor pharmacokinetic profile (Bradshaw, 2003). Recent efforts to address these concerns include sequence modifications to enhance antimicrobial activity while minimizing toxicity to eukaryotic cells, as well as development of nanoparticle-based strategies to improve AMP delivery while further reducing off-target treatment effects.

4.1 Sequence modifications

As with CPPs, the ability of an AMP to disrupt biological membranes depends primarily on its general structure and physicochemical properties rather than its specific amino acid composition. Thus, sequence modifications intended to change peptide activity often alter key peptide characteristics such as charge and hydrophobicity. Addition of cationic residues to either the N- or C-terminus of magainin II, an AMP derived from the African clawed frog, substantially improved its antibacterial activity without significantly altering its hemolytic activity (Bessalle, Haas, Goria, Shalit, & Fridkin, 1992). This effect was attributed to enhanced electrostatic interaction between the magainin derivatives and bacterial membranes, which contain high concentrations of anionic phospholipids. A similar trend was observed using cationic derivatives of the AMP V_{681}, which is derived from a fusion construct of cecropin A, an AMP isolated from the *Hyalophora cecropia* moth, and melittin, an AMP found in bee venom (Jiang et al.,

2008). However, this study also identified a charge threshold above which hemolytic activity of the V_{681} derivatives sharply increased.

The effects of hydrophobicity on antibacterial and hemolytic activity are similarly difficult to predict. Replacing two leucine residues with more hydrophobic hexafluoroleucine residues enhanced antibacterial activity without altering hemolytic activity in buforin II, an AMP derived from stomach tissue of the Asian toad (Meng & Kumar, 2007). However, similar alterations of the magainin sequence increased hemolytic activity without substantially affecting antibacterial activity. Many other sequence modifications have been applied to a wide variety of AMPs in order to enhance their selectivity for bacterial membranes or gain information regarding their mechanism of action. For a more comprehensive assessment of these efforts, the reader is referred to the review by Findlay, Zhanel, and Schweizer (2010).

Modification of an AMP sequence may also be performed to introduce properties which are beneficial for a particular application. For instance, Pan et al. (2010) recently developed a peptide linker to noncovalently attach cargo to the surface of lipid nanoparticles. This linker is generated by deleting seven amino acids from the N-terminus of the AMP melittin. The resulting peptide is substantially less cytotoxic and hemolytic than native melittin but retains the ability to stably bind and insert into lipid membranes. A fusion construct of this linker with the NEMO-binding domain peptide was used to generate perfluorocarbon nanoparticles which potently inhibited NF-κB signaling in leukemia/lymphoma cells (Pan et al., 2011). Additionally, this linker was used to create perfluorocarbon nanoparticles targeted to the inflammatory marker VCAM-1 (Pan et al., 2013). These nanoparticles accumulated in atherosclerotic plaques and breast tumors to a much greater extent than untargeted nanoparticles, making them promising vehicles for drug delivery to these sites.

While truncation of an AMP sequence permanently reduces its cytotoxicity, AMP activity may be reversibly inhibited by addition of a removable anionic blocking sequence. Desgranges et al. (2011) demonstrated this concept using a cecropin-magainin hybrid peptide containing an N-terminal polyglutamate sequence joined via a trialanine linker. This linker sequence was designed to be cleaved by neutrophil elastase, leading to activation of the cytolytic peptide only in the presence of inflammation. While this compound exhibited potent activity against *Pseudomonas aeruginosa* following incubation with recombinant neutrophil elastase, activation by bronchoalveolar lavage specimens was much less pronounced despite the presence of similar elastase concentrations (Forde, Humphreys, Greene, Fitzgerald-Hughes, & Devocelle, 2014). This effect resulted from inhibition of cytolytic activity by anionic components of the bronchoalveolar lavage samples, demonstrating the complexity of AMP development for clinical applications.

4.2 Nanoparticle delivery strategies

In addition to sequence modifications, nanoparticle delivery vehicles may be used to minimize off-target effects and improve AMP release kinetics. For instance, incorporation of melittin into the lipid monolayer of perfluorocarbon nanoparticles has been shown to reduce toxicity to sperm and vaginal epithelium (Jallouk et al., 2014) while retaining antiviral activity against HIV-1 (Hood, Jallouk, Campbell, Ratner, & Wickline, 2013). This preferential inactivation of HIV-1 virions was attributed to their relatively small surface area, high internal pressure, and inability to repair membrane defects. These findings demonstrated the potential utility of melittin-loaded nanoparticles as a topical vaginal virucide. Similarly, LL-37, an AMP derived from human cathelicidin, retained activity against *Escherichia coli* following incorporation into polymer nanoparticles (Chereddy et al., 2014). Intradermal administration of LL-37-loaded nanoparticles led to sustained LL-37 release and improved wound healing in a mouse injury model.

Like CPPs, amphipathic AMPs may self-assemble to form nanoparticles via electrostatic and hydrophobic interactions (Hamley, 2011; Zhao, Pan, et al., 2010). This self-assembly process, in turn, may be influenced through covalent addition of cationic or hydrophobic groups to the peptide sequence. Liu et al. (2009) generated one such peptide, known as CG_3R_6TAT, which consisted of cholesterol joined via a triglycine linker to a hexaarginine-modified version of Tat peptide. As shown in Fig. 3, this construct self-assembled into 150-nm nanoparticles with core–shell morphology. The presence of a CPP modified with additional cationic and hydrophobic groups enhanced the ability of these nanoparticles to disrupt and translocate across biological membranes. These nanoparticles exhibited antimicrobial activity against a wide range of bacteria, yeast, and other fungi *in vitro*, and demonstrated an antibacterial effect in a rabbit model of *Staphylococcus aureus*-induced meningitis. In a more detailed study using a rabbit model of *Cryptococcus neoformans*-induced meningitis, these nanoparticles were shown to penetrate the blood–brain barrier and appeared to have an antifungal effect similar to amphotericin B with fewer off-target effects on liver and kidney function (Wang et al., 2010).

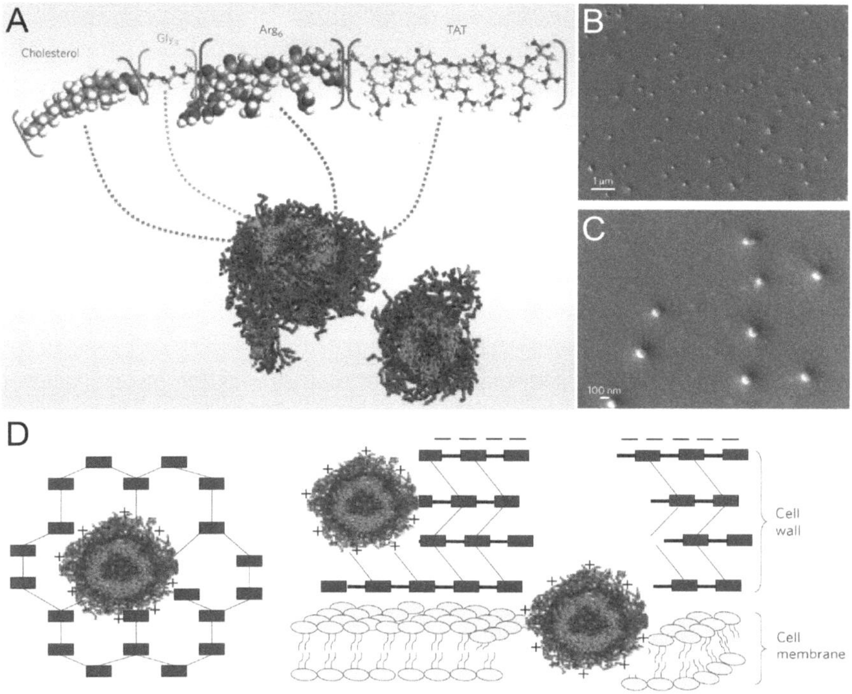

Figure 3 Self-assembly and mechanism of action of CG_3R_6TAT nanoparticles. (A) Schematic diagram of CG_3R_6TAT self-assembly into nanoparticles with core–shell morphology. (B and C) Scanning electron micrographs of CG_3R_6TAT nanoparticles. (D) Schematic diagram of CG_3R_6TAT nanoparticle antimicrobial activity. Electrostatic interaction with anionic cellular components leads to cell wall and cell membrane disruption. *All panels adapted with permission from Liu et al. (2009). Copyright 2009 Nature Publishing Group.*

5. PEPTIDE TOXINS

In addition to their primary role in defense from predators and capture of prey, animal venoms represent a potent reservoir of potentially bioactive molecules with broad therapeutic applications (Bailey & Wilce, 2001). Venom is composed of many different bioactive peptides which contribute to the pharmacologic response after delivery to the recipient (Lewis & Garcia, 2003). Due to their high potency and specificity, venom peptides remain an attractive tool for probing the structure and function of their various targets, including voltage-gated ion channels, nicotinic acetylcholine receptors, *N*-methyl-D-aspartate (NMDA) receptors, and acid-sensing ion channels (Dutertre & Lewis, 2010). Additionally, venom peptides have been used as structural templates for the development of drugs targeting these receptors, many of which have been successfully translated to the clinic. In this section, we will discuss venom peptides derived from several different organisms, as well as their potential therapeutic applications.

5.1 Mechanisms of action

The mechanism of action of individual peptide toxins varies greatly based on the organism from which they are derived and their intended biological effect. Nevertheless, many peptides derived from animal venoms are toxic to bacteria, fungi, and animal cells by virtue of their ability to disrupt biological membranes. The exact mechanism of membrane disruption is heavily debated and can vary based on the peptide being studied. However, most proposed mechanisms are based on several commonly accepted models (Shai, 1999). In the "carpet" model, peptides in an α-helical conformation coat the membrane and alter its curvature in a detergent-like fashion. Increased peptide deposition eventually leads to collapse of the membrane and leakage of its contents. In the "barrel stave" model, peptides bind electrostatically to membrane phospholipids and oligomerize, resulting in the formation of transmembrane pores and osmotic lysis. In the "toroidal pore" model, insertion of peptides into the membrane results in conformational strain leading to increased distance between phospholipid head groups, cavitation, pore formation, and osmotic lysis. This disruption of biological membranes forms the basis for therapeutic application of a number of peptide toxins.

5.2 Therapeutic applications

5.2.1 Amphibian toxins

Toxins from amphibians, particularly frogs and toads, are typically delivered as a secretion from the skin, where they play a major role in defending the host from invasion by potentially lethal bacteria and fungi. Studies of skin secretions from amphibians have isolated several potential therapeutic agents with antimicrobial properties. Like other AMPs, these toxins have been studied for their ability to disrupt bacterial membranes and induce cell death. The first of these peptides to be isolated was magainin, derived from the African clawed frog *Xenopus laevis*. Additional work has resulted in a library of more than 300 cytolytic peptides found in skin secretions, ranging from 10 to 48 amino acids in length and generally cationic due to the presence of several lysine residues (Conlon, Kolodziejek, & Nowotny, 2004). Toxins commonly studied for their antimicrobial activity include dermaseptin, dermatoxin, distinctin, phylloseptin, phylloxin, and plasticins, among others (Calderon, Silva Ade, Ciancaglini, & Stábeli, 2011). Despite their efficacy as antimicrobial agents, clinical translation of many amphibian toxins has been slow due to their nonspecific toxicity.

In addition to these AMPs, other components of amphibian secretions have been isolated for their anticancer properties, such as the bufadienolide family of compounds derived from the skin glands of Chinese toads (*Bufo melanostictus Schneider* and *Bufo bufo gargarizans Cantor*). Bufadienolides are a class of C-24 steroids consisting of a characteristic α-pyrone ring at C-17 and include bufalin, cinobufagin, and resibufogenin (Li, Wang, He, & Tang, 2008). These compounds have been shown to induce reactive oxygen species-dependent autophagy pathways in a number of cancer cell lines. Yet, their clinical translation has been limited by their cardiotoxicity and low solubility in water, which presents significant formulation challenges (Yin et al., 2012).

Several nanoparticle-based strategies have been employed to enhance the therapeutic potential of amphibian toxins. As with AMPs, nanoparticles may sequester peptide toxin activity until release at a desired therapeutic site (Calderon et al., 2011). For poorly soluble compounds such as bufadienolides, delivery and therapeutic efficacy have been improved following encapsulation in nanoparticles with hydrophobic components, including liposomes (Li et al., 2008; Li, Yang, Weng, & Tang, 2009) and polymer nanoparticles composed of pluronic polyetherimide (Hu et al., 2014) or mPEG-PLGA-PLL-cRGD block copolymers (Yin et al., 2012).

5.2.2 Bee venom

The bee venom component most commonly studied for its cytolytic properties is melittin, a peptide derived from the venom of the European honeybee *Apis mellifera*. Melittin is a cationic amphipathic peptide composed of 26 amino acids and represents approximately 50% of the dry weight of bee venom. Characterized as a host defense peptide, melittin has demonstrated potent antitumor efficacy as a result of its cytolytic activity (Gajski & Garaj-Vrhovac, 2013). The ability of melittin to disrupt lipid membranes arises from oligomerization of melittin peptides to form transmembrane channels which lead to osmotic cell lysis (Tosteson & Tosteson, 1981). Each peptide chain of melittin consists of two α-helical segments which are joined in a "bent rod" configuration due to disruption of the overall helical structure at amino acid positions 10–12 (Bechinger, 1997). Despite its great promise as an antimicrobial and antitumor agent, therapeutic applications of melittin have been severely limited by its nonspecific cytotoxicity and hemolytic activity (Hoskin & Ramamoorthy, 2008).

Nanoparticle delivery systems have been shown to reduce off-target toxicity of melittin and enhance its delivery to therapeutic sites *in vivo*. Soman et al. (2008) first demonstrated the ability of melittin to stably bind and insert into the lipid monolayer of perfluorocarbon nanoparticles. Due to the biophysical properties of the lipid monolayer and the hydrophobicity of the perfluorocarbon core, nanoparticle-bound melittin peptides did not oligomerize or disrupt nanoparticle structure in this construct. In the presence of cell membranes or other lipid nanoparticles, however, formation of a hemifusion stalk between perfluorocarbon nanoparticles and target membranes allowed for melittin transfer and subsequent membrane disruption as shown in Fig. 4. Incorporation of melittin into perfluorocarbon nanoparticles significantly reduced systemic toxicity following intravenous administration while retaining antitumor activity against established melanoma tumors and precancerous skin lesions (Soman et al., 2009). N-terminal fusion of melittin with an amphipathic α-helical peptide and subsequent incorporation into lipid nanoparticles also resulted in substantial antitumor activity with reduced nonspecific cytotoxicity (Huang et al., 2013).

5.2.3 Scorpion venom

Scorpion venom contains numerous compounds which may be adapted for cancer therapy. Of these, peptides which block ion channels have been the most heavily studied. The extent of interaction with different types of ion channels depends on the properties of each individual peptide. Longer peptides (60–70 amino acids) with multiple disulfide bridges have been shown to interact with sodium channels while shorter peptides (30–40 amino acids) primarily interact with potassium and chloride channels. The most commonly used scorpion venom toxin is the 36-amino acid peptide chlorotoxin, which is derived from the deathstalker scorpion *Leiurus quinquestriatus* and has historically been used for the study of voltage-gated chloride channels. In addition to its use as a biological research tool, chlorotoxin has been identified as a potential agent for identification and treatment of gliomas. Chlorotoxin binds reliably and specifically to gliomas and other tumors of neuroectodermal origin (Lyons, O'Neal, & Sontheimer, 2002). Fluorescently labeled chlorotoxin has been shown to stain glioma cells *in vitro*, *in situ*, and in patient tissue samples

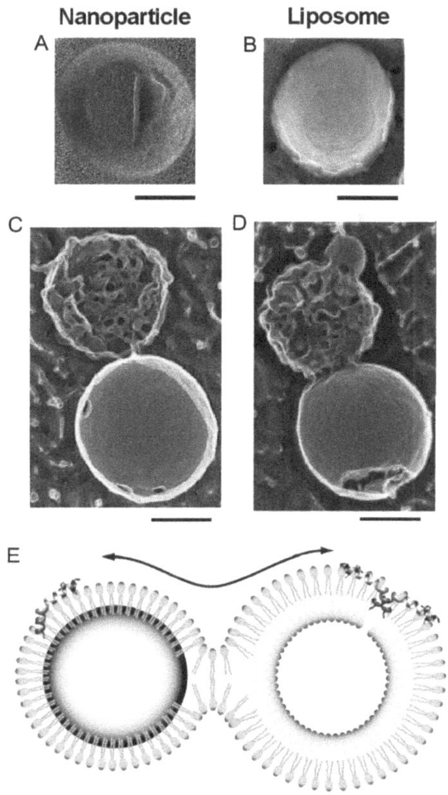

Figure 4 Interaction of melittin-loaded perfluorocarbon nanoparticles with liposomes. (A) Deep-etch platinum replica of a perfluorocarbon nanoparticle. (B) Deep-etch platinum replica of a liposome. (C and D) Deep-etch platinum replicas of perfluorocarbon nanoparticle–liposome complexes. Scale bars correspond to 100 nm. (E) Schematic diagram of hemifusion stalk formation and melittin transfer between perfluorocarbon nanoparticles and liposomes. *All panels adapted with permission from Soman et al. (2008). Copyright 2008 American Chemical Society.*

(Ding, Chua, Bay, & Gopalakrishnakone, 2014). Additionally, synthetic chlorotoxin labeled with radioactive ^{131}I is currently in clinical trials for the treatment of solid tumors and recurrent gliomas. Many different types of nanoparticles have been conjugated to chlorotoxin in order to facilitate delivery of contrast agents (Huang et al., 2011; Veiseh et al., 2010) and anticancer therapies (Veiseh et al., 2009, 2010; Wang, Gu, Xiao, Ye, & Xu, 2014) to gliomas. Furthermore, several studies have demonstrated the efficacy of chlorotoxin itself in preventing invasion and migration of glioma cells (Deshane, Garner, & Sontheimer, 2003; Soroceanu, Manning, &

Sontheimer, 1999). In addition to chlorotoxin, several other ion channel blockers have been isolated from scorpion venom, including iberiotoxin, margatoxin, and charybdotoxin, all of which could serve as potential therapeutic agents due to their antiproliferative effects on cancer cells.

5.2.4 Snake venom

Constituents of snake venom have been extensively studied for over 30 years due to their effects on a wide variety of physiological targets. Snake venom peptides and their derivatives have found particular success in the field of cardiovascular medicine. Bradykinin potentiating factors were first isolated from the venom of the lancehead viper *Bothrops jararaca* in the 1960s and demonstrated potent inhibition of angiotensin-converting enzyme (ACE) activity. Structure–function studies of these peptides identified the chemical groups most responsible for this inhibition, eventually leading to the development of the first orally active ACE inhibitor, captopril (Koh & Kini, 2012). ACE inhibitors continue to be extremely popular drugs, with one study reporting that ACE inhibitors are prescribed to 27% of all patients with hypertension (Smith & Ashiya, 2007).

Peptides derived from snake venom have had a substantial impact on the fields of thrombosis and hemostasis due to their pro- and anticoagulant activity. Factor X activators derived from elapid and viperid venoms, such as Russell's Viper Venom-X, have been primarily used as diagnostic tools in the hematology laboratory for detection of lupus anticoagulants (Takeya et al., 1992; Thiagarajan, Pengo, & Shapiro, 1986). Similarly, prothrombin and factor V activators are beneficial for diagnostic use due to the high specificity of venom components for their substrates. For instance, factor V remains the only known substrate for Russell's Viper Venom-V, allowing this compound to be developed into a specific tool for detection of factor V (Esmon & Jackson, 1973; McCleary & Kini, 2013). Ecarin, an enzyme derived from the venom of the saw-scaled viper *Echis carinatus*, promotes conversion of prothrombin to meizothrombin but is inhibited by hirudin, thereby serving as the basis for a clotting time assay for patients treated with hirudin (Nowak, 2003). Additionally, the class of enzymes known as snake venom thrombin-like enzymes (SVTLEs) has found use in diagnostic labs to defibrinogenate blood samples, as addition of SVTLEs results in formation of loose fibrin clots due to cleavage at only one of the two fibrinogen chains (Koh & Kini, 2012). Other venom components have been utilized for their procoagulant effects on platelets, where snake C-type lectin-like (snaclec) toxins have demonstrated platelet activation through binding of von Willebrand factor (e.g., botrocetin, bitiscetin) or direct binding to platelet glycoprotein receptors (e.g., convulxin; Sajevic, Leonardi, & Križaj, 2011). Platelet aggregation has also been shown to occur through activation of protease activated receptors on platelets by snake venom serine proteases such as thrombocytin (McCleary & Kini, 2013).

The major classes of anticoagulants present in snake venom include enzymes such as phospholipase A_2 (Carredano et al., 1998) and peptides such as snaclecs and "three-finger toxins" (Kini, 2006). Snaclecs, including anticoagulation factor I, bind to coagulation factors IX and X, resulting in inhibition of clotting as measured by activated partial thromboplastin time, prothrombin time, and thrombin time (Zhang et al., 2012). Similarly, three-finger toxins, which consist of four to five disulfide bridges and three β-stranded loops (Kini, 2006), have been studied for their anticoagulant effects on the extrinsic pathway of the clotting cascade (Kini, Haar, & Evans, 1988) and factor VIIa inhibition (Banerjee, Mizuguchi, Iwanaga, & Kini, 2005).

Outside the cardiovascular field, peptides derived from snake venom have been studied as anticancer therapies due to their proapoptotic properties. To this end, silica nanoparticles bearing venom from the black snake *Walterinnesia aegyptia* have been shown to enhance function of normal lymphocytes (Badr, Al-Sadoon, El-Toni, & Daghestani, 2012), as well as induce apoptosis in human prostate cancer cells (Badr, Al-Sadoon, Rabah, & Sayed, 2013) and mouse models of multiple myeloma (Al-Sadoon, Rabah, & Badr, 2013).

6. CONCLUSION

Natural peptides possess unique biological, chemical, and physical properties which underlie their application as therapeutic agents. The use of natural peptides in conjunction with nanoparticle delivery systems holds great promise for the development of new therapeutic formulations as well as novel platforms for the delivery of various cargoes.

REFERENCES

Ahn, E. H., Kim, D. W., Shin, M. J., Kim, H. R., Kim, S. M., Woo, S. J., et al. (2014). PEP-1-PEA-15 protects against toxin-induced neuronal damage in a mouse model of Parkinson's disease. *Biochimica et Biophysica Acta—General Subjects*, *1840*(6), 1686–1700. http://dx.doi.org/10.1016/j.bbagen.2014.01.004.

Allen, T. M., & Cullis, P. R. (2013). Liposomal drug delivery systems: From concept to clinical applications. *Advanced Drug Delivery Reviews*, *65*(1), 36–48. http://dx.doi.org/10.1016/j.addr.2012.09.037.

Al-Sadoon, M. K., Rabah, D. M., & Badr, G. (2013). Enhanced anticancer efficacy of snake venom combined with silica nanoparticles in a murine model of human multiple myeloma: Molecular targets for cell cycle arrest and apoptosis induction. *Cellular Immunology*, *284*(1–2), 129–138. http://dx.doi.org/10.1016/j.cellimm.2013.07.016.

Andersson, D. I., & Hughes, D. (2011). Persistence of antibiotic resistance in bacterial populations. *FEMS Microbiology Reviews*, *35*(5), 901–911. http://dx.doi.org/10.1111/j.1574-6976.2011.00289.x.

Aoshiba, K., Yokohori, N., & Nagai, A. (2003). Alveolar wall apoptosis causes lung destruction and emphysematous changes. *American Journal of Respiratory Cell and Molecular Biology*, *28*(5), 555–562. http://dx.doi.org/10.1165/rcmb.2002-0090OC.

Apte, A., Koren, E., Koshkaryev, A., & Torchilin, V. P. (2014). Doxorubicin in TAT peptide-modified multifunctional immunoliposomes demonstrates increased activity against both drug-sensitive and drug-resistant ovarian cancer models. *Cancer Biology & Therapy*, *15*(1), 69–80. http://dx.doi.org/10.4161/cbt.26609.

Babar, I. A., Cheng, C. J., Booth, C. J., Liang, X., Weidhaas, J. B., Saltzman, W. M., et al. (2012). Nanoparticle-based therapy in an *in vivo* microRNA-155 (miR-155)-dependent mouse model of lymphoma. *Proceedings of the National Academy of Sciences*, *109*(26), E1695–E1704. http://dx.doi.org/10.1073/pnas.1201516109.

Badr, G., Al-Sadoon, M. K., El-Toni, A. M., & Daghestani, M. (2012). Walterinnesia aegyptia venom combined with silica nanoparticles enhances the functioning of normal lymphocytes through PI3K/AKT, NFkappaB and ERK signaling. *Lipids in Health and Disease*, *11*(1), 27. http://dx.doi.org/10.1186/1476-511X-11-27.

Badr, G., Al-Sadoon, M. K., Rabah, D. M., & Sayed, D. (2013). Snake (Walterinnesia aegyptia) venom-loaded silica nanoparticles induce apoptosis and growth arrest in human prostate cancer cells. *Apoptosis*, *18*(3), 300–314. http://dx.doi.org/10.1007/s10495-012-0787-1.

Bailey, P., & Wilce, J. (2001). Venom as a source of useful biologically active molecules. *Emergency Medicine*, *13*(1), 28–36. http://dx.doi.org/10.1046/j.1442-2026.2001.00174.x.

Balayssac, S., Burlina, F., Convert, O., Bolbach, G., Chassaing, G., & Lequin, O. (2006). Comparison of penetratin and other homeodomain-derived cell-penetrating peptides: Interaction in a membrane-mimicking environment and cellular uptake efficiency. *Biochemistry*, *45*(5), 1408–1420. http://dx.doi.org/10.1021/bi0518390.

Banerjee, Y., Mizuguchi, J., Iwanaga, S., & Kini, R. M. (2005). Hemextin AB complex, a unique anticoagulant protein complex from Hemachatus haemachatus (African Ringhals cobra) venom that inhibits clot initiation and factor VIIa activity. *Journal of Biological Chemistry*, *280*(52), 42601–42611. http://dx.doi.org/10.1074/jbc.M508987200.

Bardag-Gorce, F., Riley, N., Nguyen, V., Montgomery, R. O., French, B. A., Li, J., et al. (2003). The mechanism of cytokeratin aggresome formation: The role of mutant ubiquitin (UBB + 1). *Experimental and Molecular Pathology*, *74*(2), 160–167. http://dx.doi.org/10.1016/S0014-4800(02)00024-2.

Bechinger, B. (1997). Structure and functions of channel-forming peptides: Magainins, cecropins, melittin and alamethicin. *The Journal of Membrane Biology*, *156*(3), 197–211. http://dx.doi.org/10.1007/s002329900201.

Bessalle, R., Haas, H., Goria, A., Shalit, I., & Fridkin, M. (1992). Augmentation of the antibacterial activity of magainin by positive-charge chain extension. *Antimicrobial Agents and Chemotherapy*, *36*(2), 313–317. http://dx.doi.org/10.1128/AAC.36.2.313.

Bradshaw, J. (2003). Cationic antimicrobial peptides: Issues for potential clinical use. *BioDrugs: Clinical Immunotherapeutics, Biopharmaceuticals and Gene Therapy*, *17*(4), 233–240.

Bush, K., Courvalin, P., Dantas, G., Davies, J., Eisenstein, B., Huovinen, P., et al. (2011). Tackling antibiotic resistance. *Nature Reviews Microbiology*, *9*(12), 894–896. http://dx.doi.org/10.1038/nrmicro2693.

Calderon, L. A., Silva, A. A. E., Ciancaglini, P., & Stábeli, R. G. (2011). Antimicrobial peptides from Phyllomedusa frogs: From biomolecular diversity to potential nanotechnologic medical applications. *Amino Acids*, *40*(1), 29–49. http://dx.doi.org/10.1007/s00726-010-0622-3.

Carredano, E., Westerlund, B., Persson, B., Saarinen, M., Ramaswamy, S., Eaker, D., et al. (1998). The three-dimensional structures of two toxins from snake venom throw light on the anticoagulant and neurotoxic sites of phospholipase A2. *Toxicon*, *36*(1), 75–92. http://dx.doi.org/10.1016/S0041-0101(97)00051-2.

Caruthers, S. D., Wickline, S. A., & Lanza, G. M. (2007). Nanotechnological applications in medicine. *Current Opinion in Biotechnology*, *18*(1), 26–30. http://dx.doi.org/10.1016/j.copbio.2007.01.006.

Chereddy, K. K., Her, C.-H., Comune, M., Moia, C., Lopes, A., Porporato, P. E., et al. (2014). PLGA nanoparticles loaded with host defense peptide LL37 promote wound healing. *Journal of Controlled Release*, *194*, 138–147. http://dx.doi.org/10.1016/j.jconrel.2014.08.016.

Chu, Q., Moellering, R. E., Hilinski, G. J., Kim, Y.-W., Grossmann, T. N., Yeh, J. T.-H., et al. (2014). Towards understanding cell penetration by stapled peptides. *Medicinal Chemistry Communications*, Advance online publication. http://dx.doi.org/10.1039/C4MD00131A.

Conlon, J. M., Kolodziejek, J., & Nowotny, N. (2004). Antimicrobial peptides from ranid frogs: Taxonomic and phylogenetic markers and a potential source of new therapeutic agents. *Biochimica et Biophysica Acta*, *1696*(1), 1–14. http://dx.doi.org/10.1016/j.bbapap.2003.09.004.

Covic, L., Gresser, A. L., Talavera, J., Swift, S., & Kuliopulos, A. (2002). Activation and inhibition of G protein-coupled receptors by cell-penetrating membrane-tethered peptides. *Proceedings of the National Academy of Sciences*, *99*(2), 643–648. http://dx.doi.org/10.1073/pnas.022460899.

Craik, D. J., Fairlie, D. P., Liras, S., & Price, D. (2013). The future of peptide-based drugs. *Chemical Biology & Drug Design*, *81*(1), 136–147. http://dx.doi.org/10.1111/cbdd.12055.

Crombez, L., Morris, M. C., Dufort, S., Aldrian-Herrada, G., Nguyen, Q., Master, G. M., et al. (2009). Targeting cyclin B1 through peptide-based delivery of siRNA prevents tumour growth. *Nucleic Acids Research*, *37*(14), 4559–4569. http://dx.doi.org/10.1093/nar/gkp451.

Danhier, F., Pourcelle, V., Marchand-Brynaert, J., Jérôme, C., Feron, O., & Préat, V. (2012). Targeting of tumor endothelium by RGD-grafted PLGA-nanoparticles. In *Methods in enzymology*. (vol. 508, pp. 157–175). Waltham, MA: Academic Press. http://europepmc.org/abstract/MED/22449925.

De Coupade, C., Fittipaldi, A., Chagnas, V., Michel, M., Carlier, S., Tasciotti, E., et al. (2005). Novel human-derived cell-penetrating peptides for specific subcellular delivery of therapeutic biomolecules. *Biochemical Journal*, *390*(2), 407. http://dx.doi.org/10.1042/BJ20050401.

De la Fuente, J. M., & Berry, C. C. (2005). Tat peptide as an efficient molecule to translocate gold nanoparticles into the cell nucleus. *Bioconjugate Chemistry*, *16*(5), 1176–1180, 10.1021/bc050033+.

Derossi, D., Joliot, A. H., Chassaing, G., & Prochiantz, A. (1994). The third helix of the Antennapedia homeodomain translocates through biological membranes. *Journal of Biological Chemistry*, *269*(14), 10444–10450.

Desai, P., Patlolla, R. R., & Singh, M. (2010). Interaction of nanoparticles and cell-penetrating peptides with skin for transdermal drug delivery. *Molecular Membrane Biology*, *27*(7), 247–259. http://dx.doi.org/10.3109/09687688.2010.522203.

Desgranges, S., Prieult, F. L., Daly, A., Lydon, J., Brennan, M., Rai, D. K., et al. (2011). *In vitro* activities of synthetic host defense propeptides processed by neutrophil elastase against cystic fibrosis pathogens. *Antimicrobial Agents and Chemotherapy*, *55*(5), 2487–2489. http://dx.doi.org/10.1128/AAC.01384-10.

Deshane, J., Garner, C. C., & Sontheimer, H. (2003). Chlorotoxin inhibits glioma cell invasion via matrix metalloproteinase-2. *Journal of Biological Chemistry*, *278*(6), 4135–4144. http://dx.doi.org/10.1074/jbc.M205662200.

Deshayes, S., Morris, M. C., Divita, G., & Heitz, F. (2005). Cell-penetrating peptides: Tools for intracellular delivery of therapeutics. *Cellular and Molecular Life Sciences*, *62*(16), 1839–1849. http://dx.doi.org/10.1007/s00018-005-5109-0.

Ding, J., Chua, P.-J., Bay, B.-H., & Gopalakrishnakone, P. (2014). Scorpion venoms as a potential source of novel cancer therapeutic compounds. *Experimental Biology and Medicine*, *239*(4), 387–393. http://dx.doi.org/10.1177/1535370213513991.

Duchardt, F., Fotin-Mleczek, M., Schwarz, H., Fischer, R., & Brock, R. (2007). A comprehensive model for the cellular uptake of cationic cell-penetrating peptides. *Traffic*, *8*(7), 848–866. http://dx.doi.org/10.1111/j.1600-0854.2007.00572.x.

Dutertre, S., & Lewis, R. J. (2010). Use of venom peptides to probe ion channel structure and function. *Journal of Biological Chemistry*, *285*(18), 13315–13320. http://dx.doi.org/10.1074/jbc.R109.076596.

Eiríksdóttir, E., Konate, K., Langel, Ü., Divita, G., & Deshayes, S. (2010). Secondary structure of cell-penetrating peptides controls membrane interaction and insertion. *Biochimica et Biophysica Acta—Biomembranes*, *1798*(6), 1119–1128. http://dx.doi.org/10.1016/j.bbamem.2010.03.005.

Elliott, G., & O'Hare, P. (1997). Intercellular trafficking and protein delivery by a herpesvirus structural protein. *Cell*, *88*(2), 223–233. http://dx.doi.org/10.1016/S0092-8674(00)81843-7.

Elmquist, A., Lindgren, M., Bartfai, T., & Langel, Ü. (2001). VE-cadherin-derived cell-penetrating peptide, pVEC, with carrier functions. *Experimental Cell Research*, *269*(2), 237–244. http://dx.doi.org/10.1006/excr.2001.5316.

El-Sayed, A., Futaki, S., & Harashima, H. (2009). Delivery of macromolecules using arginine-rich cell-penetrating peptides: Ways to overcome endosomal entrapment. *The AAPS Journal*, *11*(1), 13–22. http://dx.doi.org/10.1208/s12248-008-9071-2.

Esmon, C. T., & Jackson, C. M. (1973). The factor V activating enzyme of Russell's viper venom. *Thrombosis Research*, *2*(6), 509–524. http://dx.doi.org/10.1016/0049-3848(73)90004-2.

Findlay, B., Zhanel, G. G., & Schweizer, F. (2010). Cationic amphiphiles, a new generation of antimicrobials inspired by the natural antimicrobial peptide scaffold. *Antimicrobial Agents and Chemotherapy*, *54*(10), 4049–4058. http://dx.doi.org/10.1128/AAC.00530-10.

Forde, É., Humphreys, H., Greene, C. M., Fitzgerald-Hughes, D., & Devocelle, M. (2014). Potential of host defense peptide prodrugs as neutrophil elastase-dependent anti-infective agents for cystic fibrosis. *Antimicrobial Agents and Chemotherapy*, *58*(2), 978–985. http://dx.doi.org/10.1128/AAC.01167-13.

Frankel, A. D., & Pabo, C. O. (1988). Cellular uptake of the tat protein from human immunodeficiency virus. *Cell*, *55*(6), 1189–1193. http://dx.doi.org/10.1016/0092-8674(88)90263-2.

Futaki, S., Suzuki, T., Ohashi, W., Yagami, T., Tanaka, S., Ueda, K., et al. (2001). Arginine-rich peptides an abundant source of membrane-permeable peptides having potential as carriers for intracellular protein delivery. *Journal of Biological Chemistry*, *276*(8), 5836–5840. http://dx.doi.org/10.1074/jbc.M007540200.

Gajski, G., & Garaj-Vrhovac, V. (2013). Melittin: A lytic peptide with anticancer properties. *Environmental Toxicology and Pharmacology*, *36*(2), 697–705. http://dx.doi.org/10.1016/j.etap.2013.06.009.

Gallo, G. (2003). Making proteins into drugs: Assisted delivery of proteins and peptides into living neurons. In *Methods in cell biology*. (vol. 71, pp. 325–338). Waltham, MA: Academic Press. http://www.sciencedirect.com/science/article/pii/S0091679X03 01015X.

Gao, S., Simon, M. J., Hue, C. D., Morrison, B., & Banta, S. (2011). An unusual cell penetrating peptide identified using a plasmid display-based functional selection platform. *ACS Chemical Biology*, *6*(5), 484–491. http://dx.doi.org/10.1021/cb100423u.

Garnon, J., Lachance, C., Marco, S. D., Hel, Z., Marion, D., Ruiz, M. C., et al. (2005). Fragile X-related protein FXR1P regulates proinflammatory cytokine tumor necrosis factor expression at the post-transcriptional level. *Journal of Biological Chemistry*, *280*(7), 5750–5763. http://dx.doi.org/10.1074/jbc.M401988200.

Grdisa, M. (2011). The delivery of biologically active (therapeutic) peptides and proteins into cells. *Current Medicinal Chemistry*, *18*(9), 1373–1379. http://dx.doi.org/10.2174/092986711795029591.

Green, M., & Loewenstein, P. M. (1988). Autonomous functional domains of chemically synthesized human immunodeficiency virus tat trans-activator protein. *Cell*, *55*(6), 1179–1188. http://dx.doi.org/10.1016/0092-8674(88)90262-0.

Gros, E., Deshayes, S., Morris, M. C., Aldrian-Herrada, G., Depollier, J., Heitz, F., et al. (2006). A non-covalent peptide-based strategy for protein and peptide nucleic acid transduction. *Biochimica et Biophysica Acta—Biomembranes*, *1758*(3), 384–393. http://dx.doi.org/10.1016/j.bbamem.2006.02.006.

Gullotti, E., Park, J., & Yeo, Y. (2013). Polydopamine-based surface modification for the development of peritumorally activatable nanoparticles. *Pharmaceutical Research*, *30*(8), 1956–1967. http://dx.doi.org/10.1007/s11095-013-1039-y.

Gupta, B., Levchenko, T. S., & Torchilin, V. P. (2007). TAT peptide-modified liposomes provide enhanced gene delivery to intracranial human brain tumor xenografts in nude mice. *Oncology Research*, *16*(8), 351–359.

Hamley, I. W. (2011). Self-assembly of amphiphilic peptides. *Soft Matter*, 7(9), 4122. http://dx.doi.org/10.1039/c0sm01218a.

Hancock, R. E. W., & Sahl, H.-G. (2006). Antimicrobial and host-defense peptides as new anti-infective therapeutic strategies. *Nature Biotechnology*, *24*(12), 1551–1557. http://dx.doi.org/10.1038/nbt1267.

He, X.-H., Wang, Y., Yan, X.-T., Wang, Y.-L., Wang, C.-Y., Zhang, Z.-Z., et al. (2013). Transduction of PEP-1–heme oxygenase-1 fusion protein reduces myocardial ischemia/reperfusion injury in rats. *Journal of Cardiovascular Pharmacology*, *62*(5), 436–442. http://dx.doi.org/10.1097/FJC.0b013e3182a0b638.

Heitz, F., Morris, M. C., & Divita, G. (2009). Twenty years of cell-penetrating peptides: From molecular mechanisms to therapeutics. *British Journal of Pharmacology*, *157*(2), 195–206. http://dx.doi.org/10.1111/j.1476-5381.2009.00057.x.

Henriques, S. T., Melo, M. N., & Castanho, M. A. R. B. (2006). Cell-penetrating peptides and antimicrobial peptides: How different are they? *Biochemical Journal*, *399*(1), 1. http://dx.doi.org/10.1042/BJ20061100.

Hood, J. L., Jallouk, A. P., Campbell, N., Ratner, L., & Wickline, S. A. (2013). Cytolytic nanoparticles attenuate HIV-1 infectivity. *Antiviral Therapy*, *18*(1), 95–103. http://dx.doi.org/10.3851/IMP2346.

Hoskin, D. W., & Ramamoorthy, A. (2008). Studies on anticancer activities of antimicrobial peptides. *Biochimica et Biophysica Acta—Biomembranes*, *1778*(2), 357–375. http://dx.doi.org/10.1016/j.bbamem.2007.11.008.

Hou, K. K., Pan, H., Lanza, G. M., & Wickline, S. A. (2013). Melittin derived peptides for nanoparticle based siRNA transfection. *Biomaterials*, *34*(12), 3110–3119. http://dx.doi.org/10.1016/j.biomaterials.2013.01.037.

Hou, K. K., Pan, H., Ratner, L., Schlesinger, P. H., & Wickline, S. A. (2013). Mechanisms of nanoparticle-mediated siRNA transfection by melittin-derived peptides. *ACS Nano*, 7(10), 8605–8615. http://dx.doi.org/10.1021/nn403311c.

Hu, Q., Liang, B., Sun, Y., Guo, X.-L., Bao, Y.-J., Xie, D.-H., et al. (2014). Preparation of bufalin-loaded pluronic polyetherimide nanoparticles, cellular uptake, distribution, and

effect on colorectal cancer. *International Journal of Nanomedicine, 9*, 4035–4041. http://dx.doi.org/10.2147/IJN.S64708.
Huang, R., Han, L., Li, J., Liu, S., Shao, K., Kuang, Y., et al. (2011). Chlorotoxin-modified macromolecular contrast agent for MRI tumor diagnosis. *Biomaterials, 32*(22), 5177–5186. http://dx.doi.org/10.1016/j.biomaterials.2011.03.075.
Huang, C., Jin, H., Qian, Y., Qi, S., Luo, H., Luo, Q., et al. (2013). Hybrid melittin cytolytic peptide-driven ultrasmall lipid nanoparticles block melanoma growth *in vivo*. *ACS Nano*, 7(7), 5791–5800. http://dx.doi.org/10.1021/nn400683s.
Jallouk, A. P., Moley, K. H., Omurtag, K., Hu, G., Lanza, G. M., Wickline, S. A., et al. (2014). Nanoparticle incorporation of melittin reduces sperm and vaginal epithelium cytotoxicity. *PLoS One, 9*(4), e95411. http://dx.doi.org/10.1371/journal.pone.0095411.
Jiang, Z., Vasil, A. I., Hale, J. D., Hancock, R. E. W., Vasil, M. L., & Hodges, R. S. (2008). Effects of net charge and the number of positively charged residues on the biological activity of amphipathic α-helical cationic antimicrobial peptides. *Peptide Science, 90*(3), 369–383. http://dx.doi.org/10.1002/bip.20911.
Josephson, L., Tung, C.-H., Moore, A., & Weissleder, R. (1999). High-efficiency intracellular magnetic labeling with novel superparamagnetic-tat peptide conjugates. *Bioconjugate Chemistry, 10*(2), 186–191. http://dx.doi.org/10.1021/bc980125h.
Juliano, R., Alam, M. R., Dixit, V., & Kang, H. (2008). Mechanisms and strategies for effective delivery of antisense and siRNA oligonucleotides. *Nucleic Acids Research, 36*(12), 4158–4171. http://dx.doi.org/10.1093/nar/gkn342.
Kale, A. A., & Torchilin, V. P. (2007a). Enhanced transfection of tumor cells *in vivo* using "Smart" pH-sensitive TAT-modified pegylated liposomes. *Journal of Drug Targeting, 15*(7–8), 538–545. http://dx.doi.org/10.1080/10611860701498203.
Kale, A. A., & Torchilin, V. P. (2007b). "Smart" drug carriers: PEGylated TATp-modified pH-sensitive liposomes. *Journal of Liposome Research, 17*(3–4), 197–203. http://dx.doi.org/10.1080/08982100701525035.
Kang, M. J., Park, S. H., Kang, M. H., Park, M. J., & Choi, Y. W. (2013). Folic acid-tethered Pep-1 peptide-conjugated liposomal nanocarrier for enhanced intracellular drug delivery to cancer cells: Conformational characterization and *in vitro* cellular uptake evaluation. *International Journal of Nanomedicine, 8*, 1155–1165. http://dx.doi.org/10.2147/IJN.S39491.
Kini, R. M. (2006). Anticoagulant proteins from snake venoms: Structure, function and mechanism. *Biochemical Journal, 397*(3), 377. http://dx.doi.org/10.1042/BJ20060302.
Kini, R. M., Haar, N. C., & Evans, H. J. (1988). Non-enzymatic inhibitors of coagulation and platelet aggregation from Naja nigricollis venom are cardiotoxins. *Biochemical and Biophysical Research Communications, 150*(3), 1012–1016. http://dx.doi.org/10.1016/0006-291X(88)90729-2.
Kleemann, E., Neu, M., Jekel, N., Fink, L., Schmehl, T., Gessler, T., et al. (2005). Nanocarriers for DNA delivery to the lung based upon a TAT-derived peptide covalently coupled to PEG–PEI. *Journal of Controlled Release, 109*(1–3), 299–316. http://dx.doi.org/10.1016/j.jconrel.2005.09.036.
Ko, Y. T., Hartner, W. C., Kale, A., & Torchilin, V. P. (2008). Gene delivery into ischemic myocardium by double-targeted lipoplexes with anti-myosin antibody and TAT peptide. *Gene Therapy, 16*(1), 52–59. http://dx.doi.org/10.1038/gt.2008.135.
Koehn, F. E., & Carter, G. T. (2005). The evolving role of natural products in drug discovery. *Nature Reviews Drug Discovery, 4*(3), 206–220. http://dx.doi.org/10.1038/nrd1657.
Koh, C. Y., & Kini, R. M. (2012). From snake venom toxins to therapeutics—Cardiovascular examples. *Toxicon, 59*(4), 497–506. http://dx.doi.org/10.1016/j.toxicon.2011.03.017.
Koren, E., Apte, A., Jani, A., & Torchilin, V. P. (2012). Multifunctional PEGylated 2C5-immunoliposomes containing pH-sensitive bonds and TAT peptide for enhanced tumor cell internalization and cytotoxicity. *Journal of Controlled Release, 160*(2), 264–273. http://dx.doi.org/10.1016/j.jconrel.2011.12.002.

Lee, J.-Y., Choi, Y.-S., Suh, J.-S., Kwon, Y.-M., Yang, V. C., Lee, S.-J., et al. (2011). Cell-penetrating chitosan/doxorubicin/TAT conjugates for efficient cancer therapy. *International Journal of Cancer, 128*(10), 2470–2480. http://dx.doi.org/10.1002/ijc.25578.
Lewis, R. J., & Garcia, M. L. (2003). Therapeutic potential of venom peptides. *Nature Reviews Drug Discovery, 2*(10), 790–802. http://dx.doi.org/10.1038/nrd1197.
Li, W., Nicol, F., & Szoka, F. C., Jr. (2004). GALA: A designed synthetic pH-responsive amphipathic peptide with applications in drug and gene delivery. *Advanced Drug Delivery Reviews, 56*(7), 967–985. http://dx.doi.org/10.1016/j.addr.2003.10.041.
Li, F., Wang, T., He, H. B., & Tang, X. (2008). The properties of bufadienolides-loaded nano-emulsion and submicro-emulsion during lyophilization. *International Journal of Pharmaceutics, 349*(1–2), 291–299. http://dx.doi.org/10.1016/j.ijpharm.2007.08.011.
Li, F., Yang, R., Weng, Y., & Tang, X. (2009). Preparation and evaluation of lyophilized liposome-encapsulated bufadienolides. *Drug Development and Industrial Pharmacy, 35*(9), 1048–1058. http://dx.doi.org/10.1080/03639040902762987.
Liu, L., Guo, K., Lu, J., Venkatraman, S. S., Luo, D., Ng, K. C., et al. (2008). Biologically active core/shell nanoparticles self-assembled from cholesterol-terminated PEG–TAT for drug delivery across the blood–brain barrier. *Biomaterials, 29*(10), 1509–1517. http://dx.doi.org/10.1016/j.biomaterials.2007.11.014.
Liu, Z., Xiong, M., Gong, J., Zhang, Y., Bai, N., Luo, Y., et al. (2014). Legumain protease-activated TAT-liposome cargo for targeting tumours and their microenvironment. *Nature Communications, 5*, 4280. http://dx.doi.org/10.1038/ncomms5280.
Liu, L., Xu, K., Wang, H., Tan, P. K. J., Fan, W., Venkatraman, S. S., et al. (2009). Self-assembled cationic peptide nanoparticles as an efficient antimicrobial agent. *Nature Nanotechnology, 4*(7), 457–463. http://dx.doi.org/10.1038/nnano.2009.153.
Lyons, S. A., O'Neal, J., & Sontheimer, H. (2002). Chlorotoxin, a scorpion-derived peptide, specifically binds to gliomas and tumors of neuroectodermal origin. *Glia, 39*(2), 162–173. http://dx.doi.org/10.1002/glia.10083.
Malhotra, M., Tomaro-Duchesneau, C., & Prakash, S. (2013). Synthesis of TAT peptide-tagged PEGylated chitosan nanoparticles for siRNA delivery targeting neurodegenerative diseases. *Biomaterials, 34*(4), 1270–1280. http://dx.doi.org/10.1016/j.biomaterials.2012.10.013.
Manzoor, A. A., Lindner, L. H., Landon, C. D., Park, J.-Y., Simnick, A. J., Dreher, M. R., et al. (2012). Overcoming limitations in nanoparticle drug delivery: Triggered, intravascular release to improve drug penetration into tumors. *Cancer Research, 72*(21), 5566–5575. http://dx.doi.org/10.1158/0008-5472.CAN-12-1683.
Marks, J. R., Placone, J., Hristova, K., & Wimley, W. C. (2011). Spontaneous membrane-translocating peptides by orthogonal high-throughput screening. *Journal of the American Chemical Society, 133*(23), 8995–9004. http://dx.doi.org/10.1021/ja2017416.
Marr, A. K., Gooderham, W. J., & Hancock, R. E. (2006). Antibacterial peptides for therapeutic use: obstacles and realistic outlook. *Current Opinion in Pharmacology, 6*(5), 468–472. http://dx.doi.org/10.1016/j.coph.2006.04.006.
Marty, C., Meylan, C., Schott, H., Ballmer-Hofer, K., & Schwendener, R. A. (2004). Enhanced heparan sulfate proteoglycan-mediated uptake of cell-penetrating peptide-modified liposomes. *Cellular and Molecular Life Sciences, 61*(14), 1785–1794. http://dx.doi.org/10.1007/s00018-004-4166-0.
McCleary, R. J. R., & Kini, R. M. (2013). Snake bites and hemostasis/thrombosis. *Thrombosis Research, 132*(6), 642–646. http://dx.doi.org/10.1016/j.thromres.2013.09.031.
Meng, H., & Kumar, K. (2007). Antimicrobial activity and protease stability of peptides containing fluorinated amino acids. *Journal of the American Chemical Society, 129*(50), 15615–15622. http://dx.doi.org/10.1021/ja075373f.
Milletti, F. (2012). Cell-penetrating peptides: Classes, origin, and current landscape. *Drug Discovery Today, 17*(15–16), 850–860. http://dx.doi.org/10.1016/j.drudis.2012.03.002.

Morris, M. C., Chaloin, L., Méry, J., Heitz, F., & Divita, G. (1999). A novel potent strategy for gene delivery using a single peptide vector as a carrier. *Nucleic Acids Research*, *27*(17), 3510–3517. http://dx.doi.org/10.1093/nar/27.17.3510.

Morris, M. C., Depollier, J., Mery, J., Heitz, F., & Divita, G. (2001). A peptide carrier for the delivery of biologically active proteins into mammalian cells. *Nature Biotechnology*, *19*(12), 1173–1176. http://dx.doi.org/10.1038/nbt1201-1173.

Morris, M. C., Vidal, P., Chaloin, L., Heitz, F., & Divita, G. (1997). A new peptide vector for efficient delivery of oligonucleotides into mammalian cells. *Nucleic Acids Research*, *25*(14), 2730–2736. http://dx.doi.org/10.1093/nar/25.14.2730.

Muñoz-Morris, M. A., Heitz, F., Divita, G., & Morris, M. C. (2007). The peptide carrier Pep-1 forms biologically efficient nanoparticle complexes. *Biochemical and Biophysical Research Communications*, *355*(4), 877–882. http://dx.doi.org/10.1016/j.bbrc.2007.02.046.

Myerson, J., He, L., Lanza, G., Tollefsen, D., & Wickline, S. (2011). Thrombin-inhibiting perfluorocarbon nanoparticles provide a novel strategy for the treatment and magnetic resonance imaging of acute thrombosis. *Journal of Thrombosis and Haemostasis*, *9*(7), 1292–1300. http://dx.doi.org/10.1111/j.1538-7836.2011.04339.x.

Nakase, I., Akita, H., Kogure, K., Gräslund, A., Langel, Ü., Harashima, H., et al. (2012). Efficient intracellular delivery of nucleic acid pharmaceuticals using cell-penetrating peptides. *Accounts of Chemical Research*, *45*(7), 1132–1139. http://dx.doi.org/10.1021/ar200256e.

Nakayama, F., Yasuda, T., Umeda, S., Asada, M., Imamura, T., Meineke, V., et al. (2011). Fibroblast growth factor-12 (FGF12) translocation into intestinal epithelial cells is dependent on a novel cell-penetrating peptide domain involvement of internalization in the *in vivo* role of exogenous FGF12. *Journal of Biological Chemistry*, *286*(29), 25823–25834. http://dx.doi.org/10.1074/jbc.M110.198267.

Nguyen, L. T., Haney, E. F., & Vogel, H. J. (2011). The expanding scope of antimicrobial peptide structures and their modes of action. *Trends in Biotechnology*, *29*(9), 464–472. http://dx.doi.org/10.1016/j.tibtech.2011.05.001.

Nowak, G. (2003). The ecarin clotting time, a universal method to quantify direct thrombin inhibitors. *Pathophysiology of Haemostasis and Thrombosis*, *33*(4), 173–183. http://dx.doi.org/81505.

Ochocki, J. D., Mullen, D. G., Wattenberg, E. V., & Distefano, M. D. (2011). Evaluation of a cell penetrating prenylated peptide lacking an intrinsic fluorophore via *in situ* click reaction. *Bioorganic & Medicinal Chemistry Letters*, *21*(17), 4998–5001. http://dx.doi.org/10.1016/j.bmcl.2011.04.138.

Omata, D., Negishi, Y., Hagiwara, S., Yamamura, S., Endo-Takahashi, Y., Suzuki, R., et al. (2011). Bubble liposomes and ultrasound promoted endosomal escape of TAT-PEG liposomes as gene delivery carriers. *Molecular Pharmaceutics*, *8*(6), 2416–2423. http://dx.doi.org/10.1021/mp200353m.

Palekar, R. U., Myerson, J. W., Schlesinger, P. H., Sadler, J. E., Pan, H., & Wickline, S. A. (2013). Thrombin-targeted liposomes establish a sustained localized anticlotting barrier against acute thrombosis. *Molecular Pharmaceutics*, *10*(11), 4168–4175. http://dx.doi.org/10.1021/mp400210q.

Pan, H., Ivashyna, O., Sinha, B., Lanza, G. M., Ratner, L., Schlesinger, P. H., et al. (2011). Post-formulation peptide drug loading of nanostructures for metered control of NF-κB signaling. *Biomaterials*, *32*(1), 231–238. http://dx.doi.org/10.1016/j.biomaterials.2010.08.080.

Pan, H., Myerson, J. W., Hu, L., Marsh, J. N., Hou, K., Scott, M. J., et al. (2013). Programmable nanoparticle functionalization for *in vivo* targeting. *The FASEB Journal*, *27*(1), 255–264. http://dx.doi.org/10.1096/fj.12-218081.

Pan, H., Myerson, J. W., Ivashyna, O., Soman, N. R., Marsh, J. N., Hood, J. L., et al. (2010). Lipid membrane editing with peptide cargo linkers in cells and synthetic nanostructures. *The FASEB Journal*, *24*(8), 2928–2937. http://dx.doi.org/10.1096/fj.09-153130.

Pooga, M., Hällbrink, M., Zorko, M., & Langel, U. (1998). Cell penetration by transportan. *The FASEB Journal, 12*(1), 67–77.
Presente, A., & Dowdy, F. S. (2013). *PTD/CPP peptide-mediated delivery of siRNAs*. (November 8, 2014). http://wustl.library.ingentaconnect.com.beckerproxy.wustl.edu/content/ben/cpd/2013/00000019/00000016/art00008.
Qin, Y., Chen, H., Zhang, Q., Wang, X., Yuan, W., Kuai, R., et al. (2011). Liposome formulated with TAT-modified cholesterol for improving brain delivery and therapeutic efficacy on brain glioma in animals. *International Journal of Pharmaceutics, 420*(2), 304–312. http://dx.doi.org/10.1016/j.ijpharm.2011.09.008.
Rhee, M., & Davis, P. (2006). Mechanism of uptake of C105Y, a novel cell-penetrating peptide. *Journal of Biological Chemistry, 281*(2), 1233–1240. http://dx.doi.org/10.1074/jbc.M509813200.
Richard, J. P., Melikov, K., Vives, E., Ramos, C., Verbeure, B., Gait, M. J., et al. (2003). Cell-penetrating peptides a reevaluation of the mechanism of cellular uptake. *Journal of Biological Chemistry, 278*(1), 585–590. http://dx.doi.org/10.1074/jbc.M209548200.
Ruan, G., Agrawal, A., Marcus, A. I., & Nie, S. (2007). Imaging and tracking of Tat peptide-conjugated quantum dots in living cells: New insights into nanoparticle uptake, intracellular transport, and vesicle shedding. *Journal of the American Chemical Society, 129*(47), 14759–14766. http://dx.doi.org/10.1021/ja074936k.
Sajevic, T., Leonardi, A., & Križaj, I. (2011). Haemostatically active proteins in snake venoms. *Toxicon, 57*(5), 627–645. http://dx.doi.org/10.1016/j.toxicon.2011.01.006.
Sawant, R. R., & Torchilin, V. P. (2009). Enhanced cytotoxicity of TATp-bearing paclitaxel-loaded micelles *in vitro* and *in vivo*. *International Journal of Pharmaceutics, 374*(1–2), 114–118. http://dx.doi.org/10.1016/j.ijpharm.2009.02.022.
Schwarze, S. R., Ho, A., Vocero-Akbani, A., & Dowdy, S. F. (1999). *In vivo* protein transduction: Delivery of a biologically active protein into the mouse. *Science, 285*(5433), 1569–1572. http://dx.doi.org/10.1126/science.285.5433.1569.
Shai, Y. (1999). Mechanism of the binding, insertion and destabilization of phospholipid bilayer membranes by α-helical antimicrobial and cell non-selective membrane-lytic peptides. *Biochimica et Biophysica Acta—Biomembranes, 1462*(1–2), 55–70. http://dx.doi.org/10.1016/S0005-2736(99)00200-X.
Sharma, G., Modgil, A., Zhong, T., Sun, C., & Singh, J. (2014). Influence of short-chain cell-penetrating peptides on transport of doxorubicin encapsulating receptor-targeted liposomes across brain endothelial barrier. *Pharmaceutical Research, 31*(5), 1194–1209. http://dx.doi.org/10.1007/s11095-013-1242-x.
Simeoni, F., Morris, M. C., Heitz, F., & Divita, G. (2003). Insight into the mechanism of the peptide-based gene delivery system MPG: Implications for delivery of siRNA into mammalian cells. *Nucleic Acids Research, 31*(11), 2717–2724. http://dx.doi.org/10.1093/nar/gkg385.
Smith, R. E. T., & Ashiya, M. (2007). Antihypertensive therapies. *Nature Reviews Drug Discovery, 6*(8), 597–598. http://dx.doi.org/10.1038/nrd2354.
Soman, N. R., Baldwin, S. L., Hu, G., Marsh, J. N., Lanza, G. M., Heuser, J. E., et al. (2009). Molecularly targeted nanocarriers deliver the cytolytic peptide melittin specifically to tumor cells in mice, reducing tumor growth. *Journal of Clinical Investigation, 119*(9), 2830–2842. http://dx.doi.org/10.1172/JCI38842.
Soman, N. R., Lanza, G. M., Heuser, J. M., Schlesinger, P. H., & Wickline, S. A. (2008). Synthesis and characterization of stable fluorocarbon nanostructures as drug delivery vehicles for cytolytic peptides. *Nano Letters, 8*(4), 1131–1136. http://dx.doi.org/10.1021/nl073290r.
Soppimath, K. S., Aminabhavi, T. M., Kulkarni, A. R., & Rudzinski, W. E. (2001). Biodegradable polymeric nanoparticles as drug delivery devices. *Journal of Controlled Release, 70*(1–2), 1–20. http://dx.doi.org/10.1016/S0168-3659(00)00339-4.

Soroceanu, L., Manning, T. J., & Sontheimer, H. (1999). Modulation of glioma cell migration and invasion using Cl − and K + ion channel blockers. *The Journal of Neuroscience, 19*(14), 5942–5954.

Splith, K., & Neundorf, I. (2011). Antimicrobial peptides with cell-penetrating peptide properties and *vice versa*. *European Biophysics Journal, 40*(4), 387–397. http://dx.doi.org/10.1007/s00249-011-0682-7.

Suk, J. S., Suh, J., Choy, K., Lai, S. K., Fu, J., & Hanes, J. (2006). Gene delivery to differentiated neurotypic cells with RGD and HIV Tat peptide functionalized polymeric nanoparticles. *Biomaterials, 27*(29), 5143–5150. http://dx.doi.org/10.1016/j.biomaterials.2006.05.013.

Takeya, H., Nishida, S., Miyata, T., Kawada, S., Saisaka, Y., Morita, T., et al. (1992). Coagulation factor X activating enzyme from Russell's viper venom (RVV-X). A novel metalloproteinase with disintegrin (platelet aggregation inhibitor)-like and C-type lectin-like domains. *Journal of Biological Chemistry, 267*(20), 14109–14117.

Taylor, B. N., Mehta, R. R., Yamada, T., Lekmine, F., Christov, K., Chakrabarty, A. M., et al. (2009). Noncationic peptides obtained from azurin preferentially enter cancer cells. *Cancer Research, 69*(2), 537–546. http://dx.doi.org/10.1158/0008-5472.CAN-08-2932.

Thiagarajan, P., Pengo, V., & Shapiro, S. S. (1986). The use of the dilute Russell viper venom time for the diagnosis of lupus anticoagulants. *Blood, 68*(4), 869–874.

Torchilin, V. P., Levchenko, T. S., Rammohan, R., Volodina, N., Papahadjopoulos-Sternberg, B., & D'Souza, G. G. M. (2003). Cell transfection *in vitro* and *in vivo* with nontoxic TAT peptide-liposome–DNA complexes. *Proceedings of the National Academy of Sciences, 100*(4), 1972–1977. http://dx.doi.org/10.1073/pnas.0435906100.

Torchilin, V. P., Rammohan, R., Weissig, V., & Levchenko, T. S. (2001). TAT peptide on the surface of liposomes affords their efficient intracellular delivery even at low temperature and in the presence of metabolic inhibitors. *Proceedings of the National Academy of Sciences, 98*(15), 8786–8791. http://dx.doi.org/10.1073/pnas.151247498.

Tosteson, M. T., & Tosteson, D. C. (1981). The sting. Melittin forms channels in lipid bilayers. *Biophysical Journal, 36*(1), 109–116. http://dx.doi.org/10.1016/S0006-3495(81)84719-4.

Tran, T. D., Caruthers, S. D., Hughes, M., Marsh, J. N., Cyrus, T., Winter, P. M., et al. (2007). Clinical applications of perfluorocarbon nanoparticles for molecular imaging and targeted therapeutics. *International Journal of Nanomedicine, 2*(4), 515–526.

Tseng, Y.-L., Liu, J.-J., & Hong, R.-L. (2002). Translocation of liposomes into cancer cells by cell-penetrating peptides penetratin and Tat: A kinetic and efficacy study. *Molecular Pharmacology, 62*(4), 864–872. http://dx.doi.org/10.1124/mol.62.4.864.

Tünnemann, G., Ter-Avetisyan, G., Martin, R. M., Stöckl, M., Herrmann, A., & Cardoso, M. C. (2008). Live-cell analysis of cell penetration ability and toxicity of oligo-arginines. *Journal of Peptide Science, 14*(4), 469–476. http://dx.doi.org/10.1002/psc.968.

Veiseh, O., Gunn, J. W., Kievit, F. M., Sun, C., Fang, C., Lee, J. S. H., et al. (2009). Inhibition of tumor-cell invasion with chlorotoxin-bound superparamagnetic nanoparticles. *Small, 5*(2), 256–264. http://dx.doi.org/10.1002/smll.200800646.

Veiseh, O., Kievit, F. M., Fang, C., Mu, N., Jana, S., Leung, M. C., et al. (2010). Chlorotoxin bound magnetic nanovector tailored for cancer cell targeting, imaging, and siRNA delivery. *Biomaterials, 31*(31), 8032–8042. http://dx.doi.org/10.1016/j.biomaterials.2010.07.016.

Veldhoen, S., Laufer, S. D., Trampe, A., & Restle, T. (2006). Cellular delivery of small interfering RNA by a non-covalently attached cell-penetrating peptide: Quantitative analysis of uptake and biological effect. *Nucleic Acids Research, 34*(22), 6561–6573. http://dx.doi.org/10.1093/nar/gkl941.

Vivès, E., Brodin, P., & Lebleu, B. (1997). A truncated HIV-1 Tat protein basic domain rapidly translocates through the plasma membrane and accumulates in the cell nucleus. *Journal of Biological Chemistry, 272*(25), 16010–16017. http://dx.doi.org/10.1074/jbc.272.25.16010.

Vlieghe, P., Lisowski, V., Martinez, J., & Khrestchatisky, M. (2010). Synthetic therapeutic peptides: Science and market. *Drug Discovery Today*, *15*(1–2), 40–56. http://dx.doi.org/10.1016/j.drudis.2009.10.009.

Wagner, E., Plank, C., Zatloukal, K., Cotten, M., & Birnstiel, M. L. (1992). Influenza virus hemagglutinin HA-2 N-terminal fusogenic peptides augment gene transfer by transferrin-polylysine-DNA complexes: Toward a synthetic virus-like gene-transfer vehicle. *Proceedings of the National Academy of Sciences*, *89*(17), 7934–7938.

Wang, H., Gu, W., Xiao, N., Ye, L., & Xu, Q. (2014). Chlorotoxin-conjugated graphene oxide for targeted delivery of an anticancer drug. *International Journal of Nanomedicine*, *9*, 1433–1442. http://dx.doi.org/10.2147/IJN.S58783.

Wang, A. Z., Langer, R., & Farokhzad, O. C. (2012). Nanoparticle delivery of cancer drugs. *Annual Review of Medicine*, *63*(1), 185–198. http://dx.doi.org/10.1146/annurev-med-040210-162544.

Wang, G., Li, X., & Wang, Z. (2009). APD2: The updated antimicrobial peptide database and its application in peptide design. *Nucleic Acids Research*, *37*(Suppl. 1), D933–D937. http://dx.doi.org/10.1093/nar/gkn823.

Wang, B., Lv, L., Wang, Z., Zhao, Y., Wu, L., Fang, X., et al. (2014). Nanoparticles functionalized with Pep-1 as potential glioma targeting delivery system via interleukin 13 receptor α2-mediated endocytosis. *Biomaterials*, *35*(22), 5897–5907. http://dx.doi.org/10.1016/j.biomaterials.2014.03.068.

Wang, H., Xu, K., Liu, L., Tan, J. P. K., Chen, Y., Li, Y., et al. (2010). The efficacy of self-assembled cationic antimicrobial peptide nanoparticles against Cryptococcus neoformans for the treatment of meningitis. *Biomaterials*, *31*(10), 2874–2881. http://dx.doi.org/10.1016/j.biomaterials.2009.12.042.

Xia, H., Gao, X., Gu, G., Liu, Z., Hu, Q., Tu, Y., et al. (2012). Penetratin-functionalized PEG–PLA nanoparticles for brain drug delivery. *International Journal of Pharmaceutics*, *436*(1–2), 840–850. http://dx.doi.org/10.1016/j.ijpharm.2012.07.029.

Yin, P., Wang, Y., Qiu, Y., Hou, L., Liu, X., Qin, J., et al. (2012). Bufalin-loaded mPEG-PLGA-PLL-cRGD nanoparticles: Preparation, cellular uptake, tissue distribution, and anticancer activity. *International Journal of Nanomedicine*, 7, 3961–3969. http://dx.doi.org/10.2147/IJN.S32063.

Zasloff, M. (2002). Antimicrobial peptides in health and disease. *New England Journal of Medicine*, *347*(15), 1199–1200. http://dx.doi.org/10.1056/NEJMe020106.

Zhang, Y., Xu, X., Shen, D., Song, J., Guo, M., & Yan, X. (2012). Anticoagulation factor I, a snaclec (snake C-type lectin) from Agkistrodon acutus venom binds to FIX as well as FX: Ca2+ induced binding data. *Toxicon*, *59*(7–8), 718–723. http://dx.doi.org/10.1016/j.toxicon.2012.03.006.

Zhao, M., Kircher, M. F., Josephson, L., & Weissleder, R. (2002). Differential conjugation of Tat peptide to superparamagnetic nanoparticles and its effect on cellular uptake. *Bioconjugate Chemistry*, *13*(4), 840–844. http://dx.doi.org/10.1021/bc0255236.

Zhao, X., Pan, F., Xu, H., Yaseen, M., Shan, H., Hauser, C. A. E., et al. (2010). Molecular self-assembly and applications of designer peptide amphiphiles. *Chemical Society Reviews*, *39*(9), 3480–3498. http://dx.doi.org/10.1039/B915923C.

Zhao, P., Wang, H., Yu, M., Cao, S., Zhang, F., Chang, J., et al. (2010). Paclitaxel-loaded, folic-acid-targeted and TAT-peptide-conjugated polymeric liposomes: *In vitro* and *in vivo* evaluation. *Pharmaceutical Research*, *27*(9), 1914–1926. http://dx.doi.org/10.1007/s11095-010-0196-5.

Zhou, H., Yan, H., Pan, H., Hou, K. K., Akk, A., Springer, L. E., et al. (2014). Peptide-siRNA nanocomplexes targeting NF-κB subunit p65 suppress nascent experimental arthritis. *Journal of Clinical Investigation*, *124*(10), 4363–4374. http://dx.doi.org/10.1172/JCI75673.

CHAPTER THREE

Hybrid Protein–Synthetic Polymer Nanoparticles for Drug Delivery

Neli S. Koseva*,[1], Joanna Rydz*,†, Ekaterina V. Stoyanova*, Violeta A. Mitova*

*Institute of Polymers, Bulgarian Academy of Sciences, Sofia, Bulgaria
†Centre of Polymer and Carbon Materials, Polish Academy of Sciences, Zabrze, Poland
[1]Corresponding author: e-mail address: koseva@polymer.bas.bg

Contents

Abstract

Among the most common nanoparticulate systems, the polymeric nanocarriers have a number of key benefits, which give a great choice of delivery platforms. Nevertheless, polymeric nanoparticles possess some limitations that include use of toxic solvents in the production process, polymer degradation, drug leakage outside the diseased tissue, and polymer cytotoxicity. The combination of polymers of biological and synthetic origin is an appealing modern strategy for the production of novel nanocarriers with unprecedented properties. Proteins' interface can play an important role in determining bioactivity and toxicity and gives perspective for future development of the polymer-based nanoparticles. The design of hybrid constructs composed of synthetic polymer and biological molecules such as proteins can be considered as a straightforward tool to integrate a broad spectrum of properties and biofunctions into a single device. This review discusses hybrid protein–synthetic polymer nanoparticles with different structures and levels in complexity and functionality, in view of their applications as drug delivery systems.

Advances in Protein Chemistry and Structural Biology, Volume 98
ISSN 1876-1623
http://dx.doi.org/10.1016/bs.apcsb.2014.12.003

1. INTRODUCTION

Drug delivery via nanoparticulate systems has become a major pillar of nanomedicine. In that field, the technological innovations in the diagnostics and therapy refer to the development of useful nanovehicles or nanodevices carrying specific load (drug, imaging agent, biomolecule, etc.) to be delivered to the target site in the organism at relevant times and doses. Various nanoscale architectures have been designed—solid nanospheres or porous nanoparticles, nanotubes, nanocapsules, nanoassemblies (liposomes, micelles, and aggregates), branched macromolecules, etc. The engineered nanoparticles are smaller in size than human cells. Having dimensions less than 100 nm, they can circulate through the body, enter in the cells, and interact with the therapeutic targets. The strategic rationale for the application of drug delivery nanoconstructs arises from the findings made in genomics and proteomics providing the society with powerful tools in the combat against cancer, cardiovascular and neurodegenerative diseases, viral infectious diseases, etc.

The delivery systems can be fabricated from different materials—lipids, polymers, biological molecules, metals, and other inorganic substances. Among the most common nanoparticulate systems, the polymeric nanocarriers are widely explored as evidenced by the great number research papers and increasing number of approved drug products (Farokhzad & Langer, 2009; Shi, Votruba, Farokhzad, & Langer, 2010; Zhang et al., 2008). This is due to the fact that the polymer nanoparticles (PNPs) possess a number of key advantageous characteristics such as tissue-penetrating ability, high payload, a sustained drug release profile, drug protection from enzymatic digestion, passive targeting, etc. In addition, the availability of various polymers (synthetic and natural) and preparation methods affords a great choice of delivery platforms. Nevertheless, PNPs possess some limitations that include use of toxic solvents in the production process, polymer degradation, drug release outside the diseased tissue, and polymer cytotoxicity. The polymer carrier is responsible for the physicochemical properties of the system and drug biodistribution (Qiu & Bae, 2006). Nanoparticles interacting with cells and the extracellular environment can trigger a sequence of biological effects. PNPs are frequently taken up by the immune cells, and hence the immunocompatibility of these materials must be carefully considered (Naahidi et al., 2013). Therefore, polymer carriers were tested for inducing unacceptable toxic, immunogenic, thrombogenic, and carcinogenic responses (Donev, Koseva, Petrov, Kowalczuk, & Thome, 2011; Singh & Lillard, 2009). It was observed that, in addition to cytotoxicity, polycations could induce hemolysis and complement activation (Duncan, 2003). Though polyanions were less cytotoxic, they could display anticoagulant activity and elicit cytokine release (Duncan, 2003). The immune reaction cascade is initiated with the adsorption of opsonins to the surface of nanoparticles (Patel, 1992). The hydrophobic surfaces enhance the opsonins' adsorption and hence lead to the easy recognition of the PNPs by the macrophages of the mononuclear phagocytic system (MPS) and removal of the drug delivery devices before they can perform their therapeutic function. Therefore, size, shape, and surface characteristics are important factors for the fate of the PNPs in the biological environment. A preferred approach of masking nanoparticles from the MPS is PEGylation, i.e., modification with poly(ethylene glycol) (PEG) via covalent binding, noncovalent entrapment, or adsorption of PEG chains onto an object (Cho, Wang, Nie, Chen, & Shin, 2008; Owens & Peppas, 2006). The biological inertness and hydrophilicity are the basis of the successful *in vivo* applications of PEG as a delivery platform (Harris & Chess, 2003). PEG is an uncharged, water-soluble, nontoxic, nonimmunogenic polymer and approved by the U.S. Food and Drug Administration (FDA). PEGylation imparts many attractive properties, for instance, increased stability of the nanocolloidal system, prolonged *in vivo* circulating half-life, and altered tolerability profile, which affords more convenient dosing schedules to patients (Hamidi, Azadi, & Rafiei, 2006; Lutz & Börner, 2008; Stoyanova et al., 2013; Tzeng & Green, 2013; Veronese, 2001). Besides the obvious benefits, the permanent PEG attachment may demonstrate deficient pharmacological properties. Most of the PEGylated systems take advantage of the enhanced permeability and retention (EPR) effect (Barenholz, 2012). However, their accumulation in tumor is still modest; thus, the treatment outcome is not satisfactory. It was also observed that PEGylation significantly reduced cellular uptake of the delivery systems after they accumulate at the tumor site, which markedly impairs the *in vivo* antitumor efficiency (Yang et al., 2012). Recent investigation has indicated the formation of PEG-specific antibodies, which clear the particle (along with the drug) from the body, thus reducing its effectiveness (Rihova & Kovar, 2010).

Designing a drug carrier in view of its functions, i.e., transportation, targeting, and release, one inevitably thinks about proteins and their inherent properties perfectly matching the above-mentioned functions. Proteins are high-molar-mass and monodisperse species with precisely defined primary structure. Although their pool of building units contains only 20 L-α-amino acids, proteins display structural and functional diversity based on the different combinations of amino acid sequences arranged in linear polypeptide chains with well-defined tertiary structures. Proteins take part in the storage and the directed transport of materials, as well as in the communication channels in the biological systems. The exquisite performance of proteins is directly related to their highly organized nanoscale structures. The potential benefits of proteinaceous micro- and nanoparticles in a wide range of biomedical applications are indisputable (Lohcharoenkal, Wang, Chen, & Rojanasakul, 2014; Vassileva & Koseva, 2010). Albumin particles were commercialized as contrast agents or anticancer therapeutics (Keller, Glasheen, & Kaul, 1989). However, proteins also have limitations. Although they are biological materials, they can be toxic and induce undesired immune responses. In general, proteins are susceptible to enzymatic degradation; therefore, chemical or biological intervention is needed to enhance their capabilities for designing protein-based nanodevices (Agrawal & Hackenberger, 2013; Baslé, Joubert, & Pucheault, 2010; Kundu et al., 2014).

It is envisioned that a revolutionary advance in therapeutic nanosystems can be achieved via hybrid constructs composed of synthetic polymer and biological molecules such as proteins, polypeptides, and nucleotides. The design of polymer–protein nanoparticles can be considered as a straightforward tool to integrate a broad spectrum of properties and biofunctions into a single device. This chapter aims to outline hybrid protein–synthetic PNPs with different structures and levels in complexity and functionality and intended for drug delivery.

2. NANOPARTICLES DERIVED FROM POLYMER(S) AND COVALENTLY BOUND PROTEIN(S)

The hybrid protein/synthetic PNPs used as drug delivery systems are submicron-sized species (3–200 nm), which synthetic component can be (bio)degradable or nondegradable polymer. The combination of polymers of biological and synthetic origin is an appealing modern strategy for the production of novel biohybrid materials with unprecedented properties (Garanger & Lecommandoux, 2012; van Rijn, 2013). Interfacing artificial and biological systems at the nanoscale level is of interest for many applications. Bionano interface can play an important role in determining bioactivity and toxicity and gives perspective for future development of the polymer-based nanoparticles. Both the composition and organization of the proteins result in different properties and new functions (Mahmoudi et al., 2011). Proteins (serum proteins, enzymes, antibodies, or antigens) are generally coupled to synthetic polymers because of their biological activity, secondary structure, target specificity, diversity, and biocompatibility (Boyer, Huang, Whittaker, Bulmus, & Davis, 2011). Advanced properties like selective catalytic activity, highly specific recognition, or directional functions such as transport and the combination of various functions can so far only be achieved through the integration with native proteins with high-molar mass (Lutz & Börner, 2008). For instance, an intracellular postcontrol strategies of delivery systems based on fusion proteins which can take action only on certain cell types, although the nonselective nature in cell transduction (Huang et al., 2013).

A wide range of synthetic macromolecules plays an important complementary role in polymer bioconjugates not only due to the diversity of properties, such as stimulus-responsive functionality, biocompatibility, (bio)degradability, and desired mechanical strength, but also because they can be produced on a relatively large scale, often less expensively than biological molecules (Lutz & Börner, 2008; Wang et al., 2013). Polymer architecture, composition, backbone stability, and water solubility are important factors that determine the effectiveness of drug delivery carriers (Naahidi et al., 2013). Poly(*N*-(2-hydroxypropyl)methacrylamide) and PEG are the most commonly used nonbiodegradable synthetic polymers (Cho et al., 2008), while degradable particles are mostly obtained from poly(lactic acid) (PLA) or poly(lactic-*co*-glycolic acid) (PLGA). Depending on the method of preparation, the drug can be physically entrapped or covalently bound to the polymer matrix. Among synthetic polypeptides, poly(L-glutamic acid) (PGlu) was the first biodegradable polymer used for conjugate synthesis. Its copolymers with PEG or other synthetic chains are also referred as hybrid macromolecules (Rydz, Zawidlak-Węgrzyńska, & Christova, 2015).

2.1 Hybrid micelles derived from copolymers of PEG and PGlu

Lin et al. (2009) reported the synthesis of hybrid polymeric micelles for drug delivery applications (Fig. 1). These micelles were composed of PGlu- and PEG-mixed corona (also termed shell) and hydrophobic core of poly(propylene oxide) (PPO). The micelles were prepared via a comicellization of PGlu-*b*-PPO-*b*-PGlu and PEG-*b*-PPO in water. Under acidic conditions, PGlu, water-soluble random coils at neutral pH, is

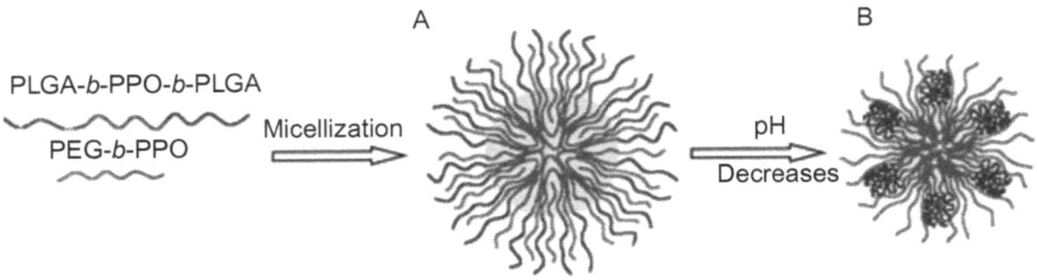

Figure 1 Schematic representation of the models proposed for the PGlu-*b*-PPO-*b*-PGlu/PEG-*b*-PPO (PGlu pointed in figure as PLGA) hybrid micelles formed under different conditions: (A) basic and neutral condition and (B) acidic condition (Lin et al., 2009). *Reprinted with permission from Elsevier.*

transformed into a water-insoluble α-helix structure. The latter transformation leads to microphase separation in the micelle shell and formation of PEG channels. These channels connect the inner core with the outer milieu, accelerating the drug diffusion out of the micelles.

The micelles were loaded with the anticancer drug doxorubicin hydrochloride. The hydrodynamic radius of the drug-loaded micelles was about 90 nm. Drug release from the carrier was studied at 37 °C in a solution with pH of 5.5 and then of 7.4. From these two sets of experiments, it became obvious that hybrid micelle structure can facilitate drug releasing only under acidic conditions, i.e., the corona containing channels formed by PEG chain aggregation can help drug diffusion from the core of micelles.

As long as the shrinkage of PPO cores at higher temperatures could lead to a quick release of the drug (Jain, Aswal, Goyal, & Bahadur, 1998; Su, Wang, & Liu, 2002), the effect of temperature on the release profile of the drug was also investigated. The authors observed that the temperature practically did not influence the release rate at pH of 7.4, but its effect became significant at pH of 5.5. This phenomenon resulted from the microphase separation which took place at lower pH and lead to PEG channel formation in the shell, i.e., a passway for the rapid drug release. Such type of hybrid micelles can be of interest for clinical applications for a couple of reasons: firstly, all blocks are generally characterized with good biocompatibility, and then, the pH range (from 7.4 to 4.0) used to tune drug releasing rate just fits that in the pathological processes. The authors infer that on this basis, a drug carrier can be designed which is able to keep the micelle structure and retain the drug in blood circulation, yet quickly release it at the tumor sites.

Nanosized drug delivery systems such as liposomes, nanoparticles, and polymeric micelles have been widely investigated for codelivery of anticancer drugs (Batist et al., 2009; Kolishetti et al., 2010; Shin et al., 2011). Nevertheless, coincorporation of drug molecules with different physicochemical properties, such as hydrophilic cisplatin and hydrophobic paclitaxel, is a rather difficult task. The simultaneous delivery of more than one drug at a controlled ratio via the same carrier to the cancer cells offers huge advantages such as spatial–temporal synchronization of drug exposure, synergistic therapeutic effects, and increased therapeutic efficacy.

A multidrug vehicle based on the functional biodegradable and biocompatible polypeptide–polymeric micelles has been recently developed (Desale, Cohen, Zhao, Kabanov, & Bronich, 2013). Using NCA-based ring-opening copolymerization, triblock copolymers composed of the blocks of ethylene glycol, glutamic acid, and phenylalanine (PEG-PGlu-PPhe) were obtained. In the synthesis of hybrid micelles consisting of PPhe hydrophobic core, cross-linked ionic PGlu intermediate shell layer, and PEG corona, the self-assembly behavior of PEG-PGlu-PPhe was exploited. The diameter of the cross-linked micelles was about 90 nm, and their ξ-potential was -20 mV. The particles were characterized by uniform (narrow size distribution) and exhibited nanogels-like behavior. As a result, a combination of drugs that differ significantly in their physical properties such as cisplatin (15 w/w% loading) and paclitaxel (9 w/w% loading) can be incorporated in the cross-linked micelles. This binary drug combination in the micelles displayed synergistic cytotoxicity against human ovarian A2780 cancer cells and exerted a superior antitumor activity in comparison with single drug-loaded micelles or free cisplatin in cancer xenograft model *in vivo*. It can be inferred that tunable composition and stability of these hybrid biodegradable micelles render them as a platform for drug combination delivery in a broad range of cancers. Altogether, this study reveals the possibility for codelivery of chemotherapeutic drugs via single well-defined and structurally tunable hybrid nanocarrier.

2.2 Hybrid nanoparticles based on (bio)degradable polyesters

Micro- and nanoparticles based on the aliphatic polyesters such as poly(ε-caprolactone) (PCL), poly(lactide), and poly(lactide-*co*-glycolide) have been approved by FDA for human clinical use. These polymers are well known as biodegradable, biocompatible, and nontoxic suitable for fabrication of drug delivery systems (Mansour, Sohn, Al-Ghananeem, & DeLuca,

2010). Nanoparticulate systems derived from aliphatic polyesters have been investigated as drug carriers for intravenous or oral administration (Danhier et al., 2012; Dinarvand, Sepehri, Manoochehri, Rouhani, & Atyabi, 2011), in cancer vaccine formulations (Silva, Videira, Gaspar, Préat, & Florindo, 2013), as topical delivery systems for the treatment of dermatological diseases (Zhang, Sun, et al., 2013; Zhang, Tsai, Ramezanli, & Michniak-Kohn, 2013), etc.

Attaching biological entities of higher complexity such as proteins to synthetic polymers has evolved rapidly in recent years. Several strategies have been developed to create PNP–protein conjugates. Recently, a review paper on conjugation strategies for linking synthetic polymers and biomolecules has been published (Börner & Lutz, 2012). The formation of a covalent bond with either the polymeric nanoparticles' core atoms or the PNP ligand molecules forms a more stable construct and allows controlling the site of attachment to the protein. Furthermore, cofunctionalization of the polymeric nanoparticles with PEG could prevent unfolding of protein (Aubin-Tam, 2013).

Glioma is the most common and deadliest of malignant primary brain tumors in adults. A drug delivery system with dual-targeting (glioma cells and blood–brain barrier, BBB) property for glioma treatment via simple design is still needed. In order to improve water solubility and bioavailability and to reduce the side effects of the anticancer drugs, polymeric nanoparticles based on polyesters grafted with PEG on the surface have been applied as a drug delivery system. Low-molar-mass methoxy-poly(ethylene glycol) (MPEG) and maleimide-poly(ethylene glycol) (MAL-PEG) as initiators for ring-opening polymerization of D,L-lactide were used to obtain MPEG–PLA and MAL-PEG–PLA block copolymers, respectively, for generating (bio)degradable polymeric nanoparticles (Zhang et al., 2014). In that case, proteins can be conjugated to the particle surface through their sulfhydryl groups reacting with the maleimide residue. The application of *N*-hydroxysuccinimide–PEG–PLA or aldehyde–PEG–PLA in the particles' formulation affords other options for coupling with proteins through their primary amine groups (Béduneau, Saulnier, & Benoit, 2007; Olivier, 2005; Olivier, Huertas, Lee, Calon, & Pardridge, 2002; Fig. 2).

For glioma treatment, fluorescence-labeled, targeted nanoparticles were obtained by *in vitro* protein binding using the enhanced green fluorescent protein–epidermal growth factor 1 (EGFP–EGF1) fusion protein (Mei et al., 2010). EGFP was used as an autofluorescent tag to label EGF1 and conjugated to nanoparticles' surface via maleimide–thiol coupling reaction.

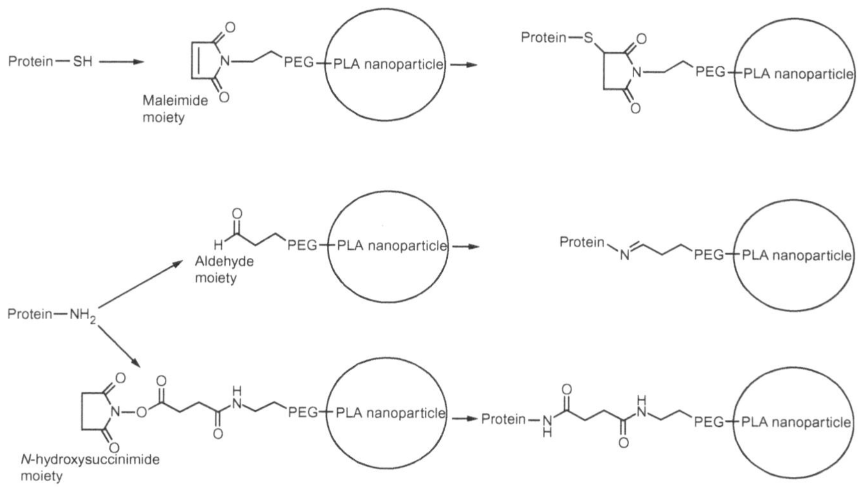

Figure 2 Coupling strategies of proteins to polyester-based nanoparticles.

EGFP–EGF1-functionalized nanoparticles have to facilitate antiglioma delivery of drug paclitaxel by targeting both neovascular and glioma cells (Zhang et al., 2014).

In a similar approach, low-density lipoprotein receptor-mediated peptide-22-conjugated nanoparticles were obtained from MPEG–PLA and HOOC–PEG–PLA block copolymers for drug delivery of paclitaxel (Zhang, Sun, et al., 2013; Zhang, Tsai, et al., 2013). Moreover, poly(ethylene glycol)-*block*-poly(ε-caprolactone) (PEG–PCL) copolymer as the matrix for delivery of paclitaxel was prepared by conjugating fluorescently labeled, activated protamine to the PEG–PCL nanoparticles (Gu et al., 2013). There are several other dual-targeting delivery systems of anticancer drug-loaded and/or fluorescence-labeled based on the PEG–polyesters that have been developed in the past few years (Caruso et al., 2011; Lu et al., 2005; Shi et al., 2012, Xin et al., 2011, 2012). Therapeutic effectiveness of the paclitaxel delivery from membrane protein-targeted polymeric nanoparticles obtained using poly(D,L-lactide-*co*-glycolide) was also evaluated. Surface functionalization of the nanoparticles with a monoclonal antibody results in a significant improvement in therapeutic efficacy and reduced local tumor recurrence in breast cancer (Bourel, Rolland, Le Verge, & Genetet, 1988; Swaminathan et al., 2013). The newly developed cascade-targeting strategy based on the fluorescence-labeled PEG–PCL nanoparticles prepared by the emulsion/solvent evaporation method was considered to be promising solution for precisely targeting of neuroglial cells (Huile et al., 2011).

Although variety of pharmaceutical agents have been developed for cancer therapy, in the absence of active transport, the hydrophobic cell membrane hinders cellular uptake, and hence it is barely possible for antitumor agents to reach their designated intracellular targets (Nichols & Bae, 2012; Park, 2013; Yan, Such, Johnston, Best, & Caruso, 2012). Insufficient internalization of chemotherapeutic agents can result in multidrug resistance of tumor cells, tumor relapse, and metastasis (Nichols & Bae, 2012; Park, 2013; Yan et al., 2012). Cell-penetrating peptides (CPPs) represent short peptides that facilitate cellular uptake of various molecular cargos—from nanosized particles to small chemical molecules and large fragments of DNA. The cargo can be associated with the peptides either through chemical linkage via covalent bonds or through noncovalent interactions. The function of the CPPs is to deliver the cargo into cells, a process that commonly occurs through endocytosis. CPPs have been used to carry therapeutics to their intracellular targets (Gao, Xiang, Meng, Liu, & Qi, 2013; Jones & Sayers, 2012; Pan, Liu, He, Wang, & Shi, 2013). The CPP which was discovered first was TATp (transactivating transcriptional activator peptide from human immunodeficiency virus 1). Due to its strong cellular penetrating ability, the peptide can be utilized to overcome multidrug resistance in tumor cells (Frankel & Pabo, 1988; Green & Loewenstein, 1988; Krpetic et al., 2011; Sethuraman & Bae, 2007; Yang et al., 2013). Nevertheless, the ability of TATp to penetrate the membrane is not cell-specific, and hence it enters both tumor cells and healthy ones (Fonseca, Pereira, & Kelley, 2009; Johnson, Cashman, Read, & Kumar-Singh, 2010; Kuai et al., 2011). Furthermore, TATp is characterized with poor stability against proteolytic enzymes presenting in bloodstream (Grunwald, Rejtar, Sawant, Wang, & Torchilin, 2009; Koren, Apte, Sawant, Grunwald, & Torchilin, 2011; Tikhonov, Ruckwardt, Berg, Hatfield, & Pauza, 2004). Therefore, intracellular delivery driven by TATp will be compromised, when TATp is cleaved by proteolytic enzymes.

The above problems were addressed via design of hybrid micelles based on the block copolymer poly(L-glutamic acid)$_n$-*b*-poly(D,L-lactic acid)$_m$ (PGlu$_n$-*b*-PLA$_m$) for intracellular delivery of doxorubicin upon activation of the TATp under tumor-acidic conditions (Wang et al., 2014; Fig. 3). In view of incorporating the CPP TATp into the micelles, a hybrid polymer was synthesized via conjugation with 1,2-distearoyl-*sn*-glycero-3-phosphoethanolamine-*b*-poly(ethylene glycol)-maleimide. The

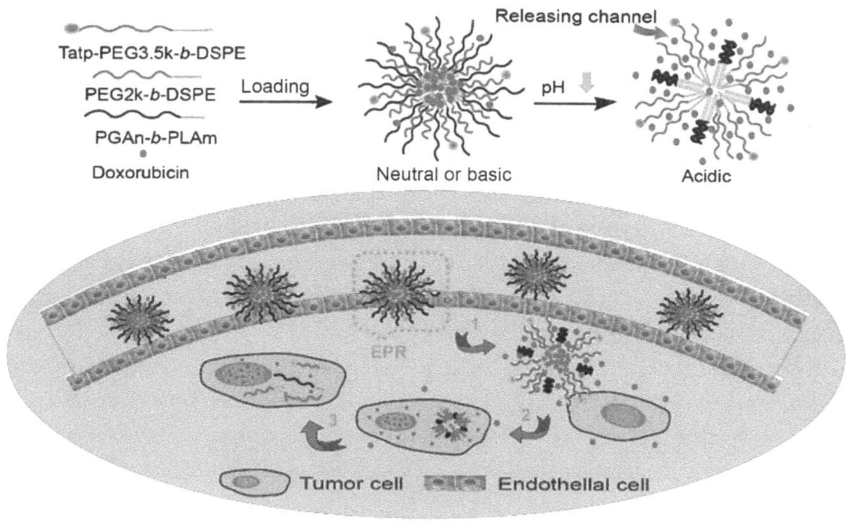

Figure 3 Illustration of hybrid micelles composed of PLA core and mixed PEG–PGlu shell with incorporated TATp intracellular delivery of doxorubicin. The accumulation of the micelles' tumor tissues is partially based on the EPR effect (1). At the tumor-acidic site, $PGlu_n$-*b*-PLA_m (PGlu pointed in figure as PGA) undergoes a conformational change which exposes TAT peptides to tumor cells and promotes cellular uptake of the micelles (2). The conformational change also produces channels that accelerate the release rate of the drug (3) (Wang et al., 2014). *Reprinted with permission from Elsevier.*

micelles demonstrated high targeting and tumor-suppressing ability against malignant melanoma *in vivo*. The conformational change of PGlu is responsible for the antitumor activity of the loaded hybrid micelles, as long as this transition leads to the formation of channels in the shell of the carrier, which in turn accelerates the release rate of chemotherapeutic agent. Moreover, $PGlu_n$-*b*-PLA_m functions to protect TATp from proteolytic cleavage in the bloodstream, yet under tumor-acidic conditions, the carrier undergoes a conformational change. As a result, it exposes TATp to cancer cells and, consequently, improves the cellular internalization of the drug-loaded micelles. As long as the hybrid micelles are composed of polypeptide-based polymers and have a small size, the EPR phenomenon has also contributed to the accumulation of the loaded particles in tumor tissues (Maeda, 2012; Matsumura & Maeda, 1986).

This study represents a rational design of bioactive delivery systems to protect degradable ligands (i.e., TATp) from the proteases during circulation and trigger their biological activity at the tumor site. In addition, the work of Wang et al. (2014) demonstrates that the micelles perform their specific biological function due to the combination of properties imparted by the components of the hybrid micelles.

Amphiphilic derivatives of poly(benzyl malate) incorporating degradable and biocompatible building blocks have been synthesized for the design of functional nanoparticles. An anticancer model drug, doxorubicin, has been successfully encapsulated into the prepared PNPs, and its release profile has been evaluated. A selected biotinylated peptide was grafted to the surfaces of the PNPs for site-specific drug delivery by targeting of biotin receptors overexpressed on the surface of several cancer cells (Loyer, Bedhouche, Huang, & Cammas-Marion, 2013).

The BBB impedes the brain drug delivery of patients suffering from brain tumor or central nervous system disorders as Parkinson's disease. Odorranalectin-conjugated PEG–PLGA nanoparticles were synthesized and blended with unmodified PEG–PLGA by double emulsion method. Novel effective, noninvasive nose-to-brain drug delivery system based on the odorranalectin-modified nanoparticles in the treatment of central nervous system disorders exhibited improved nose-to-brain drug delivery and reduced immunogenicity compared to a traditional lectin delivery system (Wen, Yan, Hu, et al., 2011). Furthermore, lactoferrin-conjugated PEG–PLA and PEG–PLGA nanoparticles were constructed as novel biodegradable brain drug delivery systems, and their delivery properties *in vitro* and *in vivo* were evaluated (Hu et al., 2009, 2011). Lectins, as odorranalectin or wheat germ agglutinin, have been widely used to conjugate with the nanoparticles in order to improve the oral absorption of the drugs (Bies, Lehr, & Woodley, 2004; Gao et al., 2006, 2007; Li et al., 2008).

The number of drug delivery formulations penetrating directly to the brain interstitium has significantly increased due to the design of polymer-based systems, functionalized with active drug-transporter proteins (Misra, Ganesh, Shahiwala, & Shah, 2003). P-glycoprotein was tested as BBB transport protein to deliver drugs to the brain parenchymal tissue via the nasal route (Graff & Pollack, 2005). Also, a recombinant Tat-β-galactosidase fusion protein showed ability to penetrate the BBB after intraperitoneal injection (Huang et al., 2013). Polymeric nanoparticles with attached large molecules such as recombinant proteins or monoclonal antibodies are potential neuropharmaceuticals, but in principle all large-molecule drugs do not cross BBB (Pardridge, 2002). Also, some difference in the drug delivery capacity can be explained with the long-term effects of interactions between PNPs and target cells which may cause toxic effects resulting from the presence of proteins attached to the surface of the PNPs.

Moreover, it is important to know the penetration capacity of various PNPs within the target cells and consequently the amount of molecules used to carry a particular compound (Caruso et al., 2011). Novel and efficient approach to deliver drug-loaded polymeric nanoparticles inside the cells for cancer therapy is based on the usage of CPP–protein complexes, which facilitate uptake across the cell membranes and thus increase the drug delivery within the cell (Torchilin, 2007).

Active targeting moieties, such as transferrin (Tf), folate ligand, glycoproteins (e.g., lactoferrin), epidermal growth factor, single-chain variable fragment, integrins, or antibody, have been conjugated to the PNPs and successfully demonstrated active targeted delivery to cancer cells (Danhier et al., 2012; Danhier, Feron, & Préat, 2010; Daniels et al., 2012; Hadinoto, Sundaresan, & Cheow, 2013; Hu et al., 2010; Wang et al., 2008; Zhang & Zhang, 2010; Zheng et al., 2010). Various examples of Tf receptor-targeted carriers used to deliver therapeutic agents to the tumor cells were studied. The efficacies of paclitaxel-loaded biodegradable PLGA, PLA–PEG, PLA–PEG–biotin, and poly[amino poly(ethylene glycol) cyanoacrylate-*co*-hexadecyl cyanoacrylate] nanoparticles conjugated to holo-Tf or biotinylated-Tf ligands were investigated on different types of tumor cells (Pulkkinen et al., 2008; Sahoo & Labhasetwar, 2005; Sahoo, Ma, & Labhasetwar, 2004; Shah, Chaudhari, Dantuluri, Murthy, & Das, 2009; Xu et al., 2005). In addition, PEG-hydroxycamptothecin conjugate-loaded poly[amino poly(ethylene glycol) cyanoacrylate-*co*-hexadecyl cyanoacrylate] and poly(hexadecyl cyanoacrylate) nanoparticles conjugated to holo-Tf ligand as well as docetaxel-loaded poly(lactide)–[D-α-tocopheryl poly(ethylene glycol) succinate] (PLA–TPGS) showed improved therapeutic efficacy (Gan & Feng, 2010; Hong et al., 2010). Other specific species have been used as active targeting protein moieties to target tumoral endothelium, e.g., endothelium-targeting ligands have to recognize such as the vascular endothelial growth factor, integrins, vascular cell adhesion molecule 1, and matrix metalloproteinases. In this strategy, targeted nanocarriers were aimed to directly kill angiogenic blood vessels and thus starving tumor cells by reduced blood supply. However, the introduction of targeting moieties often leads to increase in immunogenicity and protein adsorption (Byrne, Betancourt, & Brannon-Peppas, 2008; Cho et al., 2008; Danhier et al., 2012, 2010). Nanoparticles derived from the blend of copolymers, PLA–TPGS for hydrophobic–lipophilic balance, were designed for chemotherapy with paclitaxel. Folate ligand was conjugated to TPGS–COOH, which significantly promoted active drug targeting to the respective tumor cells and thereby increased the therapeutic effect (Pan & Feng, 2008).

Polymersomes are a new class of synthetic thin-shelled capsules based on the amphiphilic block copolymer. They are self-assembled vesicles with thicker and tougher membranes than lipids. For the first time, a drug carrier for brain delivery based on the PEG–PCL polymersomes conjugated with antibody was prepared, and its brain delivery property was evaluated (Pang et al., 2008). Furthermore, the effect of lactoferrin- and Tf-conjugated PEG–PLA polymersomes in brain targeting was evaluated (Gao et al., 2010).

Polyhydroxyalkanoates (PHAs) are aliphatic polyesters produced by a wide range of microorganisms as intracellular carbon and energy storage compounds (Lee, 1996). Due to their biocompatible and biodegradable properties, they can be successfully applied in the field of drug delivery (Madison & Huisman, 1999). Functionalization of the hydrophobic nanocarrier surface with hydrophilic active protein layer by enzymatic polymerization was a new approach to obtain semisynthetic poly[(*R*)-3-polyhydroxybutyrate] nanoparticles loaded with a model drug molecule (Lee et al., 2011). The enzyme that was used in this experiment was PHA synthase—it catalyzes polymerization of 3-hydroxybutyryl-coenzyme A to PHA. The obtained polymer chain remains covalently bound to a specific amino acid residue within the enzyme (Wodzinska et al., 1996). The covalent linkage between the polymer chain and enzyme leads to the formation of amphiphilic block copolymer, which self-assembles into micelle structure in which the protein domain faces the surrounding aqueous environment. The nanoparticles were loaded with model drug molecule, Nile red, and their surface was functionalized with tumor-specific ligand, RGD4C, fused with PHA synthase that enabled the coupling reaction (Fig. 4). The modified PHB nanoparticles demonstrated a specific affinity to MDA-MB 231 breast cancer cells. It was an indication that the tumor-specific ligand was effectively displayed on the surface of the particles, and the hybrid vehicle has the potential as delivery system for active targeting.

2.3 Hybrid nanoparticles based on nondegradable polymers

Many studies were performed to develop nondegradable polymer-based nanoparticles for drug delivery using poly(methyl methacrylate) (PMMA) or polystyrene (PS; Lee, Ryou, & Ha, 2003). It is well known that PMMA

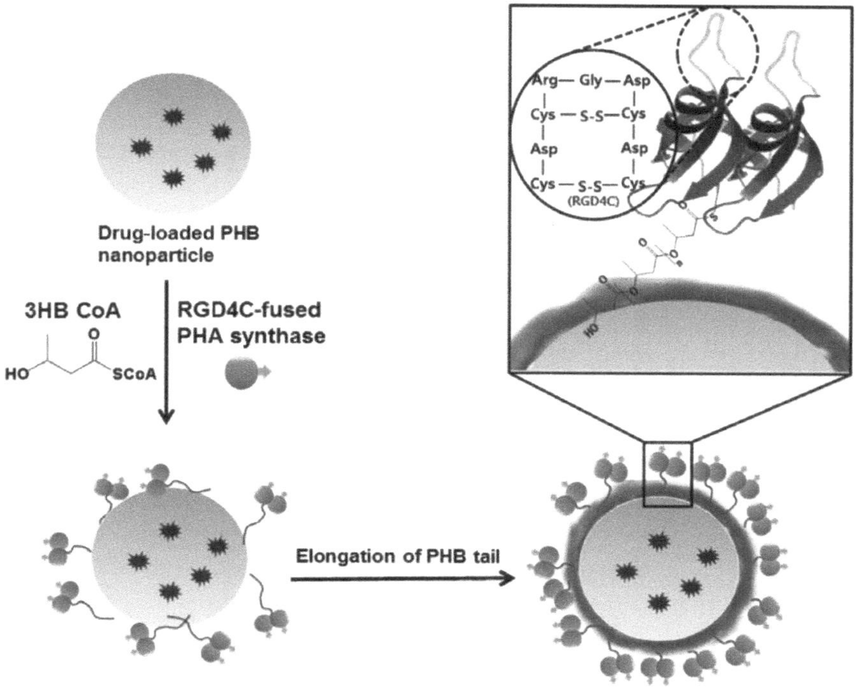

Figure 4 Surface functionalization of hydrophobic semisynthetic PHB nanoparticles through enzymatic reaction (Lee et al., 2011). *Reprinted with permission from Elsevier.*

is not biodegradable *in vivo*, but it is biocompatible and approved by FDA for medical applications. Drug delivery platform based on the self-assembled uniform core–shell spherical nanoparticles of a bovine serum albumin–poly(methyl methacrylate) (BSA–PMMA) conjugate was obtained from hydrophobic anticancer drug encapsulated in the hydrophobic PMMA core by nanoprecipitation method (Ge, Lei, & Zare, 2011; Ge, Neofytou, Lei, Beygui, & Zare, 2012) and by precipitating in supercritical CO_2 as an antisolvent (Hua & Hua, 2014). The new delivery platform based on the PNPs conjugated to albumin could selectively deliver its payload to tumor cells, while normal cells remain less affected. Proteins of albumin were considered to be specifically accumulated in tumor cells—a phenomenon based on their interaction with specific protein receptors situated in caveolae (small invaginations of the plasma membrane; Kratz, 2008; Montero, Adams, Diaz-Montero, & Glück, 2011). The anticancer drug camptothecin was encapsulated as a model chemotherapeutic agent in the BSA–PMMA nanoparticles. The hybrid nanoparticles were characterized by a tunable size and surface charge, ease of modification, excellent biocompatibility, and efficient cellular uptake. In comparison with free drug, the camptothecin-encapsulated BSA–PMMA nanoparticles demonstrated enhanced antitumor activity both *in vitro* and *in vivo*.

Jiang et al. (2014) reported the synthesis of albumin–PMMA conjugate in solution. Maleimide-terminated PMMA was prepared by the reversible addition fragmentation chain transfer polymerization in DMSO. The conjugation of albumin occurred upon slow addition of its solution in phosphate-buffered saline to the solution of the functionalized PMMA in DMSO. The system was optimized by the slow mixing of the two solutions over several hours, followed by dialysis to remove DMSO. The size of albumin-coated PMMA particles was dependent on the molar mass of the hydrophobic polymer. Changing the polymer chain length from 50 to 425 units, the particle size increased from 68 nm to 210 nm. Curcumin was employed as a model drug. An advantage of this hybrid delivery system is the fact that the nanoparticles can be prepared in one pot, while their size can be adjusted by the molar mass of the synthetic polymer. Besides the simplicity in synthesis, the presence of albumin imparts to the system specificity toward cancerous cells.

Active targeting protein moieties as a single-chain variable fragment, which aimed fibroblast activation protein commonly found in cancer cells, were conjugated to the lipid–PEG precursor, as well as a chemotherapeutic drug was conjugated to the PS-based nanoparticle core of the lipid–polymer hybrid PNPs, and active targeted delivery has been demonstrated (Messerschmidt et al., 2009).

3. HYBRID NANOPARTICLES BASED ON ELECTROSTATIC INTERACTIONS BETWEEN PROTEINS/PEPTIDES AND SYNTHETIC POLYELECTROLYTES

The combination of proteins and synthetic polymers via macromolecular assembly is a very attractive approach to the fabrication of new nanomaterials avoiding elaborate synthetic procedures. The diverse family of synthetic polymers affords wide spectrum of combinations and supramolecular structures. The assembling process is directed by versatile in nature interactions including electrostatic, hydrogen bonding, van der Waals forces, or hydrophobic interactions. The investigations of protein–polymer-interacting behavior generate new knowledge about those hybrid supramolecular constructs which aids their application as novel biohybrid materials.

Synthetic polyelectrolytes are appropriate partners since proteins are charged species. The charges on the proteins originate from the ionogenic side groups in the lysine, arginine, and histidine units (generating positive charges) and in the aspartic acid, glutamic acid, and tyrosine residues (providing negative charges) of the polypeptide chains. The net charge of the proteins can be tuned by change of the pH of the solution which provides possibility for control over the interactions with the synthetic polyelectrolytes, i.e., the delivery system with a programmed response to pH changes in the biological environment can be designed. This smart behavior can include release of drug, release of shell, or (dis)assemble of the complex structure.

The electrostatic interactions are a very convenient approach for protein–polymer assembly in aqueous media under mild conditions. They are used for obtaining hybrid particles in solution and for surface modification as well. The first approach includes mixing of the synthetic polyelectrolyte with the protein at appropriate pH of the medium which results in the formation of hybrid complex aggregates (Coué & Engbersen, 2011; Kurinomaru et al., 2012). The second approach refers to the binding of protein or polypeptide species to the oppositely charged surface of prior formed nanoparticles. The multiple deposition of oppositely charged protein and synthetic polymer in alternating fashion yields a hybrid layered surfaces applicable as biodegradable coatings and drug-releasing layers (Li, Wang, & Sun, 2012).

The most commonly used CPP is TAT peptide, derived from the protein transduction domain of the human immunodeficiency virus TAT protein. TAT peptide effectively crosses cell membranes and is able to carry macromolecular drugs (Langel et al., 2010; Liu et al., 2012; Yan et al., 2013). The peptide has been used in the design of a delivery system of insulin to the brain via the nasal-to-brain route for the treatment of Alzheimer's disease (Yan et al., 2013). The carrier presented PLGA nanoparticles modified with TAT peptide via electrostatic binding. The free carboxyl end groups of the polyester chains determine the negative charge on the surface of the PLGA nanoparticles, and the TAT modification was conducted via charge affinity. Particle coating was monitored by ξ-potential and size measurements. The size of unmodified PNPs was 189 nm, and the ξ-potential was −10.4 mV (Fig. 5A and B). After TAT attachment the particle size slightly increased, while the ξ-potential changed to positive charge, +11.3 mV (Fig. 5B).

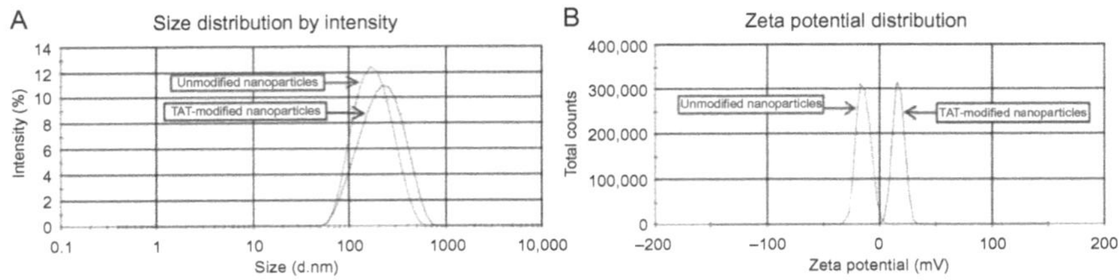

Figure 5 Characterization of the insulin-loaded PNPs: (A) particle size distribution and (B) ξ-potential (Yan et al., 2013). *Reprinted with permission from Springer.*

Drug-loading efficiency of 13.2% was achieved. The *in vitro* release experiment of insulin from the delivery system was conducted over a period of 24 h, showing a cumulative release of 23%. The TAT-modified nanoparticles displayed 4.5-fold higher intracellular delivery efficiency of insulin in comparison to the unmodified ones, and a gradient penetration across the brain compartments (olfactory bulb and cerebrum). The results obtained by Yan et al. (2012) indicate the potential of intranasal delivery of insulin to the brain, based on the use of hybrid drug delivery system, in treating neurodegenerative diseases. Moreover, the application of PLGA nanoparticles for brain delivery has been demonstrated to be safe (Semete et al., 2010; Wen, Yan, He, et al., 2011).

Pitek et al. (2012) used PS nanoparticles coated with Tf as a model system to study the interaction between nanomaterials and the biological environment. With respect to the increasing application of nanomaterials in medicine, many research efforts are focused on understanding the effect of the nanoparticle–protein interactions on the biological response *in vitro* and *in vivo*. In a biological medium, adsorption of biomolecules occurs on the surface of nanoparticles leading to the formation of a protein corona. It is accepted that the modified surface of the particles, i.e., the protein corona plays a key role for its further behavior and effective functioning as nanodevice.

The authors outlined a methodology to address the different aspects governing the formation and the function of the protein corona. PS PNPs bearing either sulfonic ($PSOSO_3H$) or carboxylic (PSCOOH) groups were used in the investigation. The PNPs were modified with Tf, applying two different immobilization approaches: physical adsorption and covalent bonding (Fig. 6). Then the stability of the hybrid species was studied in human plasma.

Holo-Tf was selected for the construction of the hybrid particles, since Tf has been largely used to functionalize nanomaterials, for its potential to

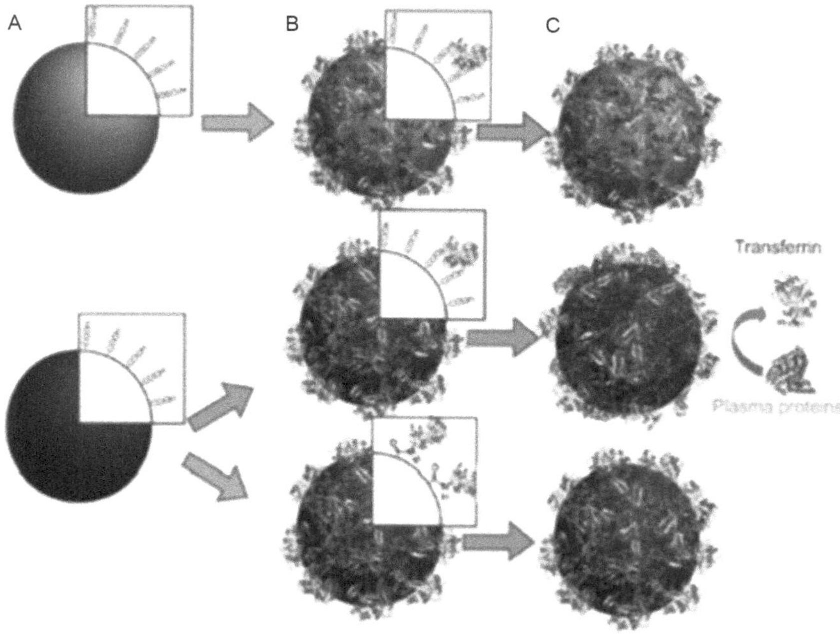

Figure 6 Strategies of PS PNPs' functionalization with Tf. Cartoon representation of 100-nm polystyrene PNPs modified with sulfonic groups ($PSOSO_3H$, upper) and carboxylic groups (PSCOOH, down) coated with Tf by either physical adsorption (Tf@$PSOSO_3H$ and Tf@PSCOOH) or covalent coupling (Tf–PSCOOH): (A) pristine PNPs, (B) Tf-coated PNPs, and (C) Tf-coated PNPs after incubation in human plasma. Both Tf@$PSOSO_3H$ and Tf–PSCOOH PNPs were stable in human plasma, while Tf on Tf@PSCOOH PNPs was easily replaced by environmental proteins (Pitek et al., 2012). *Reprinted with permission from* PLoS One.

target cancer cells (Choi, Alabi, Webster, & Davis, 2010; Zheng et al., 2010). These model systems were used to highlight the importance of the surface functional groups in defining the PNP affinity to the protein rather than the material of the core. The Tf adsorption on the sulfonated PNPs leads to the formation of a robust protein corona in the biological environment, but not for carboxylated ones. Tf coating on the sulfonated PNPs exhibited very little exchange of Tf with the environmental proteins, while Tf@PSCOOH PNPs underwent rearrangement at the bionano interface with a big loss of Tf and consequent formation of a multiprotein corona being in contact with environmental proteins. The composition of the rearranged corona was similar to that of pristine-carboxylated PNPs. Moreover, the interaction of the studied PNPs with eukaryotic protein arrays in phosphate-buffered saline and in human plasma showed that the absence of a multiprotein corona has a big impact on the extrinsic interactions of the PNPs *in situ*. It seemed that a stable coating will be able to compete with plasma proteins in binding the arrayed targets, while uncoated PS PNPs decorated by a multiprotein corona in plasma showed a higher tendency to bind the arrayed proteins. The authors stress on the importance of the methodology used to address the variety of factors governing the formation of a protein corona and to assess the behavior of engineered PNPs in a biological environment.

4. CONCLUSIONS

Progress in nanotechnology, biology, and materials' chemistry has revealed numerous opportunities for producing bioactive pharmaceutical agents capable of interacting with specific cells or subcellular components and controlling the release of therapeutic payloads. Thus, bioactive materials can be applied in the development of highly functional carriers for drug delivery. It is envisioned that a revolutionary advance in therapeutic nanosystems can be achieved via hybrid constructs composed of synthetic polymer and biological molecules such as proteins, polypeptides, and nucleotides. The design of polymer–protein nanoparticles can be considered as a straightforward tool to integrate a broad spectrum of properties and biofunctions into a single device. Nevertheless, developments of appropriate methodologies to assess the impact of the biological environment on the behavior of engineered hybrid nanoparticles are required. The accumulated knowledge in the fundamental areas such as protein engineering, polymer synthesis, particles manufacturing, medicinal chemistry in combination with the advances in diagnostic tools, and nanotechnology opens the avenue for targeted and personalized medicine.

REFERENCES

Agrawal, D., & Hackenberger, C. P. R. (2013). Site-specific chemical modifications of proteins. *Indian Journal of Chemistry Section A: Inorganic, Bio-inorganic, Physical, Theoretical and Analytical Chemistry*, *52*, 973–991.

Aubin-Tam, M. E. (2013). Conjugation of nanoparticles to proteins. *Methods in Molecular Biology*, *1025*, 19–27.

Barenholz, Y. C. (2012). Doxil®—The first FDA-approved nano-drug: Lessons learned. *Journal of Controlled Release*, *160*(2), 117–134.

Baslé, E., Joubert, N., & Pucheault, M. (2010). Protein chemical modification on endogenous amino acids. *Chemistry and Biology*, *17*(3), 213–227.

Batist, G., Gelmon, K., Chi, K., Miller, W., Chia, S., Mayer, L., et al. (2009). Safety, pharmacokinetics, and efficacy of CPX-1 liposome injection in patients with advanced solid tumors. *Clinical Cancer Research*, *15*(2), 692–700.

Béduneau, A., Saulnier, P., & Benoit, J. P. (2007). Active targeting of brain tumors using nanocarriers. *Biomaterials*, *28*(33), 4947–4967.

Bies, C., Lehr, C. M., & Woodley, J. F. (2004). Lectin-mediated drug targeting: History and applications. *Advanced Drug Delivery Reviews*, *56*(4), 425–435.

Börner, H., & Lutz, J. (2012). Synthetic–biological hybrid polymers: Synthetic designs, properties, and applications. In K. Matyjazsewski & M. Möller (Eds.), *Polymer science: A comprehensive reference* (pp. 543–585). Amsterdam, Netherlands: Elsevier B.V.

Bourel, D., Rolland, A., Le Verge, R., & Genetet, B. (1988). A new immunoreagent for cell labeling. CD3 monoclonal antibody covalently coupled to fluorescent polymethacrylic nanoparticles. *Journal of Immunological Methods*, *106*(2), 161–167.

Boyer, C., Huang, X., Whittaker, M. R., Bulmus, V., & Davis, T. P. (2011). An overview of protein–polymer particles. *Soft Matter*, 7(5), 1599–1614.

Byrne, J. D., Betancourt, T., & Brannon-Peppas, L. (2008). Active targeting schemes for nanoparticle systems in cancer therapeutics. *Advanced Drug Delivery Reviews*, *60*(15), 1615–1626.

Caruso, G., Caffo, M., Alafaci, C., Raudino, G., Cafarella, D., Lucerna, S., et al. (2011). Could nanoparticle systems have a role in the treatment of cerebral gliomas? *Nanomedicine*, 7(6), 744–752.

Cho, K., Wang, X., Nie, S., Chen, Z. G., & Shin, D. M. (2008). Therapeutic nanoparticles for drug delivery in cancer. *Clinical Cancer Research*, *14*(5), 1310–1316.

Choi, C. H., Alabi, C. A., Webster, P., & Davis, M. E. (2010). Mechanism of active targeting in solid tumors with transferrin-containing gold nanoparticles. *Proceedings of the National Academy of Sciences of the United States of America*, *107*(3), 1235–1240.

Coué, G., & Engbersen, J. F. J. (2011). Functionalized linear poly(amidoamines) are efficient vectors for intracellular protein delivery. *Journal of Controlled Release*, *152*(1), 90–98.

Danhier, F., Ansorena, E., Silva, J. M., Coco, R., Le Breton, A., & Préat, V. (2012). PLGA-based nanoparticles: An overview of biomedical applications. *Journal of Controlled Release*, *161*(2), 505–522.

Danhier, F., Feron, O., & Préat, V. (2010). To exploit the tumor microenvironment: Passive and active tumor targeting of nanocarriers for anti-cancer drug delivery. *Journal of Controlled Release*, *148*(2), 135–146.

Daniels, T. R., Bernabeu, E., Rodríguez, J. A., Patel, S., Kozman, M., Chiappetta, D. A., et al. (2012). The transferrin receptor and the targeted delivery of therapeutic agents against cancer. *Biochimica et Biophysica Acta*, *1820*(3), 291–317.

Desale, S., Cohen, S., Zhao, Y., Kabanov, A., & Bronich, T. (2013). Biodegradable hybrid polymer micelles for combination drug therapy in ovarian cancer. *Journal of Controlled Release*, *171*(3), 339–348.

Dinarvand, R., Sepehri, N., Manoochehri, S., Rouhani, H., & Atyabi, F. (2011). Polylactide-*co*-glycolide nanoparticles for controlled delivery of anticancer agents. *International Journal of Nanomedicine*, *6*, 877–895.

Donev, R., Koseva, N., Petrov, P., Kowalczuk, A., & Thome, J. (2011). Characterisation of different nanoparticles with a potential use for drug delivery in neuropsychiatric disorders. *World Journal of Biological Psychiatry*, *12*(Suppl. 1), 44–51.

Duncan, R. (2003). The dawning era of polymer therapeutics. *Nature Reviews. Drug Discovery*, *2*(5), 347–360.

Farokhzad, O. C., & Langer, R. S. (2009). Impact of nanotechnology on drug delivery. *ACS Nano*, *3*(1), 16–20.

Fonseca, S. B., Pereira, M. P., & Kelley, S. O. (2009). Recent advances in the use of cell-penetrating peptides for medical and biological applications. *Advanced Drug Delivery Reviews*, *61*(11), 953–964.

Frankel, A. D., & Pabo, C. O. (1988). Cellular uptake of the tat protein from human immunodeficiency virus. *Cell*, *55*(6), 1189–1193.

Gan, C. W., & Feng, S. S. (2010). Transferrin-conjugated nanoparticles of poly(lactide)-D-alpha-tocopheryl polyethylene glycol succinate diblock copolymer for targeted drug delivery across the blood–brain barrier. *Biomaterials*, *31*(30), 7748–7757.

Gao, H. L., Pang, Z. Q., Fan, L., Hu, K. L., Wu, B. X., & Jiang, X. G. (2010). Effect of lactoferrin- and transferrin-conjugated polymersomes in brain targeting: *In vitro* and *in vivo* evaluations. *Acta Pharmacologica Sinica*, *31*(2), 237–243.

Gao, X., Tao, W., Lu, W., Zhang, Q., Zhang, Y., Jiang, X., et al. (2006). Lectin-conjugated PEG–PLA nanoparticles: Preparation and brain delivery after intranasal administration. *Biomaterials*, *27*(18), 3482–3490.

Gao, X., Wu, B., Zhang, Q., Chen, J., Zhu, J., Zhang, W., et al. (2007). Brain delivery of vasoactive intestinal peptide enhanced with the nanoparticles conjugated with wheat germ agglutinin following intranasal administration. *Journal of Controlled Release*, *121*(3), 156–167.

Gao, W., Xiang, B., Meng, T. T., Liu, F., & Qi, X. R. (2013). Chemotherapeutic drug delivery to cancer cells using a combination of folate targeting and tumor microenvironment-sensitive polypeptides. *Biomaterials*, *34*(16), 4137–4149.

Garanger, E., & Lecommandoux, S. (2012). Towards bioactive nanovehicles based on protein polymers. *Angewandte Chemie International Edition in English*, *51*(13), 3060–3062.

Ge, J., Lei, J., & Zare, R. (2011). Bovine serum albumin–poly(methyl methacrylate) nanoparticles: An example of frustrated phase separation. *Nano Letters*, *11*(6), 2551–2554.

Ge, J., Neofytou, E., Lei, J., Beygui, R. E., & Zare, R. N. (2012). Protein–polymer hybrid nanoparticles for drug delivery. *Small*, *8*(23), 3573–3578.

Graff, C. L., & Pollack, G. M. (2005). Nasal drug administration: Potential for targeted central nervous system delivery. *Journal of Pharmaceutical Sciences*, *94*(6), 1187–1195.

Green, M., & Loewenstein, P. M. (1988). Autonomous functional domains of chemically synthesized human immunodeficiency virus tat trans-activator protein. *Cell*, *55*(6), 1179–1188.

Grunwald, J., Rejtar, T., Sawant, R., Wang, Z., & Torchilin, V. P. (2009). TAT peptide and its conjugates: Proteolytic stability. *Bioconjugate Chemistry*, *20*(8), 1531–1537.

Gu, G., Xia, H., Hu, Q., Liu, Z., Jiang, M., Kang, T., et al. (2013). PEG-*co*-PCL nanoparticles modified with MMP-2/9 activatable low molecular weight protamine for enhanced targeted glioblastoma therapy. *Biomaterials*, *34*(1), 196–208.

Hadinoto, K., Sundaresan, A., & Cheow, W. S. (2013). Lipid–polymer hybrid nanoparticles as a new generation therapeutic delivery platform: A review. *European Journal of Pharmaceutics and Biopharmaceutics*, *85*(3 Part A), 427–443.

Hamidi, M., Azadi, A., & Rafiei, P. (2006). Pharmacokinetic consequences of PEGylation. *Drug Delivery*, *13*(6), 399–409.

Harris, J., & Chess, R. (2003). Effect of PEGylation on pharmaceuticals. *Nature Reviews Drug Discovery*, *2*(3), 214–221.

Hong, M., Zhu, S., Jiang, Y., Tang, G., Sun, C., Fang, C., et al. (2010). Novel anti-tumor strategy: PEG-hydroxycamptothecin conjugate loaded transferring-PEG-nanoparticles. *Journal of Controlled Release*, *141*(1), 22–29.

Hu, C.-M. J., Kaushal, S., Tran Cao, H. S., Aryal, S., Sartor, M., Esener, S., et al. (2010). Half-antibody functionalized lipid–polymer hybrid nanoparticles for targeted drug delivery to carcinoembryonic antigen presenting pancreatic cancer cells. *Molecular Pharmaceutics*, 7(3), 914–920.

Hu, K., Li, J., Shen, Y., Lu, W., Gao, X., Zhang, Q., et al. (2009). Lactoferrin-conjugated PEG–PLA nanoparticles with improved brain delivery: *In vitro* and *in vivo* evaluations. *Journal of Controlled Release*, *134*(1), 55–61.

Hu, K., Shi, Y., Jiang, W., Han, J., Huang, S., & Jiang, X. (2011). Lactoferrin conjugated PEG–PLGA nanoparticles for brain delivery: Preparation, characterization and efficacy in Parkinson's disease. *International Journal of Pharmaceutics*, *415*(1–2), 273–283.

Hua, M., & Hua, X. (2014). Polymer nanoparticles prepared by supercritical carbon dioxide for in vivo anti-cancer drug delivery. *Nano–Micro Letters*, *6*(1), 20–23.

Huang, Y., Jiang, Y., Wang, H., Wang, J., Shin, M. C., Byun, Y., et al. (2013). Curb challenges of the "Trojan Horse" approach: Smart strategies in achieving effective yet safe cell-penetrating peptide-based drug delivery. *Advanced Drug Delivery Reviews*, *65*(10), 1299–1315.

Huile, G., Shuaiqi, P., Zhi, Y., Shijie, C., Chen, C., Xinguo, J., et al. (2011). A cascade targeting strategy for brain neuroglial cells employing nanoparticles modified with angiopep-2 peptide and EGFP–EGF1 protein. *Biomaterials*, *32*(33), 8669–8675.

Jain, N., Aswal, V., Goyal, P., & Bahadur, P. (1998). Micellar structure of an ethylene oxide–propylene oxide block copolymer: A small-angle neutron scattering study. *Journal of Physical Chemistry B*, *102*(43), 8452–8458.

Jiang, Y., Liang, M., Svejkar, D., Hart-Smith, G., Lu, H., Scarano, W., et al. (2014). Albumin-micelles via a one-pot technology platform for the delivery of drugs. *Chemical Communications*, *50*(48), 6394–6397.

Johnson, L. N., Cashman, S. M., Read, S. P., & Kumar-Singh, R. (2010). Cell penetrating peptide POD mediates delivery of recombinant proteins to retina, cornea and skin. *Vision Research*, *50*(7), 686–697.

Jones, A. T., & Sayers, E. J. (2012). Cell entry of cell penetrating peptides: Tales of tails wagging dogs. *Journal of Controlled Release*, *161*(2), 582–591.

Keller, M. W., Glasheen, W., & Kaul, S. (1989). Albunex: A safe and effective commercially produced agent for myocardial contrast echocardiography. *Journal of the American Society of Echocardiography*, *2*(1), 48–52.

Kolishetti, N., Dhar, S., Valencia, P., Lin, L., Karnik, R., Lippard, S., et al. (2010). Engineering of self-assembled nanoparticle platform for precisely controlled combination drug therapy. *Proceedings of the National Academy of Sciences*, *107*(42), 17939–17944.

Koren, E., Apte, A., Sawant, R. R., Grunwald, J., & Torchilin, V. P. (2011). Cell-penetrating TAT peptide in drug delivery systems: Proteolytic stability requirements. *Drug Delivery*, *18*(5), 377–384.

Kratz, F. (2008). Albumin as a drug carrier: Design of prodrugs, drug conjugates and nanoparticles. *Journal of Controlled Release*, *132*(3), 171–183.

Krpetic, Z., Saleemi, S., Prior, I. A., Sée, V., Qureshi, R., & Brust, M. (2011). Negotiation of intracellular membrane barriers by TAT-modified gold nanoparticles. *ACS Nano*, *5*(6), 5195–5201.

Kuai, R., Yuan, W., Li, W., Qin, Y., Tang, J., Yuan, M., et al. (2011). Targeted delivery of cargoes into a murine solid tumor by a cell-penetrating peptide and cleavable poly(ethylene glycol) comodified liposomal delivery system via systemic administration. *Molecular Pharmaceutics*, *8*(6), 2151–2161.

Kundu, B., Kurland, N. E., Bano, S., Patra, C., Engel, F. B., Yadavalli, V. K., et al. (2014). Silk proteins for biomedical applications: Bioengineering perspectives. *Progress in Polymer Science*, *39*(2), 251–267.

Kurinomaru, T., Tomita, S., Kudo, S., Ganguli, S., Nagasaki, Y., & Shiraki, K. (2012). Improved complementary polymer pair system: Switching for enzyme activity by PEGylated polymers. *Langmuir*, *28*(9), 4334–4338.

Langel, K., Lindberg, S., Copolovici, D., Arukuusk, P., Sillard, R., & Langel, Ü. (2010). Novel fatty acid modifications of transportan 10. *International Journal of Peptide Research and Therapeutics*, *16*(4), 247–255.

Lee, S. Y. (1996). Bacterial polyhydroxyalkanoates. *Biotechnology and Bioengineering, 49*(1), 1–14.

Lee, J., Jung, S. G., Park, C. S., Kim, H. Y., Batt, C. A., & Kim, Y. R. (2011). Tumor-specific hybrid polyhydroxybutyrate nanoparticle: Surface modification of nanoparticle by enzymatically synthesized functional block copolymer. *Bioorganic and Medicinal Chemistry Letters, 21*(10), 2941–2944.

Lee, W.-K., Ryou, J., & Ha, C. (2003). Retardation of enzymatic degradation of microbial. *Surface Science, 542*(3), 235–243.

Li, Y., Wang, X., & Sun, J. (2012). Layer-by-layer assembly for rapid fabrication of thick polymeric films. *Chemical Society Reviews, 41*(18), 5998–6009.

Li, J., Wu, H., Hong, J., Xu, X., Yang, H., Wu, B., et al. (2008). Odorranalectin is a small peptide lectin with potential for drug delivery and targeting. *PLoS One, 3*(6), 1–10.

Lin, J., Zhu, J., Chen, T., Lin, S., Cai, C., Zhang, L., et al. (2009). Drug releasing behavior of hybrid micelles containing polypeptide triblock copolymer. *Biomaterials, 30*(1), 108–117.

Liu, J. H., Zhao, Y. X., Guo, Q. Q., Wang, Z., Wang, H. Y., Yang, Y. X., et al. (2012). TAT-modified nanosilver for combating multidrug-resistant cancer. *Biomaterials, 33*(26), 6155–6161.

Lohcharoenkal, W., Wang, L., Chen, Y. Ch., & Rojanasakul, Y. (2014). Protein nanoparticles as drug delivery carriers for cancer therapy. *BioMed Research International* (Vols. 1–12). Article ID 180549.

Loyer, P., Bedhouche, W., Huang, Z. W., & Cammas-Marion, S. (2013). Degradable and biocompatible nanoparticles decorated with cyclic RGD peptide for efficient drug delivery to hepatoma cells in vitro. *International Journal of Pharmaceutics, 454*(2), 727–737.

Lu, W., Zhang, Y., Tan, Y. Z., Hu, K. L., Jiang, X. G., & Fu, S.-K. (2005). Cationic albumin-conjugated PEGylated nanoparticles as novel drug carrier for brain delivery. *Journal of Controlled Release, 107*(3), 428–448.

Lutz, J.-F., & Börner, H. G. (2008). Modern trends in polymer bioconjugates design. *Progress in Polymer Science, 33*, 1–39.

Madison, L. L., & Huisman, G. W. (1999). Metabolic engineering of poly (3-hydroxyalkanoates): From DNA to plastic. *Microbiology and Molecular Biology Reviews, 63*(1), 21–53.

Maeda, H. (2012). Macromolecular therapeutics in cancer treatment: The EPR effect and beyond. *Journal of Controlled Release, 164*(2), 138–144.

Mahmoudi, M., Lynch, I., Ejtehadi, M. R., Monopoli, M. P., Bombelli, F. B., & Laurent, S. (2011). Protein–nanoparticle interactions: Opportunities and challenges. *Chemical Reviews, 111*(9), 5610–5637.

Mansour, H. M., Sohn, M. J., Al-Ghananeem, A., & DeLuca, P. P. (2010). Materials for pharmaceutical dosage forms: Molecular pharmaceutics and controlled release drug delivery aspects. *International Journal of Molecular Sciences, 11*(9), 3298–3322.

Matsumura, Y., & Maeda, H. (1986). A new concept for macromolecular therapeutics in cancer chemotherapy: Mechanism of tumoritropic accumulation of proteins and the antitumor agent SMANCS. *Cancer Research, 46*(12), 6387–6392.

Mei, H., Shi, W., Pang, Z., Wang, H., Lu, W., Jiang, X., et al. (2010). EGFP–EGF1 protein-conjugated PEG–PLA nanoparticles for tissue factor targeted drug delivery. *Biomaterials, 31*(21), 5619–5626.

Messerschmidt, S. K., Musyanovych, A., Altvater, M., Scheurich, P., Pfizenmaier, K., Landfester, K., et al. (2009). Targeted lipid-coated nanoparticles: Delivery of tumor necrosis factor-functionalized particles to tumor cells. *Journal of Controlled Release, 137*(1), 69–77.

Misra, A., Ganesh, S., Shahiwala, A., & Shah, S. P. (2003). Drug delivery to the central nervous system: A review. *Journal of Pharmacy and Pharmaceutical Sciences, 6*(2), 252–573.

Montero, A., Adams, B., Diaz-Montero, C., & Glück, S. (2011). Nab-paclitaxel in the treatment of metastatic breast cancer: A comprehensive review. *Expert Review of Clinical Pharmacology*, *4*(3), 329–334.

Naahidi, S., Jafari, M., Edalat, F., Raymond, K., Khademhosseini, A., & Chen, P. (2013). Biocompatibility of engineered nanoparticles for drug delivery. *Journal of Controlled Release*, *166*(2), 182–194.

Nichols, J. W., & Bae, Y. H. (2012). Odyssey of a cancer nanoparticle: From injection site to site of action. *Nano Today*, 7(6), 606–618.

Olivier, J.-C. (2005). Drug transport to brain with targeted nanoparticles. *NeuroRx*, *2*(1), 108–119.

Olivier, J.-C., Huertas, R., Lee, H. J., Calon, F., & Pardridge, W. M. (2002). Synthesis of PEGylated immunonanoparticles. *Pharmaceutical Research*, *19*(8), 1137–1143.

Owens, D. E., 3rd., & Peppas, N. A. (2006). Opsonization, biodistribution, and pharmacokinetics of polymeric nanoparticles. *International Journal of Pharmaceutics*, *307*(1), 93–102.

Pan, J., & Feng, S. S. (2008). Targeted delivery of paclitaxel using folate-decorated poly(lactide)–vitamin E TPGS nanoparticles. *Biomaterials*, *29*(17), 2663–2672.

Pan, L., Liu, J., He, Q., Wang, L., & Shi, J. (2013). Overcoming multidrug resistance of cancer cells by direct intranuclear drug delivery using TAT-conjugated mesoporous silica nanoparticles. *Biomaterials*, *34*(11), 2719–2730.

Pang, Z., Lu, W., Gao, H., Hu, K., Chen, J., Zhang, C., et al. (2008). Preparation and brain delivery property of biodegradable polymersomes conjugated with OX26. *Journal of Controlled Release*, *128*(2), 120–127.

Pardridge, W. M. (2002). Why is the global CNS pharmaceutical market so under-penetrated? *Drug Discovery Today*, 7(1), 5–7.

Park, K. (2013). Facing the truth about nanotechnology in drug delivery. *ACS Nano*, 7(9), 7442–7447.

Patel, H. M. (1992). Serum opsonins and liposomes: Their interaction and opsonophagocytosis. *Critical Reviews in Therapeutic Drug Carrier Systems*, *9*(1), 39–90.

Pitek, A. S., Connell, D., Mahon, E., Monopoli, M. P., Bombelli, F. B., & Dawson, K. A. (2012). Transferrin coated nanoparticles: Study of the bionano interface in human plasma. *PLoS One*, 7(7), 1–14.

Pulkkinen, M., Pikkarainen, J., Wirth, T., Tarvainen, T., Haapa-aho, V., Korhonen, H., et al. (2008). Three-step tumor targeting of paclitaxel using biotinylated PLA–PEG nanoparticles and avidin–biotin technology: Formulation development and in vitro anticancer activity. *European Journal of Pharmaceutics and Biopharmaceutics*, *70*(1), 66–74.

Qiu, L., & Bae, Y. H. (2006). Polymer architecture and drug delivery. *Pharmaceutical Research*, *23*(1), 1–30.

Rihova, B., & Kovar, M. (2010). Immunogenicity and immunomodulatory properties of HPMA-based polymers. *Advanced Drug Delivery Reviews*, *62*(2), 184–191.

Rydz, J., Zawidlak-Węgrzyńska, B., & Christova, D. (2015). Degradable polymers. In M. K. Mishra (Ed.), *Encyclopedia of biomedical polymers and polymeric biomaterials*. Boca Ratón, FL, USA: CRC Press.

Sahoo, S. K., & Labhasetwar, V. (2005). Enhanced antiproliferative activity of transferrin-conjugated paclitaxel-loaded nanoparticles is mediated via sustained intracellular drug retention. *Molecular Pharmaceutics*, *2*(5), 373–383.

Sahoo, S. K., Ma, W., & Labhasetwar, V. (2004). Efficacy of transferrin-conjugated paclitaxel-loaded nanoparticles in a murine model of prostate cancer. *International Journal of Cancer*, *112*(2), 335–340.

Semete, B., Booysen, L., Lemmer, Y., Kalombo, L., Katata, L., Verschoor, J., et al. (2010). *In vivo* evaluation of the biodistribution and safety of PLGA nanoparticles as drug delivery systems. *Nanomedicine*, *6*(5), 662–671.

Sethuraman, V. A., & Bae, Y. H. (2007). TAT peptide-based micelle system for potential active targeting of anti-cancer agents to acidic solid tumors. *Journal of Controlled Release*, *118*(2), 216–224.

Shah, N., Chaudhari, K., Dantuluri, P., Murthy, R. S., & Das, S. (2009). Paclitaxel-loaded PLGA nanoparticles surface modified with transferrin and Pluronic((R))P85, an in vitro cell line and in vivo biodistribution studies on rat model. *Journal of Drug Targeting*, *17*(7), 533–542.

Shi, W., Mei, H., Deng, J., Chen, C., Wang, H., Guo, T., et al. (2012). A tissue factor targeted nanomedical system for thrombi-specific drug delivery. *Biomaterials*, *33*(30), 7643–7654.

Shi, J., Votruba, A. R., Farokhzad, O. C., & Langer, R. S. (2010). Nanotechnology in drug delivery and tissue engineering: From discovery to applications. *Nano Letters*, *10*(9), 3223–3230.

Shin, H., Alani, A., Cho, H., Bae, Y., Kolesar, J., & Kwon, G. (2011). A 3-in-1 polymeric micelle nanocontainer for poorly water-soluble drugs. *Molecular Pharmaceutics*, *8*(4), 1257–1265.

Silva, J. M., Videira, M., Gaspar, R., Préat, V., & Florindo, H. F. (2013). Immune system targeting by biodegradable nanoparticles for cancer vaccines. *Journal of Controlled Release*, *168*(2), 179–199.

Singh, R., & Lillard, J. W., Jr. (2009). Nanoparticle-based targeted drug delivery. *Experimental and Molecular Pathology*, *86*(3), 215–223.

Stoyanova, E., Mitova, V., Shestakova, P., Kowalczuk, A., Momekov, G., Momekova, D., et al. (2013). Reversibly PEGylated nanocarrier for cisplatin delivery. *Journal of Inorganic Biochemistry*, *120*, 54–62.

Su, Y., Wang, J., & Liu, H. (2002). FTIR spectroscopic study on effects of temperature and polymer composition on the structural properties of PEO–PPO–PEO block copolymer micelles. *Langmuir*, *18*(14), 5370–5374.

Swaminathan, S. K., Roger, E., Toti, U., Niu, L., Ohlfest, J. R., & Panyam, J. (2013). CD133-targeted paclitaxel delivery inhibits local tumor recurrence in a mouse model of breast cancer. *Journal of Controlled Release*, *171*(3), 280–287.

Tikhonov, I., Ruckwardt, T. J., Berg, S., Hatfield, G. S., & Pauza, C. D. (2004). Furin cleavage of the HIV-1 Tat protein. *FEBS Letters*, *565*(1–3), 89–92.

Torchilin, V. P. (2007). Targeted pharmaceutical nanocarriers for cancer therapy and imaging. *The AAPS Journal*, *9*(2), E128–E147.

Tzeng, S. Y., & Green, J. J. (2013). Therapeutic nanomedicine for brain cancer. *Therapeutic Delivery*, *4*(6), 687–704.

van Rijn, P. (2013). Polymer directed protein assemblies. *Polymers*, *5*(2), 576–599.

Vassileva, E. D., & Koseva, N. S. (2010). Sonochemically born proteinaceous micro- and nanocapsules. *Advances in Protein Chemistry and Structural Biology*, *80*, 205–252.

Veronese, F. (2001). Peptide and protein PEGylation: A review of problems and solutions. *Biomaterials*, *22*(5), 405–417.

Wang, Q. M., Gao, Z., Liu, S., Fana, B., Kang, L., Huang, W., et al. (2014). Hybrid polymeric micelles based on bioactive polypeptides as pH-responsive delivery systems against melanoma. *Biomaterials*, *35*(25), 7008–7021.

Wang, A. Z., Gu, F., Zhang, L., Chan, J. M., Radovic-Moreno, A., Shaikh, M. R., et al. (2008). Biofunctionalized targeted nanoparticles for therapeutic applications. *Expert Opinion on Biological Therapy*, *8*(8), 1063–1070.

Wang, R., Zhang, Y., Lu, D., Ge, J., Liu, Z., & Zare, R. N. (2013). Functional protein–organic/inorganic hybrid nanomaterials. *Wiley Interdisciplinary Reviews Nanomedicine and Nanobiotechnology*, *5*(4), 320–328.

Wen, Z., Yan, Z., He, R., Pang, Z., Guo, L., Qian, Y., et al. (2011). Brain targeting and toxicity study of odorranalectin-conjugated nanoparticles following intranasal administration. *Drug Delivery*, *18*(8), 555–561.

Wen, Z., Yan, Z., Hu, K., Pang, Z., Cheng, X., Guo, L., et al. (2011). Odorranalectin-conjugated nanoparticles: Preparation, brain delivery and pharmacodynamic study on Parkinson's disease following intranasal administration. *Journal of Controlled Release, 151*(2), 131–138.

Wodzinska, J., Snell, K., Rhomberg, A., Sinskey, A., Biemann, K., & Stubbe, J. (1996). Polyhydroxybutyrate synthase: Evidence for covalent catalysis. *Journal of the American Chemical Society, 118*(26), 6319–6320.

Xin, H., Jiang, X., Gu, J., Sha, X., Chen, L., Law, K., et al. (2011). Angiopep-conjugated poly(ethylene glycol)-*co*-poly(ε-caprolactone) nanoparticles as dualtargeting drug delivery system for brain glioma. *Biomaterials, 32*(18), 4293–4305.

Xin, H., Sha, X., Jiang, X., Zhang, W., Chen, L., & Fang, X. (2012). Anti-glioblastoma efficacy and safety of paclitaxel-loading angiopep-conjugated dual targeting PEG–PCL nanoparticles. *Biomaterials, 33*(32), 8167–8176.

Xu, Z., Gu, W., Huang, J., Sui, H., Zhou, Z., Yang, Y., et al. (2005). *In vitro* and *in vivo* evaluation of actively targetable nanoparticles for paclitaxel delivery. *International Journal of Pharmaceutics, 288*(2), 361–368.

Yan, Y., Such, G. K., Johnston, A. P. R., Best, J. P., & Caruso, F. (2012). Engineering particles for therapeutic delivery prospects and challenges. *ACS Nano, 6*(5), 3663–3669.

Yan, L., Wang, H., Jiang, Y., Liu, J., Wang, Z., Yang, Y., et al. (2013). Cell-penetrating peptide-modified PLGA nanoparticles for enhanced nose-to-brain macromolecular delivery. *Macromolecular Research, 21*(4), 435–441.

Yang, X. Z., Du, J. Z., Dou, S., Mao, C. Q., Long, H. Y., & Wang, J. (2012). Sheddable ternary nanoparticles for tumor acidity-targeted siRNA delivery. *ACS Nano, 6*(1), 771–781.

Yang, F., Huang, W., Li, Y., Liu, S., Jin, M., Wang, Y., et al. (2013). Anti-tumor effects in mice induced by survivin-targeted siRNA delivered through polysaccharide nanoparticles. *Biomaterials, 34*(22), 5689–5699.

Zhang, L., Gu, F. X., Chan, J. M., Wang, A. Z., Langer, R. S., & Farokhzad, O. C. (2008). Nanoparticles in medicine: Therapeutic applications and developments. *Clinical Pharmacology and Therapeutics, 83*(5), 761–769.

Zhang, B., Shen, S., Liao, Z., Shi, W., Wang, Y., Zhao, J., et al. (2014). EGFP–EGF1-conjugated nanoparticles for targeting both neovascular and glioma cells in therapy of brain glioma. *Biomaterials, 35*(13), 4133–4145.

Zhang, B., Sun, X., Mei, H., Wang, Y., Liao, Z., Chen, J., et al. (2013). LDLR-mediated peptide-22-conjugated nanoparticles for dual-targeting therapy of brain glioma. *Biomaterials, 34*(36), 9171–9182.

Zhang, Z., Tsai, P. C., Ramezanli, T., & Michniak-Kohn, B. B. (2013). Polymeric nanoparticles-based topical delivery systems for the treatment of dermatological diseases. *Wiley Interdisciplinary Reviews. Nanomedicine and Nanobiotechnology, 5*(3), 205–218.

Zhang, L., & Zhang, L. (2010). Lipid–polymer hybrid nanoparticles: Synthesis, characterization and applications. *Nano LIFE, 1*(1–2), 163–173.

Zheng, Y., Yu, B., Weecharangsan, W., Piao, L., Darby, M., Mao, Y., et al. (2010). Transferrin-conjugated lipid-coated PLGA nanoparticles for targeted delivery of aromatase inhibitor 7 alpha-APTADD to breast cancer cells. *International Journal of Pharmaceutics, 390*(2), 234–241.

CHAPTER FOUR

Efficient Delivery of Therapeutic Agents by Using Targeted Albumin Nanoparticles

Hasan Kouchakzadeh, Maryam Sadat Safavi, Seyed Abbas Shojaosadati[1]
Biotechnology Group, Chemical Engineering Faculty, Tarbiat Modares University, Tehran, Iran
[1]Corresponding author: e-mail address: shoja_sa@modares.ac.ir

Contents

Abstract

Albumin nanoparticles are one of the most important drug carriers for the delivery of therapeutic drugs, especially for the treatment of malignancies. This potential is due to their high binding capacity for both hydrophobic and hydrophilic drugs and the possibility of surface modification. Accumulation of albumin-bound drugs in the tumor interstitium occurs by the enhanced permeability and retention effect, which is also facilitated by the 60-kDa glycoprotein transcytosis pathway and binding to secreted protein, acidic and rich in cysteine located in the tumor extracellular matrix. In addition, specific ligands such as monoclonal antibodies, folic acid, transferrin, and peptides can be conjugated to the surface of albumin nanoparticles to actively target the drug to its site of action. The albumin-bound paclitaxel, Abraxane®, is one of the several therapeutic nanocarriers that have been approved for clinical use. By the development of Abraxane® that demonstrates a higher response rate and improved tolerability and therapeutic efficiency in comparison with solvent-based formulation, and with consideration of its commercial success, albumin is attracting the interest of many biotechnological and pharmaceutical companies. This chapter explores the current targeted and nontargeted albumin-based nanoparticles that are in various stages of development for the delivery of therapeutic agents in order to enhance the efficacy of cancer treatment.

Advances in Protein Chemistry and Structural Biology, Volume 98
ISSN 1876-1623
http://dx.doi.org/10.1016/bs.apcsb.2014.11.002

1. INTRODUCTION

Over recent decades, the field of nanoparticle-based medicine has received particular attention as it holds the promise of revolutionizing medical treatment with more potent, less toxic, and smarter therapeutics that can target sites of disease. Therapeutic nanoparticles can confer the ability to overcome biological barriers and effectively deliver hydrophobic, poorly water-soluble drugs, and biologics (Desai, 2012).

Currently, natural and synthetic polymers are typically used as drug delivery vectors. Among them, nanoparticles prepared with proteins (e.g., casein, albumin, and gelatin) are biocompatible, biodegradable, non-antigenic, and relatively easy to prepare. Protein nanoparticles demonstrate the ability to bind to different kinds of drugs in a relatively nonspecific manner due to their surface charge and polymeric matrix (Irache & Espuelas, 2006; Elzoghby, Samy, & Elgindy, 2012). A straightforward survey in the literature, combining the keywords "protein nanoparticles" and "drug delivery" from 2008 to 2013, clearly revealed that the number of published articles in this topic has continuously increased from ~250 to ~850 (Fig. 1A), and the number of citations has increased intensely to ~20,000 citations in 2013 (Fig. 1B).

Albumin can be commercially obtained in major quantities from egg white (ovalbumin), bovine serum (bovine serum albumin, BSA), and human serum (human serum albumin, HSA). Due to the obvious importance of HSA in clinical, metabolic, and genetic studies, it can be substituted

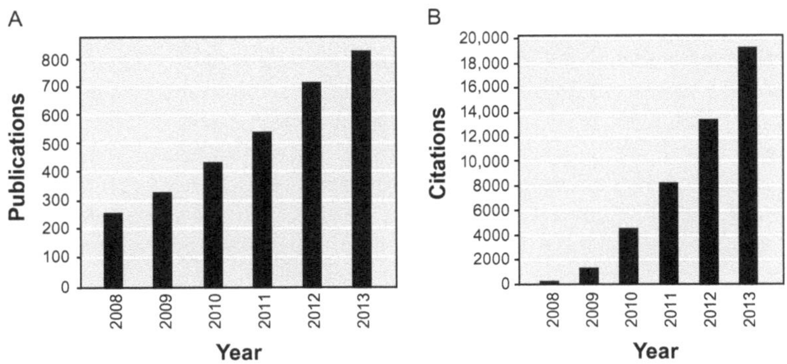

Figure 1 Literature survey. Number of publications (A) and citations (B) in Web of Knowledge combining keyword "protein nanoparticles" with "drug delivery" from 2008 to 2013.

with BSA, which is used preferably as a model protein in many *in vitro* applications. HSA, the most abundant plasma protein (35–50 g/l human serum) with a molecular weight of 66.5 kDa, is synthesized in the liver and displays an average half-life of 19 days. It shows high water solubility and is stable in the pH range of 4–9 and at 60 °C for up to 10 h (Kratz, 2008). In a physiological environment, many metabolic compounds and therapeutic molecules are transported by HSA because of its size, abundance, and binding sites available on the three-dimensional structure (Bertucci & Domenici, 2002; Carter et al., 1989; Elsadek & Kratz, 2012; Fasano et al., 2005).

Delivery systems based on albumin nanoparticles reveal several specific advantages such as stability during storage and *in vivo*, nontoxicity, reproducible preparation, and possibility of scale up during manufacturing (Elzoghby et al., 2012; Weber, Kreuter, & Langer, 2000). A large number of drugs can be effectively incorporated physically (through adsorption) or chemically (through covalent binding) into the polymeric matrix of albumin nanoparticles due to their high binding capacity (Altintas et al., 2013; Hawkins, Soon-Shiong, & Desai, 2008; Kim et al., 2011; Lanza, Langer, & Vacanti, 2011). Furthermore, a variety of targeting ligands can be attached to the surface of albumin nanoparticles due to the available superficial functional groups (i.e., amino and carboxylic) on the albumin molecule (Kouchakzadeh, Shojaosadati, Mohammadnejad, Paknejad, & Rasaee, 2012; Kouchakzadeh, Shojaosadati, Tahmasebi, & Shokri, 2013; Langer, Coester, Weber, Von Briesen, & Kreuter, 2000; Steinhauser, Langer, Strebhardt, & Spa, 2008; Ulbrich, Hekmatara, Herbert, & Kreuter, 2009).

In the following sections, after a brief review on albumin nanoparticles' fabrication methods and targeting strategies, we describe the potential of these nanoparticles for the delivery of different therapeutic agents. We also examine the pertinent developments of these systems' applications that have progressed preclinically, clinically, and in the commercial stages.

2. ALBUMIN NANOPARTICLES' FABRICATION METHODS

Various chemical- and physical-based methods can be employed for the production of albumin nanoparticles (Table 1). Although some established methods, such as desolvation, have been routinely used by many researchers and show less batch-to-batch variability, other novel methods are under further development. However, reproducibility should be kept in mind as an important factor, while the aim of these studies is to transfer

Table 1 Classification of the methods that can be used for the production of albumin nanoparticles

Chemical-based methods	Physical-based methods
Desolvation	Nanoparticle albumin-bound (NAB) technology
Emulsification	Nanospray drying
Self-assembly	Thermal gelation

the process from the laboratory to the industrial production stage and, finally, scale up. Every production process should be optimized in order to produce monodisperse nanoparticles with predictable and reproducible properties (Langer et al., 2008; von Storp, Engel, Boeker, Ploeger, & Langer, 2012).

In general, fabrication methods for albumin nanoparticles can be categorized to conventional and new methods (Elzoghby et al., 2012). By studying the production methods in the literature, it can be observed that either a chemical compound (e.g., ethanol, cottonseed oil, or β-mercaptoethanol (β-ME)) or a physical factor (e.g., heat or pressure) is employed in the preparation process to achieve albumin nanoparticles. The production methods of albumin nanoparticles, including desolvation, emulsification, self-assembly, nanoparticle albumin-bound (NAB) technology, heat denaturation, and nanospray drying, are summarized in Fig. 2.

Among the procedures, desolvation (or coacervation) is the most widely used technique applied in the preparation of albumin nanoparticles. The coacervating agent (usually ethanol or acetone) progressively and partially changes the tertiary structure of the albumin to give an increasingly hydrophobic material that tends to form submicronic aggregates of dissolved albumin. The particle size and polydispersity index (PDI) of nanoparticles produced by desolvation are controllable by varying conditions such as pH value, ionic composition, and concentration of the albumin solution (Langer et al., 2003; Weber et al., 2000; Yedomon, Fessi, & Charcosset, 2013). The desolvation process has been optimized for the fabrication of HSA nanoparticles by Weber et al. (2000) in which the lowest glutaraldehyde concentration for the cross-linking of all amine groups was reported to be 40% of the calculated amount necessary for the quantitative cross-linking of the amino groups of lysine in the HSA molecule of the particle matrix. A modified desolvation method was used by Kouchakzadeh, Shojaosadati, and Shokri (2014), which included the employment of ultrasonication on tamoxifen (Tmx)–HSA in an aqueous environment

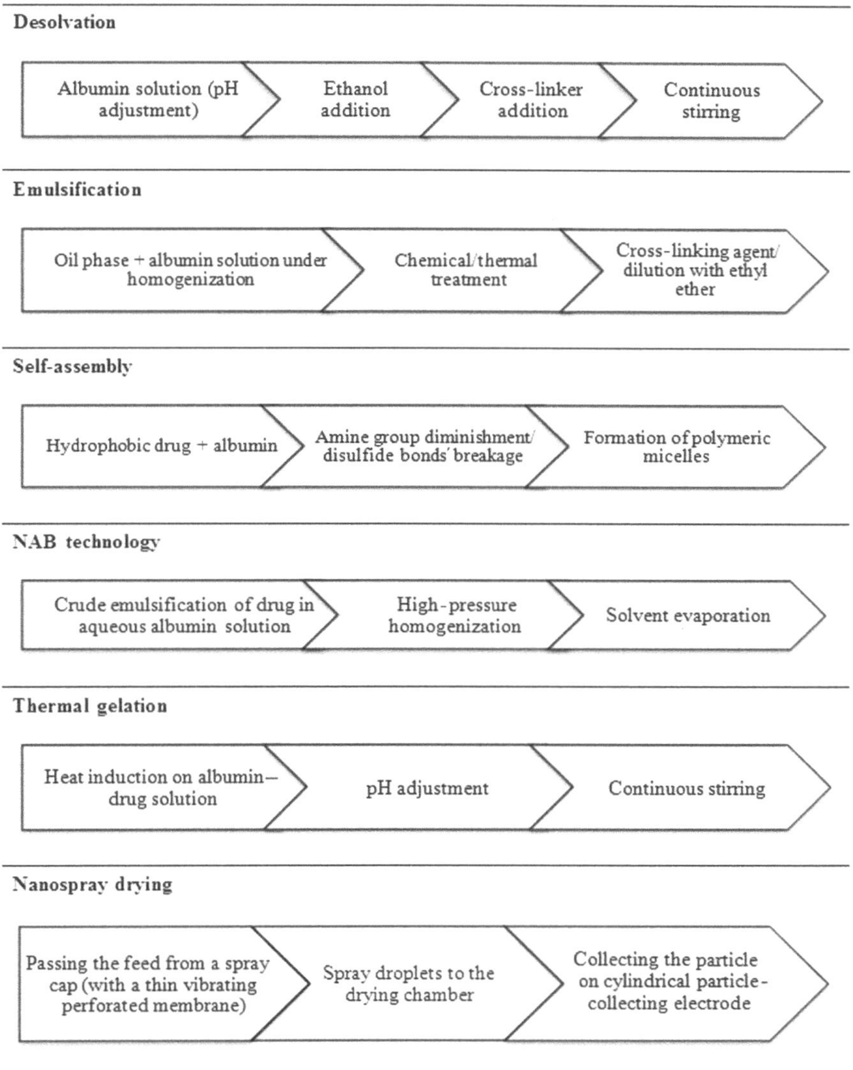

Figure 2 Methods used to prepare albumin-based nanoparticles. Production stages of albumin nanoparticles by the methods including desolvation, emulsification, self-assembly, NAB technology, thermal gelation, and nanospray drying are presented schematically. The last step for all of the methods is the separation and purification of the nanoparticles.

before the addition of a desolvating agent (ethanol) in order to achieve a higher amount of Tmx in the HSA nanoparticles. In comparison with the conventional desolvation technique, employment of the proposed method increased drug loading (DL) and drug entrapment efficiency three- and sevenfold, respectively. Efforts had been focused on scaling-up albumin desolvation as a robust and reproducible method in the laboratory-scale production of nanoparticles. Wacker et al. (2011) developed paddle-stirring systems in combination with a high performance liquid chromatography (HPLC) pump in order to find the optimal conditions for the controlled desolvation of up to 2000 mg of the protein. The paddle-stirring devices were utilized to guarantee the homogeneous distribution of the HSA molecules during the desolvation process. Results revealed no negative effects on particle size and PDI when the desolvation volume increased to 100 ml of the hydroethanolic suspension. In addition, a narrow size distribution obtained for nanoparticles after freeze-drying has been reported elsewhere (Anhorn, Mahler, & Langer, 2008; Kouchakzadeh et al., 2014; Yedomon et al., 2013). The upscale production of albumin nanoparticles was also achieved by permeating ethanol through the pores of a membrane contactor and circulating the aqueous phase tangentially to the membrane surface. It was found that BSA nanoparticles on a large scale (up to 10 g BSA) could be obtained with a mean size and PDI of around 145 nm and 0.08, respectively, by the use of a membrane contactor instead of a syringe pump (Yedomon et al., 2013). As a conclusion, the preparation of albumin nanoparticles by ethanolic desolvation is a suitable method for the large-scale production of delivery systems, and the removal of solvents without any further purification of the colloids is possible by freeze-drying.

The emulsification technique for the production of albumin nanoparticles involves the emulsification of albumin solution in an oil phase and solidification by heat denaturation or chemical cross-linking. The need to use organic solvents for the removal of the oily residues and the surfactants and the difficulty of controlling the albumin particles' size (below 500 nm) are two limitations of this method (Gallo, Hung, & Perrier, 1984; Pinto Reis, Neufeld, Ribeiro, & Veiga, 2006).

The self-assembly technique for the production of albumin nanoparticles is based on increasing the lipophilicity of albumin by the addition of a hydrophobic drug or polymer and diminishment of superficial primary amino groups on the protein molecule, which results in the self-assembly of albumin and formation of polymeric micelles in an aqueous environment (Gong et al., 2009; Jiang et al., 2014; Xu, Fisher, & Juliano, 2011). Recently, Ding et al. (2014) developed a novel method for preparing self-assembled HSA nanoparticles via the unfolding of HSA protein in a proper solution (ethanol/β-ME mixture) to expose more hydrophobic domains for the delivery of hydrophobic drugs. Difficulties with the scaling-up of the technology, insufficient storage stability, frequent instability in the body, and different solubility protocols for different drugs are the most serious problems associated with micellar carriers (Agarwal, Lvov, Sawant, & Torchilin, 2008).

Between physical approaches for the creation of albumin nanoparticles, NAB technology was successfully developed for the delivery of paclitaxel (PTX), a mitotic inhibitor drug used in cancer chemotherapy (Desai et al., 2006; Fu et al., 2009). The preparation method consists of a crude emulsion of drug in an organic phase mix to albumin solution in an aqueous phase, which is subjected to high-pressure homogenization to create a nanoemulsion. Compared to chemical-based methods, NAB technology can result in higher drug content and smaller sizes. The driving force of nanoparticle formation by NAB technology is the presence of hydrophobic drug molecules that act as glue and keep the albumin molecules in place. Therefore, more hydrophilic drug molecules and charged drugs like nucleic acids are not suitable for this technique (Yardley, 2013).

3. ALBUMIN-NANOPARTICLES-TARGETING STRATEGIES

The key point in drug delivery is to deliver and release the therapeutic agent specifically to the target site. Therefore, maintaining the loaded drug inside the delivery system before getting to the tumor site is important for all intravenously administrated formulations. After intravenous administration of the colloidal particles, their recognition by plasma protein opson and subsequent uptake by the mononuclear phagocytic system is the most limiting factor for the use of colloidal drug carrier systems (Dobrovolskaia & McNeil, 2007; Owens & Peppas, 2006; Vonarbourg, Passirani, Saulnier, & Benoit, 2006). Therefore, only a small fraction of administrated nanoparticles will reach the tumor according to their size and surface properties, and most of them become entrapped in the kidney, spleen, liver, lung, or bone marrow (He, Hu, Yin, Tang, & Yin, 2010; Khlebtsov & Dykman, 2011; Yan et al., 2012).

Generally, the delivery of drugs to tumor tissues can be achieved by either passive or active targeting. The main aims of using targeted nanoscale drug delivery systems are to decrease drug concentration in normal tissues, improve drug pharmacokinetic and pharmacodynamic profiles, improve the solubility of the drug to allow intravenous administration, release a minimal amount of the drug during transition, release a maximum of the drug at the targeted site, and increase drug stability to reduce drug degradation (Danhier, Feron, & Préat, 2010). Passive targeting of nanoparticles occurs according to the enhanced permeability and retention (EPR) effect (Fig. 3), which utilizes the tumor environment and delivery system characteristics. Nanoparticles smaller than 200 nm have low uptake by opsonization, show a longer circulation time (Langer et al., 2003), and can reach the tumor through EPR. In addition, various surface modifications such as polyethylene glycol (PEG) help to increase circulation half-life of albumin nanoparticles (Kouchakzadeh, Shojaosadati, Maghsoudi, & Vasheghani Farahani, 2010; Owens & Peppas, 2006; Vonarbourg et al., 2006), increasing the accumulation of nanoparticles in the desired site of action.

The last 30 years have revealed that the EPR effect plays an important role in the delivery of macromolecules and nanomaterials to tumors, but there is a need to study the degree of EPR variability in different patients and tumor types. Moreover, it is now clear that the physiochemical properties of nanomaterials are as important as tumor biology. Exploitation of the EPR effect in humans will involve an exact understanding and control of nanomaterial properties (Prabhakar et al., 2013).

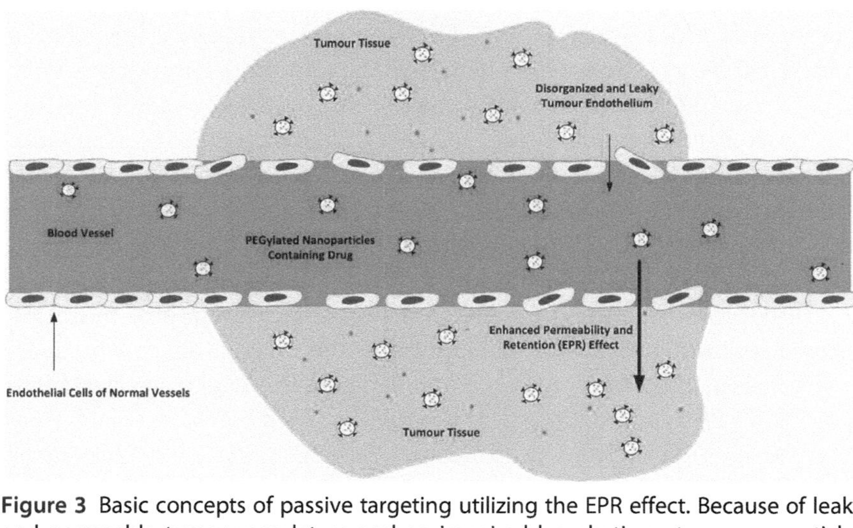

Figure 3 Basic concepts of passive targeting utilizing the EPR effect. Because of leaky and permeable tumor vasculature and an impaired lymphatic system, nanoparticles accumulate in tumors after intravenous injection.

In active targeting (ligand-targeting or receptor-mediated targeting), nanoparticles are equipped with targeting ligands. In this approach, ligands coupled to the surface of the nanoparticles through a cross-linker can result in their attachment to specific receptors at the target cells (Van der Meel, Vehmeijer, Kok, Storm, & Van Gaal, 2013). Monoclonal antibodies (mAbs; Koshkaryev, Sawant, Deshpande, & Torchilin, 2013), folic acid (Low & Kularatne, 2009), and transferrin (Daniels, Delgado, Helguera, & Penichet, 2006) are the most commonly used targeting ligands. The proteins, sugars, or lipids that are overexpressed in a diseased organ or on the surface of cells can be considered as target molecules (Bertrand, Wu, Xu, Kamaly, & Farokhzad, 2014; Yu, Tai, Xue, Lee, & Lee, 2010).

Albumin-bound drug complexes dissolve rapidly after intravenous infusion and accumulate in the tumor. Tumor uptake and distribution then occurs through a transport pathway mediated by albumin receptor, 60-kDa glycoprotein (Gp60) located on the endothelial cell surfaces (Fanciullino, Ciccolini, & Milano, 2013; Li et al., 2014; Tiruppathi, Song, Bergenfeldt, Sass, & Malik, 1997). Complexes bound to Gp60 transcytose through microvessel endothelial cells and subsequently release the drug into the subendothelial space. Accumulation of the albumin–drug complex in the tumor environment is facilitated more by albumin-binding protein SPARC (secreted protein, acidic, and rich in cysteine) after entering into the tumor interstitium (Fig. 4; Ma & Mumper, 2013). The mechanism explained for the enhanced delivery of albumin-bound drugs to the tumor site is different from the EPR effect in which stable nanoparticles smaller than 200 nm can accumulate in malignant zones due to the leaky blood vessels and lack of functional lymphatics in the tumor site, although the EPR effect may play an additional role in tumor accumulation (Li et al., 2014).

4. ALBUMIN NANOPARTICLES AS A DRUG DELIVERY CARRIER

One of the major advantages of albumin nanoparticles is their high binding capacity for various water-soluble and -insoluble drugs because of different drug binding sites available in the albumin molecule (Patil, 2003). Principally, three drug delivery approaches can be distinguished: coupling of low-molecular-weight drugs to exogenous or endogenous albumin, conjugation with bioactive proteins, and encapsulation and absorption of drugs into albumin nanoparticles (Kratz, 2008).

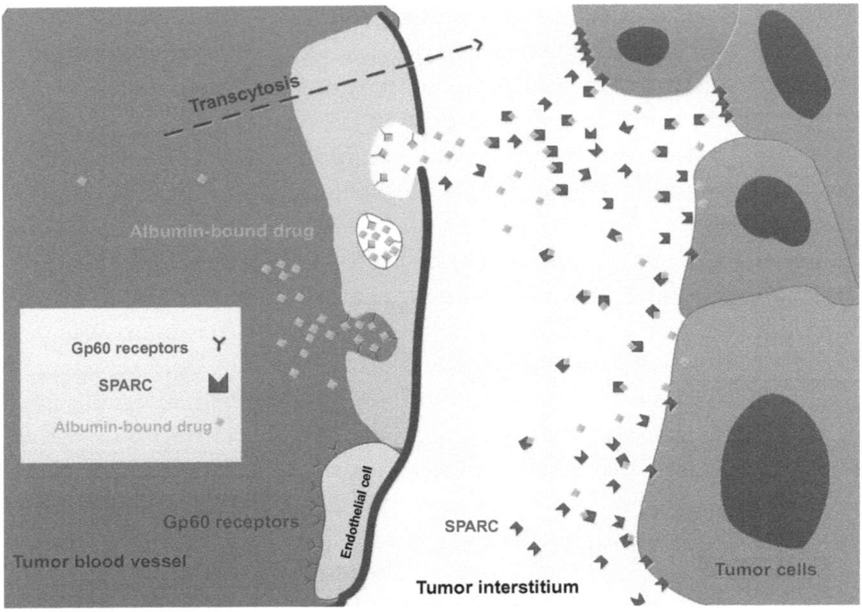

Figure 4 Albumin-bound drug endogenous pathways lead to efficient delivery of therapeutics to their site of action. Uptake of albumin into the tumor interstitium is firstly mediated by the Gp60 transcytosis pathway and then binding to SPARC in the tumor extracellular matrix, which result in the diffusion of drug into the intracellular compartment and subsequent induction of cell death.

4.1. Chemotherapeutic agents

Poor water solubility tends to be an inherent property of many drugs including potent anticancer drugs (De Vries & Willemse, 2002). Since the membrane permeability and efficacy of various drugs rises with increasing lipophilicity, their low solubility is not unexpected (Agarwal et al., 2008). On the other hand, intravenous administration of hydrophobic agents results in serious safety problems such as embolizing blood capillaries, which leads to severe side effects. Therefore, efficient and safe delivery of hydrophobic therapeutic compounds remains a serious problem for pharmaceuticals, and many promising drug candidates do not reach additional development because of the solubility problems (Desai, 2012; Lipinski, 2000). Albumin nanoparticulate delivery systems can overcome these challenges by improving solubility obstacles, which results in the modification of release rates to progress the drug's pharmacokinetics with the capability of reducing adverse side effects (Venditto & Szoka, 2013).

4.1.1 Paclitaxel

PTX, a taxane diterpenoidpseudoalkaloid obtained from the bark of *Taxus brerifolia* (pacific yew) in the late 1960s, is a mitotic inhibitor mainly used in cancer chemotherapy (Gibbs, 2000; Jordan & Wilson, 2004; Wani, Taylor, Wall, Coggon, & McPhail, 1971). More recently, PTX has been widely used in clinics for the treatment of a wide range of cancers including ovarian, breast, small and non-small-cell lung (NSCL), colon, head and neck, multiple myeloma, and Kaposi's sarcoma cancers (Rowinsky, 1993; Zhang, Mei, & Feng, 2013). In its mechanism of action, PTX acts as a microtubule-stabilizing agent that prevents cellular growth through the inhibition of cytoskeletal microtubule to free tubulin and disrupts the balance between polymerization and depolymerization, leading to termination of the cell cycle with arrest at the G2/M phase (Wang, Wang, & Soong, 2000). Its clinical application is restricted due to the poor aqueous solubility (<0.03 mg/ml; Van Zuylen, Verweij, & Sparreboom, 2001). The marketed PTX injection (Taxol®) was approved by the U.S. Food and Drug Administration for the treatment of drug-resistant ovarian cancer in 1992 and for the treatment of breast cancer in 1994. Taxol® is a solution of PTX in purified Cremophor El and dehydrated alcohol (ethanol). Due to its nonspecific, indiscriminate biodistribution in the body after intravenous administration, serious side effects including hypersensitivity reactions, nephrotoxicity, cardiotoxicity, and neurotoxicity have been reported (Van Zuylen et al., 2001).

NAB-PTX, marketed as Abraxane®, is a solvent-free form of PTX with a mean particle size of 130 nm that has been Food and Drug Administration (FDA) approved for the treatment of metastatic breast cancer, NSCL cancer, and adenocarcinoma of the pancreas in combination with gemcitabine (Ma & Mumper, 2013). NAB-PTX is obtained by high-pressure homogenization of PTX in the presence of HSA. By utilizing solvent-free PTX (Abraxane®) instead of solvent-based PTX (Taxol®), a nearly 50% higher maximum tolerated dose (300 mg/m^2 for Abraxane® vs. 175 mg/m^2 for Taxol®, every 3 weeks) and shorter infusion time (30 min for Abraxane® vs. 3–24 h for Taxol®) without the need for premedication to prevent solvent-related hypersensitivity reaction are achievable (Hawkins et al., 2008). Unstable NAB-PTX can rapidly disintegrate in the blood after intravenous injection, resulting in rapid transfer, transcytosis through endothelial cells in angiogenic tumors by Gp60, and target to SPARC. Clinical trials have shown that anticancer activity and efficacy of PTX has increased by Abraxane® in comparison with Taxol® (Desai et al., 2003; Gradishar et al., 2005). Zhao et al. (2010)

prepared a targeted delivery system of PTX-loaded BSA nanoparticles by a desolvation procedure with subsequent surface modification by folic acid. High affinity of folate-binding albumin nanoparticles in human prostate cancer cells and a continuous release profile of PTX were observed. In a study by Li et al. (2014), two versions of albumin-bound-PTX including cross-linked and noncross-linked ones were prepared, and the influence of cross-linking on the formulation properties and biological activity of albumin-bound nanoparticles were investigated. The first version was obtained by the NAB technology technique and was similar to Abraxane®. Then, one part of the resulting nanoparticles was cross-linked with glutaraldehyde to achieve the cross-linked version. Both formulations showed similar sizes and size distribution. Rapid disintegration of noncross-linked particles besides stable cross-linked ones was observed *in vitro* that can affect the PTX-release profile. The pharmacokinetics of both formulations was different, especially at the early time of administration, and the noncross-linked particles were cleared rapidly. Although it is reported that the tumor accumulation of noncross-linked particles is more than cross-linked ones, the concentration peak of PTX in heart, liver, kidney, intestine, and brain was higher for noncross-linked particles in comparison with the cross-linked ones, which may exhibit serious side effects. The surface modification of cross-linked particles with targeting ligands is a solution to increase their tumor specificity. Peptide–Abraxane® conjugates were prepared by coupling peptides (tumor-homing peptides (CREKA), a pentapeptide (cysteine–arginine–glutamic acid–lysine–alanine) and a cyclic nine-amino acid peptide (LyP-1)) to Abraxane® (Karmali et al., 2009). LyP-1–Abraxane® showed a highly significant inhibition of tumor growth compared with untargeted delivery system and the other conjugates on mice bearing MDA-MB-435 human cancer xenografts. PTX was also successfully loaded into octyl-modified bovine serum albumin (OSA) micelles by a dialysis method. Because of a synergistic effect of micellar encapsulation and binding interaction between the drug and OSA, a high PTX loading of 33.1% and an entrapment efficacy of 90.5% were obtained (Gong et al., 2009).

4.1.2 Doxorubicin

Doxorubicin (DOX), an antineoplastic anticancer drug, is largely used against breast cancer; however, its clinical use is limited by dose-dependent toxicity (myelosuppression and cardiotoxicity), the emergence of multidrug resistance, and its low specificity against cancer cells (Minotti, Menna, Salvatorelli, Cairo, & Gianni, 2004; Prados et al., 2012; Singal, Li,

Kumar, Danelisen, & Iliskovic, 2000). Moreover, DOX is rapidly eliminated by enzymes and encounters hydrolytic degradation in plasma after intravenous administration (Danesi, Fogli, Gennari, Conte, & Del Tacca, 2002). Two main mechanisms by which DOX acts on cancerous cells includes interaction with DNA and disruption of topoisomerase-II-mediated DNA repair and generation of free radicals; their damage to cellular membranes, DNA, and proteins is known (Thorn et al., 2011). Doxil® (PEGylated liposomal DOX) is an example of a DOX delivery system fabricated to reduce side effects, maintains high anticancer efficacy, and prolongs the circulation time of the drug in bloodstream (Takara et al., 2012).

DOX was loaded to the HSA nanoparticles either by adsorption to the nanoparticles' surface or by incorporation into the particle matrix by the use of a desolvation technique (Dreis et al., 2007). HSA nanoparticles of a size range between 150 and 500 nm with a DOX-loading efficiency of 70–95% were obtained. A stability study revealed no changes in HSA nanoparticles' size when stored in aqueous suspension for up to 6 months at 4 °C with maintenance of the biological activity of DOX. Furthermore, cell growth inhibition by DOX-loaded HSA nanoparticles was comparable or even better than DOX control solution on two neuroblastoma cell lines (UKF-NB3 and IMR32). The long-term storage stability of freeze-dried DOX-loaded HSA nanoparticles was investigated in the presence of different concentrations of sucrose, trehalose, and manitol. In all of the freeze-dried formulations, in the absence of cryoprotectants, particle growth was observed. The physicochemical characterization (lyophilisate appearance, particle size, and PDI) of freeze-dried DOX-loaded nanoparticles in the presence of sucrose, trahalose, and manitol gives comparable results (Anhorn et al., 2008). In order to achieve a high drug level in tumor tissue and to overcome cardiotoxic and other secondary effects, Anhorn et al. conjugated trastuzumab (a humanized antihuman epidermal growth factor receptor 2 (HER2) mAb) to the surface of DOX-loaded HSA nanoparticles prepared by the desolvation technique (Wagner et al., 2010). These trastuzumab-modified nanoparticles effectively bind to and are specifically internalized in HER2-overexpressing breast cancer cells, resulting in the release of the incorporated drug on the targeted cells. Due to the mAb-directed specificity of the nanoparticles, this system can be more efficient compared to Doxil®. Concerning the specificity and reduced drug infusion volume, a reduction of the side effects can be expected. In another study, DI17E6, an mAb directed against αv integrin, coupled to DOX-loaded HSA nanoparticles through a heterobifunctional PEG-based cross-linker (Wagner et al.,

2010). Targeted nanoparticles attached specifically to αvβ3-positive melanoma cells. DOX-loaded DI17E6-modified nanoparticles showed increased cytotoxic activity on αvβ3-positive melanoma cells in comparison with the free drug. HSA-based micelles of about 30 nm diameter and loaded with DOX were prepared (Xu et al., 2011). The surface of the micelles was conjugated to cyclic Arg-Gly-Asp (RGD) peptides for their selective delivery to cells overexpressing αvβ3 integrin. Albumin conjugated to DOX via disulfide bonds formed the inner core of the micelles. Additional DOX was also physically adsorbed into this core to obtain a high drug-loaded system. Higher uptake and longer retention of DOX was observed with the RGD-targeted micelles than the untargeted control systems or free drug when incubated with human melanoma cells (M21+) expressing the αvβ3 integrin. Consequently, the RGD-targeted micelles revealed cytotoxicity at lower doses of DOX in comparison with control micelles or free drug. Folic acid-conjugated DOX-loaded BSA nanoparticles showed incorporation into HeLa cells (tumor cells) only after 2 h incubation, whereas HeLa cells failed to incorporate albumin microspheres without surface modification after 4 h incubation (Shen, Li, Kohama, Oneill, & Bi, 2011). Folic-acid-modified nanoparticles demonstrated greater efficiency in the case of cell toxicity in comparison with free DOX and unmodified nanoparticles. As reported in the literature, and according to the findings, the combination of specific targeting with high DL in these nanoparticulate formulations can lead to an improvement in cancer therapy.

4.1.3 Other drugs

Due to the high capacity of albumin nanoparticles for water-soluble and -insoluble drugs and their possibility of surface modification, utilization of this system by academics for the effective delivery of therapeutic agents has increased. 10-Hydroxycamptothecin (HCPT) has been shown to have a strong antitumor activity against gastric carcinoma, hepatoma, leukemia, and tumors of the head and neck. However, because of its poor solubility in water and in physiologically acceptable organic solvents, the practical use of HCPT is limited (O'Leary & Muggia, 1998). BSA nanoparticles entrapped with HCPT were prepared by reformative emulsion-heat stabilization technique (Yang et al., 2007). Spherical-shaped nanoparticles with a mean size of around 600 nm, DL, and encapsulation efficiency and yields of 2.21%, 57.5%, and 90.5% were achieved, respectively, under the optimal preparation conditions with sustained *in vitro* drug release behavior. It was found that the concentration of HCPT in the liver and spleen remained relatively high even at 24 h postinjection, possibly due to the passive targeting ability of BSA nanoparticles. In order to overcome the poor aqueous solubility of curcumin (CCM), a yellow natural polyphenol extracted from turmeric (*Curcuma longa*) with potent anticancer properties, it was loaded to HSA nanoparticles using the NAB technology method for intravenous administration (Kim et al., 2011). Particles were obtained in the size range of 130–150 nm showed 300-fold water solubility in comparison with free CCM. More vascular endothelial cell binding and tumor accumulation were demonstrated by CCM-loaded HSA nanoparticles than CCM. *In vivo* antitumor investigations expressed that CCM-loaded nanoparticles had a greater therapeutic effect (50% or 66% tumor growth inhibition) than CCM (18% tumor growth inhibition) in tumor xenograft HCT116 models. The effective chemotherapeutic anticancer drug, 5-fluorouracil (5-FU), which has been used against colon, breast, ovarian, and skin cancers, was effectively loaded to BSA nanoparticles by an optimized desolvation technique (Maghsoudi, Shojaosadati, & Vasheghani Farahani, 2008). Under the optimum conditions, nanoparticles of 210 nm diameter and narrow size distribution were obtained. PEGylation of 5-FU-loaded BSA nanoparticles was performed by succinimidyl propionic acid-activated mPEG through superficial free amino groups, and the process was optimized using response surface methodology (Kouchakzadeh et al., 2010). Drug release from PEGylated nanoparticles was almost slower than non-PEGylated ones, probably due to the presence of a PEG layer around PEGylated particles, which creates additional resistance against drug diffusion. A system based on HSA nanoparticles with the aim of rapid crossing through the blood–brain barrier for the delivery of antidote obidoxime drug was fabricated (Kufleitner, Wagner, Worek, Von Briesen, & Kreuter, 2010). A maximum DL of 93.5 μg of obidoxime per mg of nanoparticles was obtained at pH 8.3 with rapid *in vitro* release of the drug from nanoparticles within 3 h. The poorly water-soluble drug, Tmx, commonly used for the treatment of breast cancer, was successfully loaded to HSA nanoparticles by a modified desolvation technique (Kouchakzadeh et al., 2014). Tmx loading and entrapment efficiency increased three- and sevenfold, respectively, by the employment of ultrasonication of the albumin–drug environment before the addition of ethanol. A more sustained Tmx release was observed from PEGylated nanoparticles in comparison with non-PEGylated ones. Long-term stability evaluation of freeze-dried Tmx-loaded HSA nanoparticles revealed almost constant physicochemical properties after 3 months of storage at room temperature. Methotrexate (MTX) is a folate antagonist and induces cancer cure in women suffering from metastatic choriocarcinoma. MTX rapidly clears from the body after intravenous administration, resulting in short tumor exposure (Bertino, 1993). MTX–HSA conjugates, functionalized by luteinizing hormone-releasing hormone (LHRH) as targeting moieties, with the aim of specifically targeting the cancer cells prepared (Taheri, Dinarvand, Ahadi, Khorramizadeh, & Atyabi, 2012; Taheri et al., 2011). LHRH-targeted MTX–HSA nanoparticles-treated group (female Balb/c mice bearing 4 T1 breast cancer) decreased the average tumor volume to 8.67% of the initial tumor volume when the number of attached LHRH molecules on nanoparticles was at the highest level. The average tumor volume in nontargeted MTX–HSA nanoparticle-treated mice grew rapidly and reached 250.7% of the initial tumor volume 7 days after the treatment. HSA nanoparticles also proved to be effective and safe carriers for the delivery of ganciclovir (an anticytomegaloviral agent) when administrated by the intravitreal route and did not induce autoimmune reactions (Kim, Robinson, & Csaky, 2009).

4.2. Biological macromolecules

Antisense oligonucleotides (ASOs), known as genetic materials, are promising agents used in the regulation of unsuitable expression of genes in pathological situations by the specific inhibition of expression of their mRNA targets (Agrawal & Kandimalla, 2000). However, the widespread use of these molecules in clinic has been limited by two major obstacles, including their enzymatic instability and poor and inadequate intracellular delivery (Beltinger et al., 1995; Lebedeva & Stein, 2001). The complexation of ASOs with albumin nanoparticles protects unmodified ASOs against enzymatic degradation and improves their nucleus accumulation and biological activity (Arnedo, Irache, González Gaitano, Valgañón, & Espuelas, 2003; Arnedo, Irache, Merodio, & Espuelas Millán, 2004). The association of ASOs with albumin carriers is simple and does not require the addition of any positive compounds (Arnedo, Espuelas, & Irache, 2002; Arnedo et al., 2004). Although albumin nanoparticles prepared by desolvation showed high ASO efficacy, the use of glutaraldehyde can lead to an inactivation of ASOs through chemical cross-linking due to its toxic entity. Steinhauser et al. (2008) developed targeted HSA nanoparticles loaded with ASOs against polo-like kinase 1 (PIK1) prepared by heat denaturation instead of using glutaraldehyde. Stable monodisperse nanoparticles with an ASOs' recovery of more than 80% could be prepared at a temperature of 105 °C for 10 min.

Significant reduction of PIK1 mRNA and protein expression was observed by ASOs-loaded trastuzumab-modified nanoparticles in BT474 breast cancer cells using real-time polymerase chain reaction (PCR) and Western blot analyses. The multikinase inhibitor 17864, a platinum-bound sunitinib analogue, was loaded to albumin nanoparticles via chemical conjugation to the albumin backbone (Altintas et al., 2013). Then, nanoparticles were surface-coated with heterobifunctional PEG and a nanobody (the single variable domain of an antibody) against epidermal growth factor receptor (EGFR). Nanobody-functionalized nanoparticles showed 40-fold higher binding to EGFR-positive 14C squamous head and neck cancer cells in comparison with PEGylated nanoparticles. Successful release of the kinase inhibitor in the target cell and inhibition of proliferation were observed from targeted nanoparticles, whereas the nontargeted formulations had no anti-proliferative effects on 14C cells.

5. ALBUMIN NANOPARTICLES MARKETED AND UNDER CLINICAL TRIAL

Abraxane®, the first albumin-bound drug delivery system approved in oncology, is an albumin-bound form of PTX. It was fabricated by NAB technology and marketed by American Bioscience in 2005 for the treatment of metastatic breast cancer in the United States, Europe, China, Russia, and several other countries. In October 2012 and September 2013, the FDA approved the use of Abraxane® for the first-line treatment of metastatic NSCL cancer and metastatic pancreatic cancer (in combination with gemcitabine), respectively (Kratz, 2014). In addition, a phase II study of Abraxane® for the treatment of metastatic melanoma showed promising results in comparison with dacarbazine, as a control (Hersh et al., 2010). It is predicted that the sale of Abraxane® will reach to $1.25 billion in 2015 and as high as $2 billion in 2017 through the use of the drug against pancreatic cancer and melanoma. In addition to Abraxane®, the NAB technology method has been successfully used to encapsulate other hydrophobic anticancer drugs including docetaxel (under phase II/III clinical trials), the mTOR inhibitor rapamycin (under phase II/III clinical trials), the heat-shock protein inhibitor Hsp90 (under phase I/II clinical trials), the tubulin polymerization/topoisomerase I dual inhibitor 5404 (in preclinical stage), and a novel taxane CY196 (in preclinical stage; Elsadek & Kratz, 2012).

6. CONCLUSIONS

Today, there is a strong focus on the application of nanotechnology to cancer. Cancer nanotechnology is a field of interdisciplinary research cutting across biology, chemistry, engineering, and medicine, aiming to lead major advances in cancer diagnosis and treatment. The poor pharmacokinetic properties of some chemotherapeutic agents can be improved using nanoparticulate drug delivery systems. Albumin-based nanoparticulate systems hold great potential to overcome some obstacles including delivery of poor water-soluble drugs and efficient targeting of diverse cell types. In addition to facilitating an advantageous therapeutic index, drug-loaded albumin nanoparticles may also afford benefit in terms of pharmacokinetic profile. The challenge over further development of commercial albumin-based nanomedicines before the preclinical and clinical trials will include the possibility of process scale-up with the employment of a simple and reproducible production method. As reviewed here, although ligand-modified albumin nanoparticles have not been approved or even reached clinical trials, the reported results, especially for mAb-conjugated drug-loaded nanoparticles, support a potential delivery system for targeted drug transport into specific cells.

REFERENCES

Agarwal, A., Lvov, Y., Sawant, R., & Torchilin, V. (2008). Stable nanocolloids of poorly soluble drugs with high drug content prepared using the combination of sonication and layer-by-layer technology. *Journal of Controlled Release*, *128*(3), 255–260.

Agrawal, S., & Kandimalla, E. R. (2000). Antisense therapeutics: Is it as simple as complementary base recognition? *Molecular Medicine Today*, *6*(2), 72–81.

Altintas, I., Heukers, R., Van der Meel, R., Lacombe, M., Amidi, M., Van Bergen En Henegouwen, P. M. P., et al. (2013). Nanobody-albumin nanoparticles (NANAPs) for the delivery of a multikinase inhibitor 17864 to EGFR overexpressing tumor cells. *Journal of Controlled Release*, *165*(2), 110–118.

Anhorn, M. G., Mahler, H.-C., & Langer, K. (2008). Freeze drying of human serum albumin (HSA) nanoparticles with different excipients. *International Journal of Pharmaceutics*, *363*(1–2), 162–169.

Arnedo, A., Espuelas, S., & Irache, J. (2002). Albumin nanoparticles as carriers for a phosphodiester oligonucleotide. *International Journal of Pharmaceutics*, *244*(1–2), 59–72.

Arnedo, A., Irache, J. M., González Gaitano, G., Valgañón, M., & Espuelas, S. (2003). Bovine serum albumin modified the intracellular distribution and improved the antiviral activity of an oligonucleotide. *Journal of Drug Targeting*, *11*(4), 197–204.

Arnedo, A., Irache, J. M., Merodio, M., & Espuelas Millán, M. S. (2004). Albumin nanoparticles improved the stability, nuclear accumulation and anticytomegaloviral activity of a phosphodiester oligonucleotide. *Journal of Controlled Release*, *94*(1), 217–227.

Beltinger, C., Saragovi, H. U., Smith, R. M., LeSauteur, L., Shah, N., DeDionisio, L., et al. (1995). Binding, uptake, and intracellular trafficking of phosphorothioate-modified oligodeoxynucleotides. *The Journal of Clinical Investigation*, *95*(4), 1814–1823.

Bertino, J. R. (1993). Ode to methotrexate. *Journal of Clinical Oncology*, *11*(1), 5–14.

Bertrand, N., Wu, J., Xu, X., Kamaly, N., & Farokhzad, O. C. (2014). Cancer nanotechnology: The impact of passive and active targeting in the era of modern cancer biology. *Advanced Drug Delivery Reviews*, *66*, 2–25.

Bertucci, C., & Domenici, E. (2002). Reversible and covalent binding of drugs to human serum albumin: Methodological approaches and physiological relevance. *Current Medicinal Chemistry*, *9*(15), 1463–1481.

Carter, D. C., He, X. M., Munson, S. H., Twigg, P. D., Gernert, K. M., Broom, M. B., et al. (1989). Three-dimensional structure of human serum albumin. *Science (New York, NY)*, *244*(4909), 1195–1198.

Danesi, R., Fogli, S., Gennari, A., Conte, P., & Del Tacca, M. (2002). Pharmacokinetic–pharmacodynamic relationships of the anthracycline anticancer drugs. *Clinical Pharmacokinetics*, *41*(6), 431–444.

Danhier, F., Feron, O., & Préat, V. (2010). To exploit the tumor microenvironment: Passive and active tumor targeting of nanocarriers for anti-cancer drug delivery. *Journal of Controlled Release*, *148*(2), 135–146.

Daniels, T. R., Delgado, T., Helguera, G., & Penichet, M. L. (2006). The transferrin receptor. Part II: Targeted delivery of therapeutic agents into cancer cells. *Clinical Immunology*, *121*(2), 159–176.

De Vries, E. G. E., & Willemse, P. H. B. (2002). Cancer chemotherapy and biotherapy: Principles and practice. *Annals of Oncology*, *13*(8), 1325.

Desai, N. (2012). Challenges in development of nanoparticle-based therapeutics. *The AAPS Journal*, *14*(2), 282–295.

Desai, N., Trieu, V., Yao, Z., Louie, L., Ci, S., Yang, A., et al. (2006). Increased antitumor activity, intratumor paclitaxel concentrations, and endothelial cell transport of cremophor-free, albumin-bound paclitaxel, ABI-007, compared with cremophor-based paclitaxel. *Clinical Cancer Research*, *12*(4), 1317–1324.

Desai, N., Trieu, V., Yao, R., et al. (2003). Evidence of greater antitumor activity of ceremophorefree nanoparticle albumin-bound (nab) paclitaxel (Abraxane) compared to Taxol: Role of a novel albumin transporter mechanism. In *26th annual San Antonio Breast Cancer Symposium*. USA: San Antonio.

Ding, D., Tang, X., Cao, X., Wu, J., Yuan, A., Qiao, Q., et al. (2014). Novel self-assembly endows human serum albumin nanoparticles with an enhanced antitumor efficacy. *AAPS PharmSciTech*, *15*(1), 213–222.

Dobrovolskaia, M. A., & McNeil, S. E. (2007). Immunological properties of engineered nanomaterials. *Nature Nanotechnology*, *2*(8), 469–478.

Dreis, S., Rothweiler, F., Michaelis, M., Cinati, J., Jr., Kreuter, J., & Langer, K. (2007). Preparation, characterisation and maintenance of drug efficacy of doxorubicin-loaded human serum albumin (HSA) nanoparticles. *International Journal of Pharmaceutics*, *341*, 207–214.

Elsadek, B., & Kratz, F. (2012). Impact of albumin on drug delivery—New applications on the horizon. *Journal of Controlled Release*, *157*(1), 4–28.

Elzoghby, A. O., Samy, W. M., & Elgindy, N. a. (2012). Albumin-based nanoparticles as potential controlled release drug delivery systems. *Journal of Controlled Release*, *157*(2), 168–182.

Fanciullino, R., Ciccolini, J., & Milano, G. (2013). Challenges, expectations and limits for nanoparticles-based therapeutics in cancer: A focus on nano-albumin-bound drugs. *Critical Reviews in Oncology/Hematology*, *88*, 504–513.

Fasano, M., Curry, S., Terreno, E., Galliano, M., Fanali, G., Narciso, P., et al. (2005). The extraordinary ligand binding properties of human serum albumin. *IUBMB Life*, *57*(12), 787–796.

Fu, Q., Sun, J., Zhang, W., Sui, X., Yan, Z., & He, Z. (2009). Nanoparticle albumin-bound (NAB) technology is a promising method for anti-cancer drug delivery. *Recent Patents on Anti-Cancer Drug Discovery*, *4*(3), 262–272.

Gallo, J. M., Hung, C. T., & Perrier, D. G. (1984). Analysis of albumin microsphere preparation. *International Journal of Pharmaceutics*, *22*(1), 63–74.

Gibbs, J. B. (2000). Mechanism-based target identification and drug discovery in cancer research. *Science*, *287*(5460), 1969–1973.

Gong, J., Huo, M., Zhou, J., Zhang, Y., Peng, X., Yu, D., et al. (2009). Synthesis, characterization, drug-loading capacity and safety of novel octyl modified serum albumin micelles. *International Journal of Pharmaceutics*, *376*(1–2), 161–168.

Gradishar, W. J., Tjulandin, S., Davidson, N., Shaw, H., Desai, N., Bhar, P., et al. (2005). Phase III trial of nanoparticle albumin-bound paclitaxel compared with polyethylated castor oil-based paclitaxel in women with breast cancer. *Journal of Clinical Oncology*, *23*(31), 7794–7803.

Hawkins, M. J., Soon-Shiong, P., & Desai, N. (2008). Protein nanoparticles as drug carriers in clinical medicine. *Advanced Drug Delivery Reviews*, *60*(8), 876–885.

He, C., Hu, Y., Yin, L., Tang, C., & Yin, C. (2010). Effects of particle size and surface charge on cellular uptake and biodistribution of polymeric nanoparticles. *Biomaterials*, *31*(13), 3657–3666.

Hersh, E. M., O'Day, S. J., Ribas, A., Samlowski, W. E., Gordon, M. S., Shechter, D. E., et al. (2010). A phase II clinical trial of nab-paclitaxel in previously treated and chemotherapy-naive patients with metastatic melanoma. *Cancer*, *116*(1), 155–163.

Irache, J. M., & Espuelas, S. (2006). Albumin nanoparticles. In: C. S. S. R. Kumar (Ed.), *Biological and pharmaceutical nanomaterials, Nanotechnologies for the life sciences: Vol. 2* (pp. 185–218). Wiley-VCH.

Jiang, Y., Liang, M., Svejkar, D., Hart-Smith, G., Lu, H., Scarano, W., et al. (2014). Albumin-micelles via a one-pot technology platform for the delivery of drugs. *Chemical Communications (Cambridge, England)*, *50*(48), 6394–6397.

Jordan, M. A., & Wilson, L. (2004). Microtubules as a target for anticancer drugs. *Nature Reviews Cancer*, *4*(4), 253–265.

Karmali, P. P., Kotamraju, V. R., Kastantin, M., Black, M., Missirlis, D., Tirrell, M., et al. (2009). Targeting of albumin-embedded paclitaxel nanoparticles to tumors. *Nanomedicine: Nanotechnology, Biology, and Medicine*, *5*(1), 73–82.

Khlebtsov, N., & Dykman, L. (2011). Biodistribution and toxicity of engineered gold nanoparticles: A review of *in vitro* and *in vivo* studies. *Chemical Society Reviews*, *40*(3), 1647–1671.

Kim, T. H., Jiang, H. H., Youn, Y. S., Park, C. W., Tak, K. K., Lee, S., et al. (2011). Preparation and characterization of water-soluble albumin-bound curcumin nanoparticles with improved antitumor activity. *International Journal of Pharmaceutics*, *403*(1–2), 285–291.

Kim, H., Robinson, S. B., & Csaky, K. G. (2009). Investigating the movement of intravitreal human serum albumin nanoparticles in the vitreous and retina. *Pharmaceutical Research*, *26*(2), 329–337.

Koshkaryev, A., Sawant, R., Deshpande, M., & Torchilin, V. (2013). Immunoconjugates and long circulating systems: Origins, current state of the art and future directions. *Advanced Drug Delivery Reviews*, *65*(1), 24–35.

Kouchakzadeh, H., Shojaosadati, S. A., Maghsoudi, A., & Vasheghani Farahani, E. (2010). Optimization of PEGylation conditions for BSA nanoparticles using response surface methodology. *AAPS PharmSciTech*, *11*(3), 1206–1211.

Kouchakzadeh, H., Shojaosadati, S. A., Mohammadnejad, J., Paknejad, M., & Rasaee, M. J. (2012). Attachment of an anti-MUC1 monoclonal antibody to 5-FU loaded BSA nanoparticles for active targeting of breast cancer cells. *Human Antibodies*, *21*(3–4), 49–56.

Kouchakzadeh, H., Shojaosadati, S. A., & Shokri, F. (2014). Efficient loading and entrapment of tamoxifen in human serum albumin based nanoparticulate delivery system by a modified desolvation technique. *Chemical Engineering Research and Design*, *92*(9), 1681–1692.

Kouchakzadeh, H., Shojaosadati, S. A., Tahmasebi, F., & Shokri, F. (2013). Optimization of an anti-HER2 monoclonal antibody targeted delivery system using PEGylated human serum albumin nanoparticles. *International Journal of Pharmaceutics*, *447*(1–2), 62–69.

Kratz, F. (2008). Albumin as a drug carrier: Design of prodrugs, drug conjugates and nanoparticles. *Journal of Controlled Release*, *132*(3), 171–183.

Kratz, F. (2014). A clinical update of using albumin as a drug vehicle—A commentary. Press release, *Journal of Controlled Release*, *190*, 331–336.

Kufleitner, J., Wagner, S., Worek, F., Von Briesen, H., & Kreuter, J. (2010). Adsorption of obidoxime onto human serum albumin nanoparticles: Drug loading, particle size and drug release. *Journal of Microencapsulation*, *27*(6), 506–513.

Langer, K., Anhorn, M. G., Steinhauser, I., Dreis, S., Celebi, D., Schrickel, N., et al. (2008). Human serum albumin (HSA) nanoparticles: Reproducibility of preparation process and kinetics of enzymatic degradation. *International Journal of Pharmaceutics*, *347*(1–2), 109–117.

Langer, K., Balthasar, S., Vogel, V., Dinauer, N., Von Briesen, H., & Schubert, D. (2003). Optimization of the preparation process for human serum albumin (HSA) nanoparticles. *International Journal of Pharmaceutics*, *257*(1–2), 169–180.

Langer, K., Coester, C., Weber, C., Von Briesen, H., & Kreuter, J. (2000). Preparation of avidin-labeled protein nanoparticles as carriers for biotinylated peptide nucleic acid. *European Journal of Pharmaceutics and Biopharmaceutics*, *49*(3), 303–307.

Lanza, R., Langer, R., & Vacanti, J. (Eds.), (2011). *Principles of tissue engineering*. China: Elsevier Academic Press.

Lebedeva, I., & Stein, C. A. (2001). Antisense oligonucleotides: Promise and reality. *Annual Review of Pharmacology and Toxicology*, *41*, 403–419.

Li, C., Li, Y., Gao, Y., Wei, N., Zhao, X., Wang, C., et al. (2014). Direct comparison of two albumin-based paclitaxel-loaded nanoparticle formulations: Is the crosslinked version more advantageous? *International Journal of Pharmaceutics*, *468*(1–2), 15–25.

Lipinski, C. A. (2000). Drug-like properties and the causes of poor solubility and poor permeability. *Journal of Pharmacological and Toxicological Methods*, *44*(1), 235–249.

Low, P. S., & Kularatne, S. A. (2009). Folate-targeted therapeutic and imaging agents for cancer. *Current Opinion in Chemical Biology*, *13*(3), 256–262.

Ma, P., & Mumper, R. J. (2013). Paclitaxel nano-delivery systems: A comprehensive review. *Journal of Nanomedicine and Nanotechnology*, *4*(2), 1000164.

Maghsoudi, A., Shojaosadati, S. A., & Vasheghani Farahani, E. (2008). 5-Fluorouracil-loaded BSA nanoparticles: Formulation optimization and *in vitro* release study. *AAPS PharmSciTech*, *9*(4), 1092–1096.

Minotti, G., Menna, P., Salvatorelli, E., Cairo, G., & Gianni, L. (2004). Anthracyclines: Molecular advances and pharmacologic developments in antitumor activity and cardiotoxicity. *Pharmacological Reviews*, *56*(2), 185–229.

O'Leary, J., & Muggia, F. M. (1998). Camptothecins: A review of their development and schedules of administration. *European Journal of Cancer*, *34*(10), 1500–1508.

Owens, D. E., & Peppas, N. A. (2006). Opsonization, biodistribution, and pharmacokinetics of polymeric nanoparticles. *International Journal of Pharmaceutics*, *307*(1), 93–102.

Patil, G. V. (2003). Biopolymer albumin for diagnosis and in drug delivery. *Drug Development Research*, *58*(3), 219–247.

Pinto Reis, C., Neufeld, R. J., Ribeiro, A. J., & Veiga, F. (2006). Nanoencapsulation I. Methods for preparation of drug-loaded polymeric nanoparticles. *Nanomedicine: Nanotechnology, Biology, and Medicine*, *2*(1), 8–21.
Prabhakar, U., Maeda, H., Jain, R. K., Sevick-Muraca, E. M., Zamboni, W., Farokhzad, O. C., et al. (2013). Challenges and key considerations of the enhanced permeability and retention effect for nanomedicine drug delivery in oncology. *Cancer Research*, *73*(8), 2412–2417.
Prados, J., Melguizo, C., Ortiz, R., Vélez, C., Alvarez, P. J., Arias, J. L., et al. (2012). Doxorubicin-loaded nanoparticles: New advances in breast cancer therapy. *Anti-Cancer Agents in Medicinal Chemistry*, *12*(9), 1058–1070.
Rowinsky, E. K. (1993). Clinical pharmacology of Taxol. *Journal of the National Cancer Institute Monographs*, (15), 25–37.
Shen, Z., Li, Y., Kohama, K., Oneill, B., & Bi, J. (2011). Improved drug targeting of cancer cells by utilizing actively targetable folic acid-conjugated albumin nanospheres. *Pharmacological Research*, *63*(1), 51–58.
Singal, P. K., Li, T., Kumar, D., Danelisen, I., & Iliskovic, N. (2000). Adriamycin-induced heart failure: Mechanism and modulation. *Molecular and Cellular Biochemistry*, *207*(1–2), 77–86.
Steinhauser, I. M., Langer, K., Strebhardt, K. M., & Spa, B. (2008). Effect of trastuzumab-modified antisense oligonucleotide-loaded human serum albumin nanoparticles prepared by heat denaturation. *Biomaterials*, *29*, 4022–4028.
Taheri, A., Dinarvand, R., Ahadi, F., Khorramizadeh, M. R., & Atyabi, F. (2012). The *in vivo* antitumor activity of LHRH targeted methotrexate-human serum albumin nanoparticles in 4 T1 tumor-bearing Balb/c mice. *International Journal of Pharmaceutics*, *431*(1–2), 183–189.
Taheri, A., Dinarvand, R., Atyabi, F., Ahadi, F., Nouri, F. S., Ghahremani, M. H., et al. (2011). Enhanced anti-tumoral activity of methotrexate-human serum albumin conjugated nanoparticles by targeting with luteinizing hormone-releasing hormone (LHRH) peptide. *International Journal of Molecular Sciences*, *12*(7), 4591–4608.
Takara, K., Hatakeyama, H., Kibria, G., Ohga, N., Hida, K., & Harashima, H. (2012). Size-controlled, dual-ligand modified liposomes that target the tumor vasculature show promise for use in drug-resistant cancer therapy. *Journal of Controlled Release*, *162*(1), 225–232.
Thorn, C. F., Oshiro, C., Marsh, S., Hernandez-Boussard, T., McLeod, H., Klein, T. E., et al. (2011). Doxorubicin pathways: Pharmacodynamics and adverse effects. *Pharmacogenetics and Genomics*, *21*(7), 440–446.
Tiruppathi, C., Song, W., Bergenfeldt, M., Sass, P., & Malik, A. B. (1997). Gp60 activation mediates albumin transcytosis in endothelial cells by tyrosine kinase-dependent pathway. *The Journal of Biological Chemistry*, *272*(41), 25968–25975.
Ulbrich, K., Hekmatara, T., Herbert, E., & Kreuter, J. (2009). Transferrin- and transferrin-receptor-antibody-modified nanoparticles enable drug delivery across the blood–brain barrier (BBB). *European Journal of Pharmaceutics and Biopharmaceutics*, *71*(2), 251–256.
Van der Meel, R., Vehmeijer, L. J. C., Kok, R. J., Storm, G., & Van Gaal, E. V. B. (2013). Ligand-targeted particulate nanomedicines undergoing clinical evaluation: Current status. *Advanced Drug Delivery Reviews*, *65*(10), 1284–1298.
Van Zuylen, L., Verweij, J., & Sparreboom, A. (2001). Role of formulation vehicles in taxane pharmacology. *Investigational New Drugs*, *19*(2), 125–141.
Venditto, V. J., & Szoka, F. C. (2013). Cancer nanomedicines: So many papers and so few drugs!. *Advanced Drug Delivery Reviews*, *65*(1), 80–88.
von Storp, B., Engel, A., Boeker, A., Ploeger, M., & Langer, K. (2012). Albumin nanoparticles with predictable size by desolvation procedure. *Journal of Microencapsulation*, *29*(2), 138–146.

Vonarbourg, A., Passirani, C., Saulnier, P., & Benoit, J.-P. (2006). Parameters influencing the stealthiness of colloidal drug delivery systems. *Biomaterials*, *27*(24), 4356–4373.
Wacker, M., Zensi, A., Kufleitner, J., Ruff, A., Schütz, J., Stockburger, T., et al. (2011). A toolbox for the upscaling of ethanolic human serum albumin (HSA) desolvation. *International Journal of Pharmaceutics*, *414*(1–2), 225–232.
Wagner, S., Rothweiler, F., Anhorn, M. G., Sauer, D., Riemann, I., Weiss, E. C., et al. (2010). Enhanced drug targeting by attachment of an anti αv integrin antibody to doxorubicin loaded human serum albumin nanoparticles. *Biomaterials*, *31*, 2388–2398.
Wang, T. H., Wang, H. S., & Soong, Y. K. (2000). Paclitaxel-induced cell death: Where the cell cycle and apoptosis come together. *Cancer*, *88*(11), 2619–2628.
Wani, M. C., Taylor, H. L., Wall, M. E., Coggon, P., & McPhail, A. T. (1971). Plant antitumor agents. VI. Isolation and structure of taxol, a novel antileukemic and antitumor agent from Taxus brevifolia. *Journal of the American Chemical Society*, *93*(9), 2325–2327.
Weber, C., Kreuter, J., & Langer, K. (2000). Desolvation process and surface characteristics of HSA-nanoparticles. *International Journal of Pharmaceutics*, *196*(2), 197–200.
Xu, R., Fisher, M., & Juliano, R. L. (2011). Targeted albumin-based nanoparticles for delivery of amphipathic drugs. *Bioconjugate Chemistry*, *22*(5), 870–878.
Yan, H., Wang, L., Wang, J., Weng, X., Lei, H., Wang, X., et al. (2012). Two-order targeted brain tumor imaging by using an optical/paramagnetic nanoprobe across the blood brain barrier. *ACS Nano*, *6*(1), 410–420.
Yang, L., Cui, F., Cun, D., Tao, A., Shi, K., & Lin, W. (2007). Preparation, characterization and biodistribution of the lactone form of 10-hydroxycamptothecin (HCPT)-loaded bovine serum albumin (BSA) nanoparticles. *International Journal of Pharmaceutics*, *340*(1–2), 163–172.
Yardley, D. a. (2013). nab-Paclitaxel mechanisms of action and delivery. *Journal of Controlled Release*, *170*(3), 365–372.
Yedomon, B., Fessi, H., & Charcosset, C. (2013). Preparation of bovine serum albumin (BSA) nanoparticles by desolvation using a membrane contactor: A new tool for large scale production. *European Journal of Pharmaceutics and Biopharmaceutics*, *85*, 398–405.
Yu, B., Tai, H. C., Xue, W., Lee, L. J., & Lee, R. J. (2010). Receptor-targeted nanocarriers for therapeutic delivery to cancer. *Molecular Membrane Biology*, *27*(7), 286–298.
Zhang, Z., Mei, L., & Feng, S.-S. (2013). Paclitaxel drug delivery systems. *Expert Opinion on Drug Delivery*, *10*(3), 325–340.
Zhao, D., Zhao, X., Zu, Y., Li, J., Zhang, Y., Jiang, R., et al. (2010). Preparation, characterization, and *in vitro* targeted delivery of folate-decorated paclitaxel-loaded bovine serum albumin nanoparticles. *International Journal of Nanomedicine*, *5*, 669–677.

CHAPTER FIVE

Vasoactive Intestinal Peptide (VIP) Nanoparticles for Diagnostics and for Controlled and Targeted Drug Delivery

Rebecca Klippstein*,†, David Pozo*,†,1

*CABIMER-Andalusian Center for Molecular Biology and Regenerative Medicine (CSIC-University of Seville-UPO), Seville, Spain

†Department of Medical Biochemistry, Molecular Biology and Immunology, The University of Seville Medical School, Seville, Spain

[1]Corresponding author: e-mail address: david.pozo@cabimer.es

Contents

Abstract

Neuropeptides are potentially valuable tools for clinical applications as they offer many distinct advantages over other bioactive molecules like proteins and monoclonal antibodies due to their reduced side effects and simple chemical modifications. Despite such advantages, the difficulty with neuropeptides often relies on their poor metabolic stability and reduced biological activity intervals. Among the neuropeptides, VIP has been identified as a potentially bioactive agent for inflammatory, neurodegenerative, and cancer-related diseases. However, the effective translation of preclinical studies related to VIP to clinical realities faces several major challenges, most of which are commonplace for other neuropeptides. Here, we present recent studies aimed at developing nanostructured organic and inorganic systems either for the appropriate delivery of VIP or for VIP targeting. These technologies stand as an alternative starting point for chemical manipulations of the neuropeptides in order to improve potency, selectivity, or pharmacokinetic parameters.

Advances in Protein Chemistry and Structural Biology, Volume 98
ISSN 1876-1623
http://dx.doi.org/10.1016/bs.apcsb.2014.11.006

1. INTRODUCTION

The technological leap associated to nanotechnology is now able to control materials at a nanoscale level and has driven developments of new applications in modern molecular medicine by using nanodevices, such as nanoparticles (NPs). NPs are defined as particles with sizes between 1 and 100 nm approximately that show properties not found in bulk samples of the same material (Auffan et al., 2009). In recent years, we have seen the emergence of nanoscience and nanotechnology as cutting-edge research areas that are being rapidly developed. In particular, engineered NPs constitute an extensive field of research due to the translational potential for biomedical applications (Gendelman et al., 2014; Klippstein & Pozo, 2010; Quintana, 2013; West & Halas, 2000). The efforts are mainly related to the application of nanoengineering methods and materials to develop new diagnostic platforms and more effective therapies for human diseases. Currently, there is a growing interest in different nanosystems and their biological features as nanocarriers of chemotherapeutic agents and other compounds, known as "smart drug delivery" products. Drug delivery nanosystems are being studied to allow more selective and potent treatments as well as to improve the efficacy/toxicity ratio of our current and future therapeutic arsenal (Malam, Lim, & Seifalian, 2011). Moreover, recently developed methodologies related to surface functionalization provide materials with the final properties required for a desired application (Friedman, Claypool, & Liu, 2013). Nowadays, nanotechnology has brought an unprecedented variety of revolutionary approaches for molecular medicine. Some examples of what engineered NPs can offer to modern clinical practice include the detection of biological analytes (reducing sample size, reagent volumes, and faster generation of results) (Rosi & Mirkin, 2005), specific targeting to the action site (Drechsler, Erdogan, & Rotello, 2004), efficient drug delivery into the target cell (Drechsler et al., 2004), or *in vivo* real-time monitoring of cellular events (Morawski, Lanza, & Wickline, 2005).

Extensive libraries of NPs, composed of an assortment of different sizes, shapes, and materials, and with various chemical and surface properties, have already been constructed (Schutz, Juillerat-Jeanneret, Mueller, Lynch, & Riediker, 2013). Moreover, they can be multitask by combining different functionalities in a single stable construct, for example, incorporating a drug, a contrast agent and a targeting molecule within the same system (Fig. 1). Such combinatorial NPs may eventually provide the means to achieve

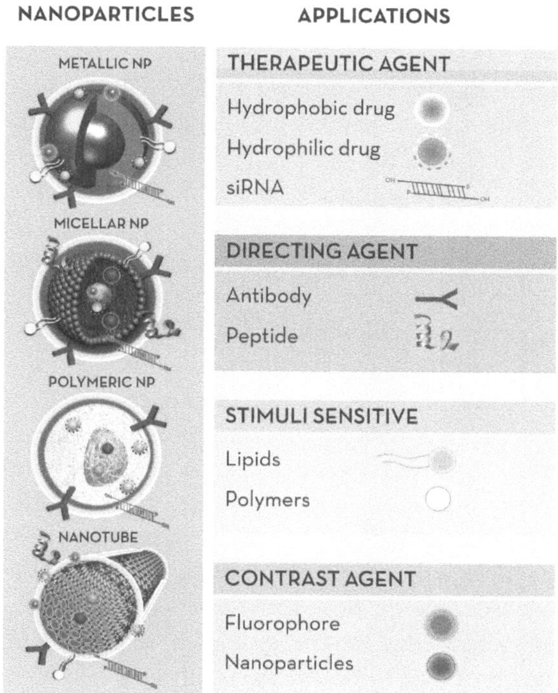

Figure 1 Schematic representation of different types of nanoparticles used for a variety of applications. Some examples of NPs are shown above, such as micelles, nanotubes, polymeric, and metallic NPs. Different molecules, such as therapeutic, directing, or contrast agents (e.g., drugs, antibodies, peptides, or fluorophores) can be used to further modify and confer novel properties for different applications.

"personalized medicine" by tailoring NP functionality to individual responses.

Basically, NPs are classified into two main groups depending on the type of material, organic or inorganic, and both groups have been used for vasoactive intestinal peptide (VIP)-based nanostructuration. Organic NPs are generally composed of proteins, lipids, or polysaccharides in order to synthesize liposomes, micelles, or dendrimers particles among others. These NPs are commonly recognized as safe due to their biodegradable and biocompatible properties (Johansson & Bradley, 2012). In addition, they also exhibit a good potential for surface modification and provide an excellent pharmacokinetic control. In the case of inorganic NPs, they are composed of silica, alumina, metals, metal oxides, and metal sulfides to produce a myriad of nanostructures with varying size, shape, and porosity. Some of their benefits include high stability over broad ranges of temperature and pH, high potential for surface functionalization, and their intrinsic optical/electrical properties (Huang, Jain, El-Sayed, & El-Sayed, 2007). However, inorganic NPs raise many safety questions, especially for long-term administration (Zuzana, Alessandra, Lise, & Maria, 2011).

Unfortunately, there is a general need in terms of comprehensive characterization of the published NP systems, which requires an additional effort toward interdisciplinary research at the nanoscience/pharmaceutical interface. While it is beyond the capability of many research groups to perform a detailed physicochemical characterization, tracking analyses, pharmacokinetics, and circulatory half-life studies, the scientific community critically needs methods of standardization for physicochemical characterization and biological evaluation which is required for the clinical translation of the field, not only in clinical tests but also to reduce the number of conflicting data already published in the scientific literature (Howard & Peer, 2013). In this work, we will discuss recent developments in the use of nanotechnology-based approaches as a tool for VIP-controlled drug delivery, an excellent example on how nano-enabled methods can fill the gap between current knowledge on the biological role of neuropeptides and their translational potential to healthcare solutions.

2. STRUCTURAL BASIS FOR THE NEUROPEPTIDE VIP-TARGETED DRUG DELIVERY

Up to date, there are over 50 neuropeptides identified. Neuropeptides are endogenous biologically active peptides secreted by neurons or nonneural cells (in particular, we pay special attention to immunocompetent cells) that bind to specific receptors in either neurons or nonneural target cells. Thus, neuropeptides are part of the common biochemical language used by the immune and neuroendocrine systems that build a bidirectional network of major relevance in health and disease. Among them, the 28-amino acid VIP neuropeptide is a molecule (Fig. 2) that has evolved from being considered a mere neuropeptide/hormone into a novel agent for modifying immune function and, possibly, as a cytokine-like molecule (Gonzalez-Rey, Delgado-Maroto, Souza Moreira, & Delgado, 2010; Pozo & Delgado, 2004; Pozo et al., 2007).

Originally identified by Said and Mutt in the late 1960s, VIP was originally isolated as a vasodilator and hypotensive peptide (Piper, Said, & Vane, 1970; Said & Mutt, 1970). Subsequently, its biochemistry was elucidated

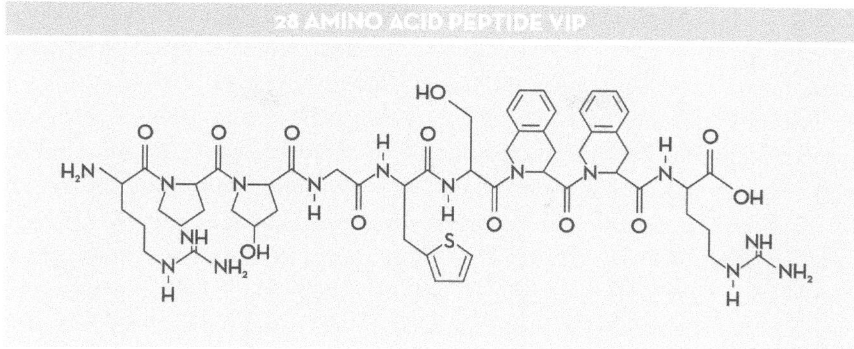

Figure 2 The 28-amino acid composition of the neuropeptide vasoactive intestinal peptide (VIP).

and within the first decade its signature features as a neuropeptide/neurotransmitter became consolidated: indeed, it is currently known to act as a neuromodulator in many organs and tissues, including heart, lung, thyroid gland, kidney, immune system, urinary tract, and genital organs (Henning & Sawmiller, 2001).

The amino acid sequence of VIP contains homologies with many gastrointestinal hormones such as pituitary adenylate cyclase-activating peptide (PACAP) and also shares some of its receptors (Miyata et al., 1989, 1990). VIP is remarkably well conserved across species and is identical in human, cow, pig, rat, dog, and goat (Mutt, 1988). Even across species, amino acid substitutions are conservative and usually do not result in changes in bioactivity (Singh, Jagannathan, Khandelwal, Abraham, & Poddar, 2013).

Two heterotrimeric G protein-coupled receptors (GPCRs) mediate the actions of VIP, namely, VPAC1 and VPAC2, according to the International Union of Basic and Clinical Pharmacology Committee on Receptor Nomenclature and Drug Classification, each of them with a unique expression pattern (Harmar et al., 1998). VIP and PACAP share receptors VPAC1 and VPAC2, but PAC1 receptor is specific for another peptide called PACAP (Harmar et al., 2012). Each of these receptors is coupled primarily to G proteins, and activate adenylyl cyclase and protein kinase A, but other pathways are often activated or inhibited in some cells in parallel or downstream of cyclic adenosine monophosphate (cAMP), including pathways involving exchange proteins activated by cAMP (Ster et al., 2007), nitric oxide (Murthy, Zhang, Jin, Grider, & Makhlouf, 1993), phospholipase C (Spengler et al., 1993), phosphatidylinositol 3-kinase (Straub & Sharp,

1996), MAP kinases (Barrie, Clohessy, Buensuceso, Rogers, & Allen, 1997; Lelievre et al., 1998), Jak/STAT, and nuclear factor kappa-light-chain-enhancer of activated B cells (NFκβ) (Delgado & Ganea, 2000).

The physical sites of interaction between VIP and its VPAC1 receptor remained elusive until the development of extensive photoaffinity experiments showing that the side chains of VIP in positions 6, 22, 24, and 28 are in direct contact with different amino acids of the receptor VPAC1 N-ted (Ceraudo et al., 2008; Tan, Couvineau, & Laburthe, 2004; Tan, Couvineau, Van Rampelbergh, & Laburthe, 2003). The N-ted structure contains two antiparallel β-sheets and is stabilized by three disulphide bonds between residues Cys50 and Cys72, Cys63 and Cys105 and Cys86. Elucidation of VIP structure by nuclear magnetic resonance and circular dichroism revealed that most of the 28-amino acid peptide displays an α-helical structure (sequence 11–26) and two β-bends (residues 2–5 and 1–10) at the N-terminus (N-ted) (Fry et al., 1989). The development of a structural model of the VPAC1 receptor N-ted has made it possible to localize the binding site of VIP. This finding is in good agreement with a speculative, but largely accepted, mechanism for peptide–ligand interaction with class B GPCRs, which is referred to as a "two-site" binding model (Hoare, 2005). In this model, the central and C-terminal α-helical segments of the ligand are trapped by the N-ted domain of the receptor. In this respect, the integrity of the α-helical conformation seems crucial for binding to their receptors. Although numerous structure and activity studies have elucidated the molecular interactions between VIP and their receptors, the determination of the structure of VIP receptors is still a challenge and more information will provide and facilitate the design of VIP-engineered NPs able to interact with VIP receptors, either triggering the effector–receptor system or just as a way of specific targeting, depending on the desired effects for particular applications.

3. FUNCTIONAL BASES FOR THE NEUROPEPTIDE VIP AS AN ENDOGENOUS LEAD COMPOUND

The physiological roles of VIP and the biological/pharmacological properties that have been reported in several animal models, among others, for neurodegenerative and inflammatory diseases, point to its therapeutically useful potential (Morell, Souza-Moreira, & Gonzalez-Rey, 2012).

In this sense, the recent use of VIP receptor and peptide knockout animals has contributed to important insights into their importance in a number of areas. These include their roles in metabolism, obesity, control of insulin release, various gastrointestinal disorders, and in CNS, particularly in regard to roles in regulation of circadian rhythm, neuroprotection, and more recently certain possible human disorders such as schizophrenia and depression (Morell et al., 2012).

In particular, due to its potent action on the CNS, VIP and its receptors could be promising therapeutic targets for the treatment of various neurological disorders. VIP is also thought to play a role in neurodevelopment and in neuroprotection following injury to the CNS. For example, VIP has been shown to be protective against excitotoxin-induced white matter lesions in neonatal mice (Gressens et al., 1997; Rangon et al., 2005), probably acting through VPAC2 receptors. VPAC2 receptors have also been implicated in the control of astrocyte proliferation (Zupan et al., 1998). Moreover, VPAC2 receptors have also been implicated in the VIP-induced expression of the neuroprotective protein activity-dependent neuroprotective protein (ADNP) in astrocytes (Zusev & Gozes, 2004) and davunetide, an active fragment of ADNP, is in clinical development for the treatment of neurodegenerative disorders (Gozes, 2011) Gozes ALS 2013 (Jouroukhin et al., 2013).

Two recent publications from independent groups have found associations between copy number variation in the gene encoding the VPAC2 receptor and susceptibility to developing schizophrenia (Levinson et al., 2011; Piggins, 2011). These findings have generated some excitement in the field because they may imply that VIP receptors are a potential target for the development of new antipsychotic drugs or for the treatment of other major neurological disorders, as is the case of Alzehimer's disease.

Furthermore, VIP has been shown to protect against neuronal cell death and to act as a modulator of the inflammatory immune response. For this reason, it has been proposed as a viable treatment option for Parkinson's disease (PD). PD is a common neurodegenerative disease of the CNS which primarily impairs an individual's motor skills and speech. It has been demonstrated by Delgado and Ganea (2003) that VIP could protect dopaminergic cells from bacterial endotoxin lipopolysaccharide (LPS)-induced inflammation in PD mouse embryonic neurons. The treatment with the bacterial endotoxin LPS results in neuronal cell loss and is thought to be mediated by increased microglial activation. In another study using a mouse model of PD, VIP treatment showed a decrease in dopaminergic neuronal loss in the substantia nigra (Delgado & Ganea, 2003; Delgado, Varela, & Gonzalez-Rey, 2008). In this scenario, VIP intervention most likely prevents neurodegeneration by deactivating microglia and the production of proinflammatory mediators via binding to VPAC1. Taken together, these findings have shown that VIP presents good potential as a therapeutic agent for neurodegenerative disorders of the CNS that affect so many people worldwide.

Alongside, VIP has certain characteristics that make it attractive for the control of the mechanisms associated to immune tolerance (Gonzalez-Rey, Anderson, & Delgado, 2007; Pozo, Anderson, & Gonzalez-Rey, 2009). In this sense, VIP restores tolerance in autoimmune disorders by acting at multiple levels (Carrion et al., 2014; Jimeno et al., 2014; Martinez et al., 2014; Pozo et al., 2009). Therefore, the capacity of VIP to regulate a wide spectrum of inflammatory factors and to switch the Th1/Th2 balance in favor of Th2 immunity makes it an attractive therapeutic candidate for the treatment of inflammatory disorders and/or Th1-type autoimmune diseases.

The targeted delivery of therapeutic agents to tumor cells remains a challenge because most of the chemotherapeutic agents distribute throughout the whole body, resulting in general toxicity and poor tolerance by patients. Recent advances have opened the way to site-specific targeting and drug delivery by NPs (Brannon-Peppas & Blanchette, 2004). Besides its relevance as potential therapeutic agent, VIP has attracted attention as a lead compound also for diagnostic purposes, particularly in the field of molecular oncology. Indeed, in the last few years there has been a great interest in the research and development of peptide-functionalized NPs for diagnostic imaging and cancer therapy (Delehanty et al., 2010). This emergence is attributable to the higher expression of hormone receptors for GPCRs in many tumor cells as compared to normal tissues (Reubi, 2003; Reubi et al., 2000). In particular, this phenomenon is of high relevance in the case of VIP receptors.

Based on this high occurrence of tumor VIP receptors, a number of potential clinical applications have been evaluated. First, it has been demonstrated that selected tumors, in particular, the VIP receptor-positive cancers, can be visualized in the patient by means of *in vitro* and *in vivo* VIP receptor scintigraphy (Reubi, Waser, Schmassmann, & Laissue, 1999; Virgolini et al., 1994). For example, Raderer et al. developed scintigraphy by using $^{[123]}$I-VIP as a radioligand and compared the results with computerized tomography (CT). In this study, the VIP scan indicated the presence of disease earlier than CT in four patients (Raderer et al., 1998). On the other hand, several studies have reported an effect of VIP and PACAP analogues on tumor growth in animal tumor models related to lung cancer mediated by specific receptors (Maruno, Absood, & Said, 1998; Moody et al., 1993).

Thus, different VIP analogues may be of relevance both for diagnosis and therapeutic interventions in cancer management.

Recently, a high incidence of PAC1 has been found in human gliomas, neuroblastomas, and pituitary adenomas (Robberecht et al., 1993, 1994; Vertongen et al., 1996), whereas VPAC1 has been identified in pancreatic, colorectal, prostate and breast cancers (Jiang, Kopras, McMichael, Bell, & Ulrich, 1997; Reubi et al., 2000).

In particular, human prostate carcinomas have been analyzed in order to better understand the differences between human normal and malignant prostate tissue. A study from Collado et al. (2005) showed that the level of expression of VIP receptors was about twofold higher in adenocarcinoma samples than in normal tissue, measured by a polymerase chain reaction method and by enzyme immune analysis. In addition, Garcia-Fernandez et al. detected VPAC1, VPAC2, and PAC1 proteins in solubilized human normal and tumoral prostate membranes by immunoblot and analyzed by immunohistochemistry the presence of VPAC1, VPAC2, and PAC1 (Garcia-Fernandez et al., 2003). The receptors were mainly located in the epithelial layer of prostate glands and also the presence of PAC1 was observed in some dispersed cells of the stroma (Garcia-Fernandez et al., 2003). The functionality of these receptors was also confirmed in both, normal and tumoral tissues by VIP stimulation of adenylyl cyclase activity. Moreover, Reubi performed radio-labeled binding experiments in prostate tumors with $^{[125]}$I-VIP and showed 2.5 times more VIP binding sites during malignant transformation (Reubi, 1996). The studies mentioned above support the notion that this family of receptors is highly expressed in cancerous tissues and that it is maintained in a functional condition during malignant transformation in the human prostate gland. Therefore, they could be considered as a convenient option for effective diagnosis and targeted drug delivery. A flowchart depicting the VIP-based strategy for targeted cancer that exploits the fact that tumors leaky vasculature is commonly used for nanoparticle passive targeting, whereas VIP NPs can use this approach and in addition take advantage of the effects of active targeting to cells that overexpress VIP receptors and as a result increase the specificity of treatments (Fig. 3). Actually, this is a healthcare demand of high priority because there is currently no consensus on the optimal management of high-risk prostate cancer, mainly because there are different primary modalities available, such as surgery or radiation. In addition, timing of different therapies is unstandardized which makes comparisons of efficacy problematic. Thus, other alternatives are being explored, and among them, VIP-engineered

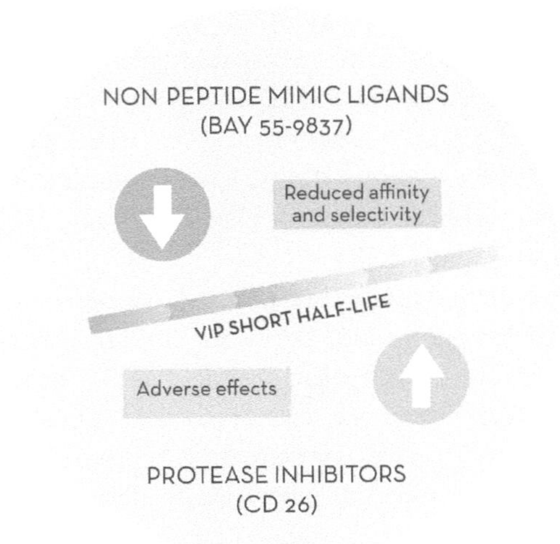

Figure 3 Flowchart depicting the VIP-based approach for targeted cancer. Tumors leaky vasculature is commonly used for NP passive targeting, whereas VIP NPs can use this approach and, in addition, take advantage of the effects of active targeting to cells that overexpress VIP receptors and, as a result, increase the specificity of treatments. In contrast, cells with a low level of VIP receptor expression would receive limited targeting and, as an overall result, the treatment would have lower side effects.

NPs are a feasible option due to high sensitivity and specificity to detect cancer cells offering the potential for early diagnostic and treatment.

4. NANOTECHNOLOGY HELPS TO SOLVE KEY VIP'S DRUGGABILITY PROBLEMS

VIP is an excellent working paradigm to explain some of the issues that impede the full development of neuropeptides toward actual market applications in the clinical setting. For example, many efforts have been made to obtain highly potent VIP analogues (Igarashi et al., 2005) focusing on the improvement of stability to create drug candidates for the treatment of several diseases including asthma (Bolin, Michalewsky, Wasserman, & O'Donnell, 1995), neurodegenerative diseases (Gozes, 2008), impotence (Gozes, Reshef, Salah, Rubinraut, & Fridkin, 1994), septic shock (Delgado & Ganea, 2001; Lv, Tang, Chen, & Xiao, 2009), diabetes (Yung et al., 2003), or as a tumor imaging agent (Kothari et al., 2007). However, none of these analogues has yet reached the clinical stage. Different

VIP ligand modifications have been performed in order to affect the binding to the cognate receptors (VPAC1 and VPAC2). Tandem extensions and additional branching methodologies have been carried out on the C-terminal domain, in an attempt to amplify VPAC1 binding. However, these alterations at the C-terminus showed no significant difference in the activation of the VPAC1 receptor to cAMP production when compared to unmodified VIP (Dangoor, Rubinraut, Fridkin, & Gozes, 2007). Nonetheless, manipulations at the C-terminal region of VIP have been shown to affect discrimination between VPAC1 and VPAC2 receptors (Caraglia et al., 2008), while modifications of the N-terminus have shown increased affinity for the VPAC2 receptor (Langer et al., 2004).

On the other hand, the stability problem has also been approached by an alternative strategy involving the use of covalent and noncovalent reversible inhibitors of specific proteases. A representative example is the case of inhibitors of dipeptidyl peptidase IV (CD26) that inactivates incretin neuropeptides aimed at the treatment of type 2 diabetes or cancer progress (Karagiannis, Paschos, Paletas, Matthews, & Tsapas, 2012; Nisal, Kela, Khunti, & Davies, 2012). Unfortunately, although in advanced phase clinical trials, this approach may show safety issues (kidney impairment, immune defects, skin lesions, etc.) or limited efficacy which result in contradictory approvals from the US or EU regulatory agencies (Drucker, Sherman, Bergenstal, & Buse, 2011; Monami, Dicembrini, Martelli, & Mannucci, 2011). These precautions regarding long-term treatments with specific proteases inhibitors are reflecting the increasing and complex world of the protease degradome where a given protease is often involved in different key physiological processes increasing the chances of adverse effects (Quesada, Ordonez, Sanchez, Puente, & Lopez-Otin, 2009; Turk, Turk du, & Turk, 2012). As an overall conclusion, the use of VIP analogues, either shorter or nonpeptide, as a strategy to increase the short half-life of the cognate VIP molecule has limited advantages while the application of protease inhibitors could lead to undesired side effects (Fig. 4).

The fact that VIP is so attractive for therapeutic use by itself leads to the study of nano-applications such as a peptide delivery system which may solve the problem of drug breakdown by digestive acids and enzymes before they reach their targets (Domschke et al., 1978). It is worth to note that the half-life of VIP needs to be substantially prolonged in biological fluids to be employed in therapeutics with increased effectiveness, as VIP-based drug design is hampered by the instability of the peptide and has limited bioavailability.

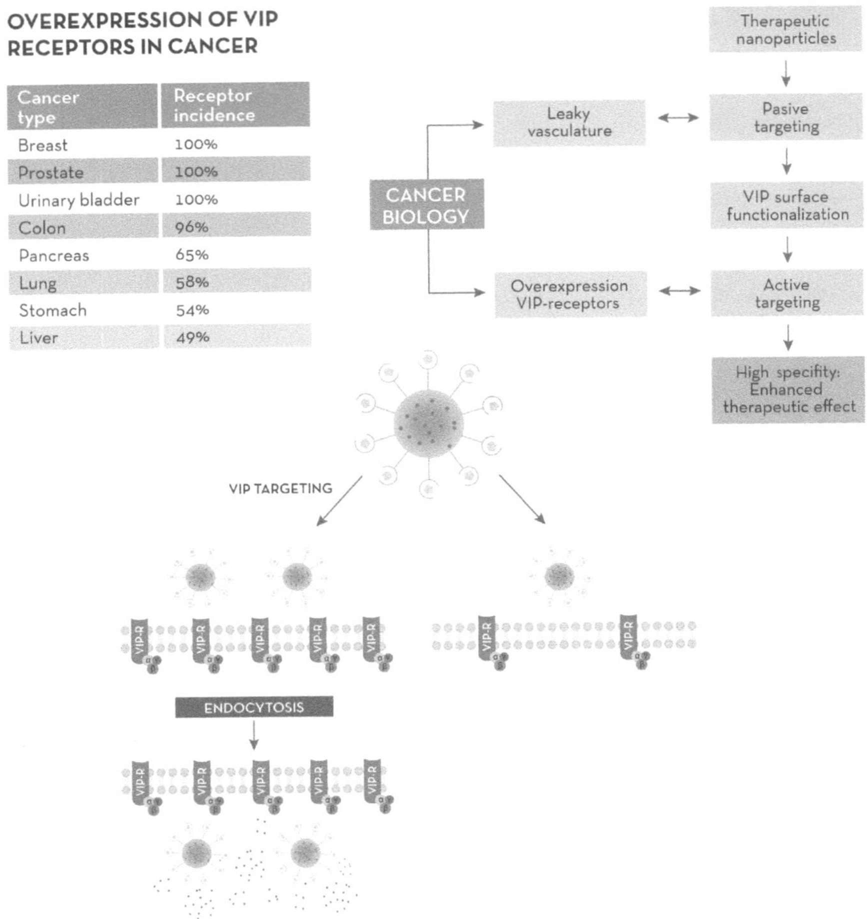

Figure 4 VIP intervention as a therapeutic approach. The neuropeptide VIP has a short circulation half-life (≈2–4 min). Several approaches have been investigated in order to solve the limited circulation half-life of the neuropeptides. An important research trend is focused on the designs from the lead compound of nonpeptide mimic ligands or analogues. Unfortunately, these molecules have shown reduced affinity and selectivity compared to its endogenous ligand counterpart. A recent line of research takes advantage of new protease inhibitors. In this case, several studies have shown different adverse effects, for example, kidney impairment or other effects due to the physiological roles of the proteases, a field of recent awareness. This scenario makes the field of engineered NPs highly attractive for alternative experimental approaches.

5. TYPES OF VIP-ENGINEERED NANOPARTICLES

Two main applications can be developed with VIP-based nanoparticles: (a) VIP encapsulated in NPs or attached to its surface in order to act as a drug itself for therapeutic applications and (b) VIP surface functionalized NPs as a targeting agent to transport drugs to a specific tissue site.

5.1. Nanoparticles designed for VIP drug delivery

Drug delivery plays a crucial role in the improvement of therapeutic agents since many drugs have unfavorable drawbacks if applied directly. As well as the chemical and metabolic issues, there is a possible drawback that the systemic administration of VIP or other VIP receptor agonists could cause moderate hypotension and/or other adverse effects depending on the rate of infusion (Morice, Unwin, & Sever, 1983). Therefore, developing NPs, which can work as carriers for a controlled delivery of a peptide, is a promising option to improve its availability by reducing its side effects and enhancing its efficacy. Several studies have been developed to establish whether the encapsulation of VIP into NPs enhances its effects and whether its attachment to a surface leads to an appropriate circulating transport.

Different NPs have been used to encapsulate VIP, such as liposomes (Gao, Noda, Rubinstein, & Paul, 1994; Stark et al., 2008), biodegradable protamine oligonucleotide NPs (Wernig et al., 2008), or polyethylene glycol–poly(lactic acid) NPs (Gao et al., 2007). Depending on the location where the peptide needs to be delivered, the nature of the nanoparticle and/or surface ligands are different (Klippstein & Pozo, 2011). Liposomes are widely used for drug delivery due to their unique properties (Lasic, 1998). They are biodegradable, typically made from natural non-immunogenic molecules, which makes them even more attractive in view of their ability to encapsulate a variety of molecules, including VIP. Stark and coworkers demonstrated that the encapsulation of VIP into liposomes protects the peptide from proteolytic degradation while maintaining its biological activity (Stark et al., 2008). Furthermore, it has been reported that VIP liposome encapsulation enhances its vasoactive effects on systemic arterial blood pressure (Gao et al., 1994) while another study showed the delayed release of VIP when administered inside liposomes within hyaluronic acid gel for uveitis (Lajavardi et al., 2009). However, although liposomes can be highly effective as VIP carriers in some cases, this is not the case of VIP transport to the brain. For this reason, other options have been engineered, such as glucose-targeted niosomes–nonionic surfactant-based liposomes allowing an efficient delivery of intact VIP to the brain by crossing the blood–brain barrier following intravenous administration (Dufes et al., 2004) or polyethylene glycol–poly(lactic acid) NPs modified with wheat germ agglutinin that enhanced VIP transport to the brain by intranasal administration (Gao et al., 2007). A lipophilic VIP analogue (stearyl-norleucine^{17}) has been proposed for such therapeutic applications (Gozes et al., 1996). However, like most endogenic peptides, its potential therapeutic applications are limited by its susceptibility to endopeptidases and poor passage across the blood–brain barrier (Dogrukol-Ak, Banks, Tuncel, & Tuncel, 2003; McCulloch & Edvinsson, 1980).

Other biodegradable NPs have been used as a drug delivery system to enhance protection against proteolytic cleavage as well as cellular uptake; this is the case of protamine–oligonucleotide NPs, which achieve an appropriate binding efficiency of VIP as well as nanoparticle stability (Wernig et al., 2008). Here, VIP encapsulation occurs during self-assembly of the components and the pharmacological VIP response of encapsulated VIP is investigated using an *ex vivo* lung arterial model system. The authors achieved high encapsulation efficiency (up to 80%) and a modified VIP response on pulmonary arteries, noting differences in the profile of artery relaxation followed by prolonged vasodilation when compared to aqueous VIP.

In addition, VIP poly(lactide-co-glycolide) nanospheres have been formulated for the treatment of airway inflammatory diseases such as asthma and chronic obstructive pulmonary disease due to VIP's potent bronchodilating and anti-inflammatory effects (Onoue et al., 2012). Onoue et al. developed a respiriable sustained-release formulation and studied the peptide stability during the formulation process. The rationale for the development of this formulation was to maximize drug concentrations, in this case VIP in the airway systems, while minimizing systemic exposure and associated toxicity. Their results show that they were able to attenuate antigen-evoked inflammatory symptoms and neutrophilia in the respiratory system.

As we have discussed, the use of biodegradable nanoparticles for encapsulating VIP offers the advantage of sustained release through its diffusion into the NP matrix and subsequent degradation. This permits a higher or lower release depending on the composition of the nanoparticle with the possibility of designing specific nanosystems according to the treatment and the therapeutic agent. The advantage of VIP encapsulation relies on the facts of peptide protection and sustained release, thus maintaining its activity and increasing its effectiveness for treatments.

Other nanoparticles have been used to enhance the therapeutic/diagnostic potential of VIP with the valuable properties of metal NPs in the field of biosensing and medical imaging owing to their plasmonic properties (Caro, Castillo, Klippstein, Pozo, & Zaderenko, 2010). This is the case of silver-protected VIP NPs that retain the functional activity of VIP, assayed by the inhibition of activated microglia under inflammatory conditions (Fernandez-Montesinos et al., 2009).

A great number of techniques are described in the literature for the preparation of peptide encapsulated NPs or peptide-functionalized NPs, but few of them make direct reference to VIP. As explained above, VIP can be encapsulated as well as attached to the surface of molecules in order to act like a drug itself or to direct another compound to a specific site. There is an invention that uses these two features of VIP to merge them in a single nanoparticle (Paul, Yasuko, & Rubinstein, 1998). This invention describes a method of delivering VIP to the surface and to the intracellular compartment of a target tissue by producing a liposome where VIP is located both on and inside the liposome, i.e., VIP is exposed to the outside solvent and to the luminal space. Other inventions describe the encapsulation of VIP but with the aim of a controlled release. This is the case of a water-soluble peptide encapsulation method in order to prepare polymeric microspheres or nanospheres that encapsulate VIP with high efficiency (Ignatious, 2001). The author describes a protocol for reducing the aqueous solubility of the drug, without sacrificing its potency, which is useful in case of high peptide loading and heterogeneous distribution of the drug particles, leading to a nonpredictable release profile. This offers an improved encapsulation method when predictable release profiles are needed.

5.2. Smart-delivery nanoparticles designed for VIP targeting

In other applications, VIP is used as a surface ligand for targeted delivery. In this case, VIP needs to be efficiently attached to the nanoparticle surface and for this purpose liposomes are commonly used. They offer the possibility of attaching a ligand to the lipid head group or to the distal end of the polyethylene-glycerol (PEG) chain on PEG-grafted liposomes and can evade the reticulo-endothelial system as well as have prolonged circulation time in blood. Notably, these nano-applications that use VIP as surface ligands and not as a cargo take advantage of the fact that VIP can be used for targeting specific cells, for example, human cancers, and has clinical impact at both diagnostic and therapeutic levels. An example of these nanosystems could be phospholipid nanomicelles grafted with VIP, which have been reported to actively target breast cancer and act as carriers for water-insoluble anticancer drugs (Onyuksel, Jeon, & Rubinstein, 2009; Rubinstein, Soos, & Onyuksel, 2008), e.g., paclitaxel or an antibiotic derivative named 17-allylamino-17-demethoxy geldanamycin (Onyuksel, Mohanty, & Rubinstein, 2009). In summary, the authors demonstrated that they could solubilize these drugs at therapeutically relevant concentrations in actively targeted VIP surface-grafted micelles. The cytotoxicity of these micelles to human breast cancer cells was retained implying high affinity to VIP receptors, which are overexpressed on these cells and also mediate, in part, their intracellular uptake, thereby amplifying the drug potency. Pioneer research teams led by Rubinstein and Onyuksel also characterized VIP grafted liposome for targeted breast cancer imaging and could demonstrate that there was significantly more accumulation of VIP liposomes than similar liposomes without VIP, therefore indicating a successful use of VIP receptors for active, rather than passive, molecular targeting (Dagar, Krishnadas, Rubinstein, Blend, & Onyuksel, 2003).

Not only targeting to cancer tissues but also targeting to other cells such as activated T-lymphocytes and macrophages in rheumatoid arthritis have been studied (Koo, Rubinstein, & Onyuksel, 2011). May Yue Koo et al. formulated camptothecin sterically stabilized micelles conjugated to VIP to actively bind to overexpressed VPAC2 receptors in the cells mentioned previously. This approach of actively targeting long circulating micelles to the effector cells in the arthritic joint enhanced the efficacy of the drug and diminished systemic toxicity. They could observe that a single, subcutaneous injection of low-dose camptothecin formulation mitigated joint inflammation for at least 32 days after treatment without systemic toxicity in collagen-induced arthritis mice (Koo et al., 2011).

A major issue for the development of clinical applications based on engineered NPs pertains to toxicology, pharmacokinetic, and biodistribution studies (Longmire, Choyke, & Kobayashi, 2008; Malam et al., 2011). In this sense, there are few data on VIP-engineered NPs, and mainly limited to animal models. Dagar and collaborators have demonstrated, with breast tumor-bearing rats, that VIP covalently attached to sterically stabilized liposomes encapsulating a radionuclide did not alter the pharmacokinetic parameters of the liposomes, as half-lives data calculated from blood radioactivity indicated (Rubinstein, Patel, Ikezaki, Dagar, & Onyuksel, 1999). Therefore, VIP could be used to decorate labeled, long circulating small liposomes to develop new targeted imaging agents for enhanced, and thus, early detection of several cancer types. Although biodistribution studies using liposomes have shown increased brain delivery of VIP (Dufes et al., 2004; Gao et al., 2007), the authors do not perform comprehensive pharmacokinetic studies, and only indirect data indicate the existence of VIP-engineered NPs clearance by cerebrospinal fluid draining (Dufes et al., 2004). Precise pharmacokinetic analyses of the entire plasma profile, including absorption, distribution, and elimination, should be included in coming studies and, whenever possible, comparison with VIP engineered NPs intravenous routes of administration should be performed.

6. CONCLUDING REMARKS

The effective translation of preclinical studies related to VIP into clinical realities faces several major challenges; most of them are commonplace for other neuropeptides. Thus, one of the major issues for the use of VIP as a therapeutic agent is that once it has been released into the body it is quickly degraded by enzymes, leaving the peptide with a very short half-life. While new peptide derivatives specifically targeting VPAC receptor subtypes are now available, their short half-life and the inconvenience related to their administration routes make them difficult to use in human therapy. Classical approaches peptide PEGylation has been limited by the propensity of the large polymers to interfere with peptide function. In this context, nanotechnological approaches could be a useful and attractive option. VIP functionalization to NPs could increase their half-life, increasing its efficacy and decreasing the dose needed for treatments. Such molecules would be of considerable interest in the therapy of human diseases, in particular inflammatory, neurodegenerative and cancer-related conditions. However, we should be cautious too in this sense, as excessive circulating times of agents known to increase the intracellular levels of cAMP might lead to side effects. In any case, the adjustability of engineered NPs should provide tools for optimal fine-tuning of VIP-based NPs circulating times. VIP-based NPs show also great potential in terms as diagnostic agents and it could also be envisage the use of VIP-engineered NPs as theranostics agents.

ACKNOWLEDGMENTS

Financial support was provided by the Carlos III Institute of Health of Spain (Spanish Ministry of Economy and Competitiveness) according to the "Strategic Action in Health" (PI14-01600) with co-funding by FEDER funds, the Andalusian Ministry of Economy, Science and Innovation (P10-CTS-6928 and P11-CTS-8161), and the PAIDI Program from the Andalusian Government (CTS-677). R. K. was supported by a research contract (associated to CTS6928) and currently holds a Marie Curie research fellowship from FP7-ITN Marie-Curie Network programme RADDEL (290023) at The Institute of Pharmaceutical Sciences, King's College London, UK.

REFERENCES

Auffan, M., Rose, J., Bottero, J. Y., Lowry, G. V., Jolivet, J. P., & Wiesner, M. R. (2009). Towards a definition of inorganic nanoparticles from an environmental, health and safety perspective. *Nature Nanotechnology*, *4*, 634–641.

Barrie, A. P., Clohessy, A. M., Buensuceso, C. S., Rogers, M. V., & Allen, J. M. (1997). Pituitary adenylyl cyclase-activating peptide stimulates extracellular signal-regulated kinase 1 or 2 (ERK1/2) activity in a Ras-independent, mitogen-activated protein kinase/ERK kinase 1 or 2-dependent manner in PC12 cells. *The Journal of Biological Chemistry*, *272*, 19666–19671.

Bolin, D. R., Michalewsky, J., Wasserman, M. A., & O'Donnell, M. (1995). Design and development of a vasoactive intestinal peptide analog as a novel therapeutic for bronchial asthma. *Biopolymers*, *37*, 57–66.

Brannon-Peppas, L., & Blanchette, J. O. (2004). Nanoparticle and targeted systems for cancer therapy. *Advanced Drug Delivery Reviews*, *56*, 1649–1659.

Caraglia, M., Carteni, M., Dicitore, A., Cassese, D., De Maria, S., Ferranti, P., et al. (2008). Experimental study on vasoactive intestinal peptide (VIP) and its diaminopropane bound (VIP-DAP) analog in solution. *Amino Acids*, *35*(2), 275–281.

Caro, C., Castillo, P. M., Klippstein, R., Pozo, D., & Zaderenko, A. P. (2010). Silver nanoparticles. In D. Pozo (Ed.), *Silver nanoparticles: Sensing and imaging applications* (pp. 201–223). Vienna: In Tech.

Carrion, M., Juarranz, Y., Seoane, I. V., Martinez, C., Gonzalez-Alvaro, I., Pablos, J. L., et al. (2014). VIP modulates IL-22R1 expression and prevents the contribution of rheumatoid synovial fibroblasts to IL-22-mediated joint destruction. *Journal of Molecular Neuroscience*, *52*, 10–17.

Ceraudo, E., Murail, S., Tan, Y. V., Lacapere, J. J., Neumann, J. M., Couvineau, A., et al. (2008). The vasoactive intestinal peptide (VIP) alpha-helix up to C terminus interacts with the N-terminal ectodomain of the human VIP/pituitary adenylate cyclase-activating peptide 1 receptor: Photoaffinity, molecular modeling, and dynamics. *Molecular Endocrinology*, *22*, 147–155.

Collado, B., Carmena, M. J., Sanchez-Chapado, M., Ruiz-Villaespesa, A., Bajo, A. M., Fernandez-Martinez, A. B., et al. (2005). Expression of vasoactive intestinal peptide and functional VIP receptors in human prostate cancer: Antagonistic action of a growth-hormone-releasing hormone analog. *International Journal of Oncology*, *26*, 1629–1635.

Dagar, S., Krishnadas, A., Rubinstein, I., Blend, M. J., & Onyuksel, H. (2003). VIP grafted sterically stabilized liposomes for targeted imaging of breast cancer: In vivo studies. *Journal of Controlled Release*, *91*, 123–133.

Dangoor, D., Rubinraut, S., Fridkin, M., & Gozes, I. (2007). Novel analogs of VIP with multiple C-terminal domains. *Peptides*, *28*, 1622–1630.

Delehanty, J. B., Boeneman, K., Bradburne, C. E., Robertson, K., Bongard, J. E., & Medintz, I. L. (2010). Peptides for specific intracellular delivery and targeting of nanoparticles: Implications for developing nanoparticle-mediated drug delivery. *Therapeutic Delivery*, *1*, 411–433.
Delgado, M., & Ganea, D. (2000). Inhibition of IFN-gamma-induced janus kinase-1-STAT1 activation in macrophages by vasoactive intestinal peptide and pituitary adenylate cyclase-activating polypeptide. *Journal of Immunology*, *165*, 3051–3057.
Delgado, M., & Ganea, D. (2001). Inhibition of endotoxin-induced macrophage chemokine production by vasoactive intestinal peptide and pituitary adenylate cyclase-activating polypeptide in vitro and in vivo. *Journal of Immunology*, *167*, 966–975.
Delgado, M., & Ganea, D. (2003). Vasoactive intestinal peptide prevents activated microglia-induced neurodegeneration under inflammatory conditions: Potential therapeutic role in brain trauma. *The FASEB Journal*, *17*, 1922–1924.
Delgado, M., Varela, N., & Gonzalez-Rey, E. (2008). Vasoactive intestinal peptide protects against beta-amyloid-induced neurodegeneration by inhibiting microglia activation at multiple levels. *Glia*, *56*, 1091–1103.
Dogrukol-Ak, D., Banks, W. A., Tuncel, N., & Tuncel, M. (2003). Passage of vasoactive intestinal peptide across the blood–brain barrier. *Peptides*, *24*, 437–444.
Domschke, S., Domschke, W., Bloom, S. R., Mitznegg, P., Mitchell, S. J., Lux, G., et al. (1978). Vasoactive intestinal peptide in man: Pharmacokinetics, metabolic and circulatory effects. *Gut*, *19*, 1049–1053.
Drechsler, U., Erdogan, B., & Rotello, V. M. (2004). Nanoparticles: Scaffolds for molecular recognition. *Chemistry*, *10*, 5570–5579.
Drucker, D. J., Sherman, S. I., Bergenstal, R. M., & Buse, J. B. (2011). The safety of incretin-based therapies—Review of the scientific evidence. *The Journal of Clinical Endocrinology and Metabolism*, *96*, 2027–2031.
Dufes, C., Gaillard, F., Uchegbu, I. F., Schatzlein, A. G., Olivier, J. C., & Muller, J. M. (2004). Glucose-targeted niosomes deliver vasoactive intestinal peptide (VIP) to the brain. *International Journal of Pharmaceutics*, *285*, 77–85.
Fernandez-Montesinos, R., Castillo, P. M., Klippstein, R., Gonzalez-Rey, E., Mejias, J. A., Zaderenko, A. P., et al. (2009). Chemical synthesis and characterization of silver-protected vasoactive intestinal peptide nanoparticles. *Nanomedicine (London, England)*, *4*, 919–930.
Friedman, A. D., Claypool, S. E., & Liu, R. (2013). The smart targeting of nanoparticles. *Current Pharmaceutical Design*, *19*, 6315–6329.
Fry, D. C., Madison, V. S., Bolin, D. R., Greeley, D. N., Toome, V., & Wegrzynski, B. B. (1989). Solution structure of an analogue of vasoactive intestinal peptide as determined by two-dimensional NMR and circular dichroism spectroscopies and constrained molecular dynamics. *Biochemistry*, *28*, 2399–2409.
Gao, X. P., Noda, Y., Rubinstein, I., & Paul, S. (1994). Vasoactive intestinal peptide encapsulated in liposomes: Effects on systemic arterial blood pressure. *Life Sciences*, *54*, PL247–PL252.
Gao, X., Wu, B., Zhang, Q., Chen, J., Zhu, J., Zhang, W., et al. (2007). Brain delivery of vasoactive intestinal peptide enhanced with the nanoparticles conjugated with wheat germ agglutinin following intranasal administration. *Journal of Controlled Release*, *121*, 156–167.
Garcia-Fernandez, M. O., Solano, R. M., Carmena, M. J., Busto, R., Bodega, G., Ruiz-Villaespesa, A., et al. (2003). Expression of functional PACAP/VIP receptors in human prostate cancer and healthy tissue. *Peptides*, *24*, 893–902.
Gendelman, H. E., Balogh, L. P., Bawa, R., Bradbury, M., Chang, E. H., Chiu, W., et al. (2014). The 4th annual meeting of the American society for nanomedicine. *Journal of Neuroimmune Pharmacology*, *9*(Suppl. 1), S1–S3.

Gonzalez-Rey, E., Anderson, P., & Delgado, M. (2007). Emerging roles of vasoactive intestinal peptide: A new approach for autoimmune therapy. *Annals of the Rheumatic Diseases*, *66*, 70–76.

Gonzalez-Rey, E., Delgado-Maroto, V., Souza Moreira, L., & Delgado, M. (2010). Neuropeptides as therapeutic approach to autoimmune diseases. *Current Pharmaceutical Design*, *16*, 3158–3172.

Gozes, I. (2008). VIP, from gene to behavior and back: Summarizing my 25 years of research. *Journal of Molecular Neuroscience*, *36*, 115–124.

Gozes, I. (2011). NAP (davunetide) provides functional and structural neuroprotection. *Current Pharmaceutical Design*, *17*, 1040–1044.

Gozes, I., Bardea, A., Reshef, A., Zamostiano, R., Zhukovsky, S., Rubinraut, S., et al. (1996). Neuroprotective strategy for Alzheimer disease: Intranasal administration of a fatty neuropeptide. *Proceedings of the National Academy of Sciences of the United States of America*, *93*, 427–432.

Gozes, I., Reshef, A., Salah, D., Rubinraut, S., & Fridkin, M. (1994). Stearyl-norleucine-vasoactive intestinal peptide (VIP): A novel VIP analog for noninvasive impotence treatment. *Endocrinology*, *134*(5), 2121–2125.

Gressens, P., Marret, S., Hill, J. M., Brenneman, D. E., Gozes, I., Fridkin, M., et al. (1997). Vasoactive intestinal peptide prevents excitotoxic cell death in the murine developing brain. *The Journal of Clinical Investigation*, *100*, 390–397.

Harmar, A. J., Arimura, A., Gozes, I., Journot, L., Laburthe, M., Pisegna, J. R., et al. (1998). International Union of Pharmacology. XVIII. Nomenclature of receptors for vasoactive intestinal peptide and pituitary adenylate cyclase-activating polypeptide. *Pharmacological Reviews*, *50*, 265–270.

Harmar, A. J., Fahrenkrug, J., Gozes, I., Laburthe, M., May, V., Pisegna, J. R., et al. (2012). Pharmacology and functions of receptors for vasoactive intestinal peptide and pituitary adenylate cyclase-activating polypeptide: IUPHAR review 1. *British Journal of Pharmacology*, *166*, 4–17.

Henning, R. J., & Sawmiller, D. R. (2001). Vasoactive intestinal peptide: Cardiovascular effects. *Cardiovascular Research*, *49*, 27–37.

Hoare, S. R. (2005). Mechanisms of peptide and nonpeptide ligand binding to Class B G-protein-coupled receptors. *Drug Discovery Today*, *10*, 417–427.

Howard, K. A., & Peer, D. (2013). Providing the full picture: A mandate for standardizing nanoparticle-based drug delivery. *Nanomedicine (London, England)*, *8*, 1031–1033.

Huang, X., Jain, P. K., El-Sayed, I. H., & El-Sayed, M. A. (2007). Gold nanoparticles: Interesting optical properties and recent applications in cancer diagnostics and therapy. *Nanomedicine (London, England)*, *2*, 681–693.

Igarashi, H., Ito, T., Mantey, S. A., Pradhan, T. K., Hou, W., Coy, D. H., et al. (2005). Development of simplified vasoactive intestinal peptide analogs with receptor selectivity and stability for human vasoactive intestinal peptide/pituitary adenylate cyclase-activating polypeptide receptors. *The Journal of Pharmacology and Experimental Therapeutics*, *315*, 370–381.

Ignatious, F. (2001). Encapsulation of water soluble, peptides. US006270700B1.

Jiang, S., Kopras, E., McMichael, M., Bell, R. H., Jr., & Ulrich, C. D., 2nd. (1997). Vasoactive intestinal peptide (VIP) stimulates in vitro growth of VIP-1 receptor-bearing human pancreatic adenocarcinoma-derived cells. *Cancer Research*, *57*, 1475–1480.

Jimeno, R., Leceta, J., Martinez, C., Gutierrez-Canas, I., Carrion, M., Perez-Garcia, S., et al. (2014). Vasoactive intestinal peptide maintains the nonpathogenic profile of human Th17-polarized cells. *Journal of Molecular Neuroscience*, *54*, 512–525.

Johansson, E. M., & Bradley, M. (2012). Biocompatible polymer nanoparticles for intracellular applications. *Chimia*, *66*, 237–240.

Jouroukhin, Y., Ostritsky, R., Assaf, Y., Pelled, G., Giladi, E., & Gozes, I. (2013). NAP (davunetide) modifies disease progression in a mouse model of severe neurodegeneration: Protection against impairments in axonal transport. *Neurobiology of Disease, 56*, 79–94.

Karagiannis, T., Paschos, P., Paletas, K., Matthews, D. R., & Tsapas, A. (2012). Dipeptidyl peptidase-4 inhibitors for treatment of type 2 diabetes mellitus in the clinical setting: Systematic review and meta-analysis. *British Medical Journal, 344*, e1369. http://dx.doi.org/10.1136/bmj.e1369.

Klippstein, R., & Pozo, D. (2010). Nanotechnology-based manipulation of dendritic cells for enhanced immunotherapy strategies. *Nanomedicine, 6*, 523–529.

Klippstein, R., & Pozo, D. (2011). Engineered nanoparticles for improved vasoactive intestinal peptide (VIP) biomedical applications. *Recent Patents on Nanomedicine, 1*, 55–59.

Koo, O. M., Rubinstein, I., & Onyuksel, H. (2011). Actively targeted low-dose camptothecin as a safe, long-acting, disease-modifying nanomedicine for rheumatoid arthritis. *Pharmaceutical Research, 28*, 776–787.

Kothari, K., Prasad, S., Korde, A., Mukherjee, A., Mathur, A., Jaggi, M., et al. (2007). 99mTc(CO)3-VIP analogues: Preparation and evaluation as tumor imaging agent. *Applied Radiation and Isotopes, 65*, 382–386.

Lajavardi, L., Camelo, S., Agnely, F., Luo, W., Goldenberg, B., Naud, M. C., et al. (2009). New formulation of vasoactive intestinal peptide using liposomes in hyaluronic acid gel for uveitis. *Journal of Controlled Release, 139*, 22–30.

Langer, I., Gregoire, F., Nachtergael, I., De Neef, P., Vertongen, P., & Robberecht, P. (2004). Hexanoylation of a VPAC2 receptor-preferring ligand markedly increased its selectivity and potency. *Peptides, 25*, 275–278.

Lasic, D. D. (1998). Novel applications of liposomes. *Trends in Biotechnology, 16*, 307–321.

Lelievre, V., Pineau, N., Du, J., Wen, C. H., Nguyen, T., Janet, T., et al. (1998). Differential effects of peptide histidine isoleucine (PHI) and related peptides on stimulation and suppression of neuroblastoma cell proliferation. A novel VIP-independent action of PHI via MAP kinase. *The Journal of Biological Chemistry, 273*, 19685–19690.

Levinson, D. F., Duan, J., Oh, S., Wang, K., Sanders, A. R., Shi, J., et al. (2011). Copy number variants in schizophrenia: Confirmation of five previous findings and new evidence for 3q29 microdeletions and VIPR2 duplications. *The American Journal of Psychiatry, 168*, 302–316.

Longmire, M., Choyke, P. L., & Kobayashi, H. (2008). Clearance properties of nano-sized particles and molecules as imaging agents: Considerations and caveats. *Nanomedicine (London, England), 3*, 703–717.

Lv, B., Tang, Y., Chen, F., & Xiao, X. (2009). Vasoactive intestinal peptide and pituary adenylate cyclase-activating polypeptide inhibit tissue factor expression in monocyte in vitro and in vivo. *Shock, 31*, 185–191.

Malam, Y., Lim, E. J., & Seifalian, A. M. (2011). Current trends in the application of nanoparticles in drug delivery. *Current Medicinal Chemistry, 18*, 1067–1078.

Martinez, C., Ortiz, A. M., Juarranz, Y., Lamana, A., Seoane, I. V., Leceta, J., et al. (2014). Serum levels of vasoactive intestinal peptide as a prognostic marker in early arthritis. *PLoS One, 9*, e85248. http://dx.doi.org/10.1371/journal.pone.0085248.

Maruno, K., Absood, A., & Said, S. I. (1998). Vasoactive intestinal peptide inhibits human small-cell lung cancer proliferation in vitro and in vivo. *Proceedings of the National Academy of Sciences of the United States of America, 95*, 14373–14378.

McCulloch, J., & Edvinsson, L. (1980). Cerebral circulatory and metabolic effects of vasoactive intestinal polypeptide. *The American Journal of Physiology, 238*, H449–H456.

Miyata, A., Arimura, A., Dahl, R. R., Minamino, N., Uehara, A., Jiang, L., et al. (1989). Isolation of a novel 38 residue-hypothalamic polypeptide which stimulates adenylate

cyclase in pituitary cells. *Biochemical and Biophysical Research Communications, 164*, 567–574.
Miyata, A., Jiang, L., Dahl, R. D., Kitada, C., Kubo, K., Fujino, M., et al. (1990). Isolation of a neuropeptide corresponding to the N-terminal 27 residues of the pituitary adenylate cyclase activating polypeptide with 38 residues (PACAP38). *Biochemical and Biophysical Research Communications, 170*, 643–648.
Monami, M., Dicembrini, I., Martelli, D., & Mannucci, E. (2011). Safety of dipeptidyl peptidase-4 inhibitors: A meta-analysis of randomized clinical trials. *Current Medical Research and Opinion, 27*, 57–64.
Moody, T. W., Zia, F., Draoui, M., Brenneman, D. E., Fridkin, M., Davidson, A., et al. (1993). A vasoactive intestinal peptide antagonist inhibits non-small cell lung cancer growth. *Proceedings of the National Academy of Sciences of the United States of America, 90*, 4345–4349.
Morawski, A. M., Lanza, G. A., & Wickline, S. A. (2005). Targeted contrast agents for magnetic resonance imaging and ultrasound. *Current Opinion in Biotechnology, 16*, 89–92.
Morell, M., Souza-Moreira, L., & Gonzalez-Rey, E. (2012). VIP in neurological diseases: More than a neuropeptide. *Endocrine, Metabolic & Immune Disorders: Drug Targets, 12*, 323–332.
Morice, A., Unwin, R. J., & Sever, P. S. (1983). Vasoactive intestinal peptide causes bronchodilatation and protects against histamine-induced bronchoconstriction in asthmatic subjects. *Lancet, 2*, 1225–1227.
Murthy, K. S., Zhang, K. M., Jin, J. G., Grider, J. R., & Makhlouf, G. M. (1993). VIP-mediated G protein-coupled Ca2+ influx activates a constitutive NOS in dispersed gastric muscle cells. *The American Journal of Physiology, 265*, G660–G671.
Mutt, V. (1988). Vasoactive intestinal polypeptide and related peptides. Isolation and chemistry. *Annals of the New York Academy of Sciences, 527*, 1–19.
Nisal, K., Kela, R., Khunti, K., & Davies, M. J. (2012). Comparison of efficacy between incretin-based therapies for type 2 diabetes mellitus. *BMC Medicine, 10*, 152.
Onoue, S., Matsui, T., Kuriyama, K., Ogawa, K., Kojo, Y., Mizumoto, T., et al. (2012). Inhalable sustained-release formulation of long-acting vasoactive intestinal peptide derivative alleviates acute airway inflammation. *Peptides, 35*, 182–189.
Onyuksel, H., Jeon, E., & Rubinstein, I. (2009). Nanomicellar paclitaxel increases cytotoxicity of multidrug resistant breast cancer cells. *Cancer Letters, 274*, 327–330.
Onyuksel, H., Mohanty, P. S., & Rubinstein, I. (2009). VIP-grafted sterically stabilized phospholipid nanomicellar 17-allylamino-17-demethoxy geldanamycin: A novel targeted nanomedicine for breast cancer. *International Journal of Pharmaceutics, 365*, 157–161.
Paul, S., Yasuko, N., & Rubinstein, I. (1998). Method of delivering a vasoactive intestinal polypeptide, an encapsulated vasoactive intestinal polypeptide, and a method of making the encapsulated vasoactive intestinal polypeptide. US005770570A.
Piggins, H. D. (2011). Schizophrenia: Zooming in on a gene. *Nature, 471*, 455–456.
Piper, P. J., Said, S. I., & Vane, J. R. (1970). Effects on smooth muscle preparations of unidentified vasoactiv peptides from intestine and lung. *Nature, 225*, 1144–1146.
Pozo, D., Anderson, P., & Gonzalez-Rey, E. (2009). Induction of alloantigen-specific human T regulatory cells by vasoactive intestinal peptide. *Journal of Immunology, 183*, 4346–4359.
Pozo, D., & Delgado, M. (2004). The many faces of VIP in neuroimmunology: A cytokine rather a neuropeptide? *The FASEB Journal, 18*, 1325–1334.
Pozo, D., Gonzalez-Rey, E., Chorny, A., Anderson, P., Varela, N., & Delgado, M. (2007). Tuning immune tolerance with vasoactive intestinal peptide: A new therapeutic approach for immune disorders. *Peptides, 28*, 1833–1846.
Quesada, V., Ordonez, G. R., Sanchez, L. M., Puente, X. S., & Lopez-Otin, C. (2009). The Degradome database: Mammalian proteases and diseases of proteolysis. *Nucleic Acids Research, 37*, D239–D243.

Quintana, F. J. (2013). Nanoparticles for the induction of antigen-specific Tregs. *Immunotherapy*, *5*(5), 437–440.
Raderer, M., Kurtaran, A., Yang, Q., Meghdadi, S., Vorbeck, F., Hejna, M., et al. (1998). Iodine-123-vasoactive intestinal peptide receptor scanning in patients with pancreatic cancer. *Journal of Nuclear Medicine*, *39*, 1570–1575.
Rangon, C. M., Goursaud, S., Medja, F., Lelievre, V., Mounien, L., Husson, I., et al. (2005). VPAC2 receptors mediate vasoactive intestinal peptide-induced neuroprotection against neonatal excitotoxic brain lesions in mice. *The Journal of Pharmacology and Experimental Therapeutics*, *314*, 745–752.
Reubi, J. C. (1996). In vitro identification of VIP receptors in human tumors: Potential clinical implications. *Annals of the New York Academy of Sciences*, *805*, 753–759.
Reubi, J. C. (2003). Peptide receptors as molecular targets for cancer diagnosis and therapy. *Endocrine Reviews*, *24*, 389–427.
Reubi, J. C., Laderach, U., Waser, B., Gebbers, J. O., Robberecht, P., & Laissue, J. A. (2000). Vasoactive intestinal peptide/pituitary adenylate cyclase-activating peptide receptor subtypes in human tumors and their tissues of origin. *Cancer Research*, *60*, 3105–3112.
Reubi, J. C., Waser, B., Schmassmann, A., & Laissue, J. A. (1999). Receptor autoradiographic evaluation of cholecystokinin, neurotensin, somatostatin and vasoactive intestinal peptide receptors in gastro-intestinal adenocarcinoma samples: Where are they really located? *International Journal of Cancer*, *81*, 376–386.
Robberecht, P., Vertongen, P., Velkeniers, B., de Neef, P., Vergani, P., Raftopoulos, C., et al. (1993). Receptors for pituitary adenylate cyclase activating peptides in human pituitary adenomas. *The Journal of Clinical Endocrinology and Metabolism*, 77, 1235–1239.
Robberecht, P., Woussen-Colle, M. C., Vertongen, P., De Neef, P., Hou, X., Salmon, I., et al. (1994). Expression of pituitary adenylate cyclase activating polypeptide (PACAP) receptors in human glial cell tumors. *Peptides*, *15*, 661–665.
Rosi, N. L., & Mirkin, C. A. (2005). Nanostructures in biodiagnostics. *Chemical Reviews*, *105*, 1547–1562.
Rubinstein, I., Patel, M., Ikezaki, H., Dagar, S., & Onyuksel, H. (1999). Conformation and vasoreactivity of VIP in phospholipids: Effects of calmodulin. *Peptides*, *20*, 1497–1501.
Rubinstein, I., Soos, I., & Onyuksel, H. (2008). Intracellular delivery of VIP-grafted sterically stabilized phospholipid mixed nanomicelles in human breast cancer cells. *Chemico-Biological Interactions*, *171*, 190–194.
Said, S. I., & Mutt, V. (1970). Polypeptide with broad biological activity: Isolation from small intestine. *Science*, *169*, 1217–1218.
Schutz, C. A., Juillerat-Jeanneret, L., Mueller, H., Lynch, I., & Riediker, M. (2013). Therapeutic nanoparticles in clinics and under clinical evaluation. *Nanomedicine (London, England)*, *8*, 449–467.
Singh, D. K., Jagannathan, R., Khandelwal, P., Abraham, P. M., & Poddar, P. (2013). In situ synthesis and surface functionalization of gold nanoparticles with curcumin and their antioxidant properties: An experimental and density functional theory investigation. *Nanoscale*, *5*, 1882–1893.
Spengler, D., Waeber, C., Pantaloni, C., Holsboer, F., Bockaert, J., Seeburg, P. H., et al. (1993). Differential signal transduction by five splice variants of the PACAP receptor. *Nature*, *365*, 170–175.
Stark, B., Andreae, F., Mosgoeller, W., Edetsberger, M., Gaubitzer, E., Koehler, G., et al. (2008). Liposomal vasoactive intestinal peptide for lung application: Protection from proteolytic degradation. *European Journal of Pharmaceutics and Biopharmaceutics*, *70*, 153–164.
Ster, J., De Bock, F., Guerineau, N. C., Janossy, A., Barrere-Lemaire, S., Bos, J. L., et al. (2007). Exchange protein activated by cAMP (EPAC) mediates cAMP activation of

p38 MAPK and modulation of Ca^{2+}-dependent K^{+} channels in cerebellar neurons. *Proceedings of the National Academy of Sciences of the United States of America*, *104*, 2519–2524.
Straub, S. G., & Sharp, G. W. (1996). A wortmannin-sensitive signal transduction pathway is involved in the stimulation of insulin release by vasoactive intestinal polypeptide and pituitary adenylate cyclase-activating polypeptide. *The Journal of Biological Chemistry*, *271*, 1660–1668.
Tan, Y. V., Couvineau, A., & Laburthe, M. (2004). Diffuse pharmacophoric domains of vasoactive intestinal peptide (VIP) and further insights into the interaction of VIP with the N-terminal ectodomain of human VPAC1 receptor by photoaffinity labeling with [Bpa6]-VIP. *The Journal of Biological Chemistry*, *279*, 38889–38894.
Tan, Y. V., Couvineau, A., Van Rampelbergh, J., & Laburthe, M. (2003). Photoaffinity labeling demonstrates physical contact between vasoactive intestinal peptide and the N-terminal ectodomain of the human VPAC1 receptor. *The Journal of Biological Chemistry*, *278*(38), 36531–36536.
Turk, B., Turk du, S. A., & Turk, V. (2012). Protease signalling: The cutting edge. *The EMBO Journal*, *31*, 1630–1643.
Vertongen, P., Devalck, C., Sariban, E., De Laet, M. H., Martelli, H., Paraf, F., et al. (1996). Pituitary adenylate cyclase activating peptide and its receptors are expressed in human neuroblastomas. *Journal of Cellular Physiology*, *167*, 36–46.
Virgolini, I., Raderer, M., Kurtaran, A., Angelberger, P., Banyai, S., Yang, Q., et al. (1994). Vasoactive intestinal peptide-receptor imaging for the localization of intestinal adenocarcinomas and endocrine tumors. *The New England Journal of Medicine*, *331*, 1116–1121.
Wernig, K., Griesbacher, M., Andreae, F., Hajos, F., Wagner, J., Mosgoeller, W., et al. (2008). Depot formulation of vasoactive intestinal peptide by protamine-based biodegradable nanoparticles. *Journal of Controlled Release*, *130*, 192–198.
West, J. L., & Halas, N. J. (2000). Applications of nanotechnology to biotechnology commentary. *Current Opinion in Biotechnology*, *11*, 215–217.
Yung, S. L., Dela Cruz, F., Hamren, S., Zhu, J., Tsutsumi, M., Bloom, J. W., et al. (2003). Generation of highly selective VPAC2 receptor agonists by high throughput mutagenesis of vasoactive intestinal peptide and pituitary adenylate cyclase-activating peptide. *The Journal of Biological Chemistry*, *278*, 10273–10281.
Zupan, V., Hill, J. M., Brenneman, D. E., Gozes, I., Fridkin, M., Robberecht, P., et al. (1998). Involvement of pituitary adenylate cyclase-activating polypeptide II vasoactive intestinal peptide 2 receptor in mouse neocortical astrocytogenesis. *Journal of Neurochemistry*, *70*, 2165–2173.
Zusev, M., & Gozes, I. (2004). Differential regulation of activity-dependent neuroprotective protein in rat astrocytes by VIP and PACAP. *Regulatory Peptides*, *123*, 33–41.
Zuzana, M., Alessandra, R., Lise, F., & Maria, D. (2011). Safety assessment of nanoparticles cytotoxicity and genotoxicity of metal nanoparticles in vitro. *Journal of Biomedical Nanotechnology*, 7, 20–21.

CHAPTER SIX

Implications of Protein- and Peptide-Based Nanoparticles as Potential Vehicles for Anticancer Drugs

Ahmed O. Elzoghby[*,†,1], **Mayada M. Elgohary**[*,†], **Nayra M. Kamel**[*,†]
[*]Cancer Nanotechnology Research Laboratory (CNRL), Faculty of Pharmacy, Alexandria University, Alexandria, Egypt
[†]Department of Industrial Pharmacy, Faculty of Pharmacy, Alexandria University, Alexandria, Egypt
[1]Corresponding author: e-mail address: dr_ahmedelzoghby@yahoo.com

Contents

Advances in Protein Chemistry and Structural Biology, Volume 98
ISSN 1876-1623
http://dx.doi.org/10.1016/bs.apcsb.2014.12.002

Abstract

Protein-based nanocarriers have gained considerable attention as colloidal carrier systems for the delivery of anticancer drugs. Protein nanocarriers possess various advantages including their low cytotoxicity, abundant renewable sources, high drug-binding capacity, and significant uptake into the targeted tumor cells. Moreover, the unique protein structure offers the possibility of site-specific drug conjugation and tumor targeting using various ligands modifying the surface of protein nanocarriers. In this chapter, we highlight the most important applications of protein nanoparticles (NPs) for the delivery of anticancer drugs. We examine the various techniques that have been utilized for the preparation of anticancer drug-loaded protein NPs. Finally, the current chapter also reviews the major outcomes of the *in vitro* and *in vivo* investigations of surface-modified tumor-targeted protein NPs.

1. INTRODUCTION

Cancer nanotherapeutics are progressing at a steady rate; research and development in the field has experienced an exponential growth since early 2000s. Nanosystems have been reported to target and deliver the drug *in situ* to selectively kill cancer cells, decreasing toxicity on healthy organs and tissues as well as side effects. Nanoparticles (NPs) offer the possibility to encapsulate poorly soluble anticancer drugs, protect unstable therapeutic molecules, and modify their blood circulation and tissue distribution. Furthermore, some nanosystems have been reported to overcome tumor resistance. Over the years, several nanosystems have been proposed to diagnose and treat cancers, such as dendrimers, polymeric micelles, superparamagnetic iron oxide cores, gold NPs, and liposomes (Bertrand, Wu, Xu, Kamaly, & Farokhzad, 2014; Egusquiaguirre, Igartua, Hernández, & Pedraz, 2012).

Novel drug delivery systems were developed to overcome the poor aqueous solubility and enhance the antitumor efficacy of anticancer drugs (Elgindy, Elkhodairy, Molokhia, & Elzoghby, 2010, 2011a, 2011b, 2011c, 2011d). Over the past few decades, there is a growing interest in developing protein nanocarriers as GRAS (generally regarded as safe) drug delivery devices due to their exceptional characteristics, namely biodegradability, nonantigenicity, high nutritional value, abundant renewable sources, and extraordinary binding capacity of various drugs. Proteins have the possibility of less opsonization by the reticuloendothelial system (RES) through an aqueous steric barrier in addition to their excellent functional properties including emulsification, gelation, and water-binding capacity (Elzoghby, Helmy, Samy, & Elgindy, 2013b; Elzoghby, Samy, & Elgindy, 2012a). Moreover, protein NPs can be easily prepared and scaled-up during manufacture. Owing to multiple functional groups in the primary sequences of polypeptides, protein NPs can be exploited to create different interactions with therapeutic compounds and subsequently form three-dimensional networks offering a variety of possibilities for reversible binding of active molecules, protecting them in a matrix, as well as specific targeting to the site of action (Hawkins, Soon-Shiong, & Desai, 2008; Maham, Tang, Wu, Wang, & Lin, 2009). Furthermore, protein NPs possess acceptability as metabolizable naturally occurring components. These properties are attractive in oncology in order to encapsulate cytotoxics exhibiting wide-ranging toxicities and physicochemical properties. They have been successfully synthesized from various proteins including hydrophilic water-soluble proteins (e.g., albumin, gelatin (GEL), and milk proteins) and hydrophobic water-insoluble proteins (e.g., zein and gliadin). In this chapter, we will discuss the application of various protein-based NPs as vehicles for anticancer drugs featuring their significant contribution to the enhancement of the antitumor efficacy and minimizing the toxicity of anticancer therapeutics.

2. PREPARATION TECHNIQUES

NPs were successfully prepared via different techniques from animal- and plant-derived proteins for the delivery of anticancer drugs.

2.1 Desolvation

In desolvation process, NPs are obtained when a desolvating agent (e.g., alcohol or acetone) is added dropwise to an aqueous solution of protein under stirring in order to dehydrate the protein resulting in conformational change from stretched to coil conformation. Next, coacervates were hardened by cross-linking the amino moieties in the protein (Langer et al., 2003). Albumin and gelatin NPs (GNPs) loaded with anticancer drugs are commonly prepared using desolvation method. More recently, silk sericin NPs loaded with plasmid DNA were also prepared via desolvation technique. In some cases, e.g., GNPs, a second desolvation step is required for the formation of smaller and uniform NPs where the high-molecular-weight (HMW) GEL was precipitated in the first desolvation step to remove the nondesolvated low-molecular-weight GEL then HMW GEL is redissolved and desolvated again (Coester, Langer, van Briesen, & Kreuter, 2000). Although extensively used to prepare GNPs, desolvation technique has two major drawbacks: the use of organic solvents and the use of toxic cross-linkers.

2.2 Coacervation–phase separation

In coacervation process, when a salt is added to protein solution, liquid–liquid phase separation occurred, giving rise to a polymer-rich dense phase at the bottom and a transparent solution above resulting in the desired NPs (Lu, Yeh, Tsai, Au, & Wientjes, 2004). Paclitaxel-loaded GNPs (600–1000 nm) were successfully prepared by slow addition of sodium sulfate to aqueous GEL solution containing surfactant (Tween 20) followed by the addition of isopropanol.

2.3 Emulsification–solvent evaporation

Bovine serum albumin (BSA) NPs loaded with the anticancer drug, 10-hydroxycamptothecin (HCPT), were prepared by W/O emulsification method in order to improve its stability mainly in its active lactone form (Yang et al., 2007). In this study, the aqueous alkali solution of HCPT and BSA was added to castor oil containing Span-80 and emulsified with high-speed homogenizer followed by adding the W/O emulsion to castor oil at 140 °C for thermal cross-linking. Similarly, methotrexate-loaded GNPs were prepared via emulsification of the aqueous phase of GEL and drug into an organic solution of polymethylmethacrylate (Cascone, Lazzeri, Carmignani, & Zhu, 2002). In our laboratory, flutamide (FLT)-loaded casein (CN) NPs were fabricated by an O/W emulsification method where the organic solution of the drug in dichloromethane was emulsified into aqueous CN solution via homogenization followed by cross-linking (Elzoghby et al., 2013b; Elzoghby, Saad, Helmy, Samy, & Elgindy, 2013). Although effective for entrapping hydrophobic anticancer drug into hydrophilic proteins, the residual organic solvent is a common limitation of this technique.

2.4 Nanoprecipitation

Nanoprecipitation is a rapid and easy technique enabling the production of small NPs with narrow unimodal distribution. GNPs of 251 nm with unimodal polydispersity of 0.096 were prepared by adding the aqueous protein solution to ethanol as the nonsolvent phase containing poloxamer as a stabilizer followed by cross-linking (Lee, Khan, Park, & Lim, 2012). This technique is particularly useful for the fabrication of drug-loaded NPs from hydrophobic proteins, zein and gliadin. The protein solution in 70–80% aqueous alcohol containing the drug is added to water containing a stabilizer, e.g., sodium caseinate, Tween 80, or poloxamer, under vigorous stirring. When the alcohol concentration decreases below a level required for dissolving protein, protein becomes insoluble and precipitates to form NPs. Zein NPs loaded with hydrophobic anticancer drugs, e.g., curcumin (Patel, Hu, Tiwari, & Velikov, 2010) and quercetin (Patel, Heussen, Hazekamp, Drost, & Velikov, 2012), were successfully prepared using this technique. This process is advantageous over conventional emulsification methods that require strong shear forces to reduce droplet sizes to submicrometers.

2.5 Nano spray-drying

Spray-drying is a well-established method commonly used in the pharmaceutical industry for producing a dry powder from a liquid phase. Unlike conventional spray dryers that use rotary atomizers and pressure nozzles for forming the spray droplets, the new Nano Spray-Dryer utilizes a vibrating mesh technology for fine droplets generation (Elzoghby, Samy, & Elgindy, 2012b). Lee, Bae, Kim, Nam, and Park (2011) and Lee, Heng, Ng, Chan, and Tan (2011) utilized a feasible approach to prepare BSA NPs 460 nm using the novel Nano Spray-Dryer (equipped with a vibrating mesh spray technology and an electrostatic particle collector). Nano spray-drying technique offers a new, simple, and alternative approach for the production of NPs. It offers the advantage of drying and particle formation in a single-step continuous and scalable process.

2.6 Nab technology

NPs prepared by NP albumin-bound technology (nab technology) were designed to be safe and suitable for intravenous administration of poorly soluble anticancer drugs. The drug is mixed with human serum albumin (HSA) in an aqueous solvent and passed under high pressure through a jet to form drug–albumin NPs in the size range of 100–200 nm (Desai, 2007). Abraxane® (nab–paclitaxel; paclitaxel–albumin NP) with an approximate diameter of 130 nm is the first FDA-approved nanotechnology-based chemotherapeutic that has shown significant benefit in the treatment of metastatic breast cancer. Similarly, Kim et al. (2011) utilized nab technology to prepare curcumin-loaded HSA NPs with a size range of 130–150 nm for intravenous administration. HSA was dissolved in water saturated with chloroform. Separately, curcumin was dissolved in chloroform saturated with water. These two solutions were then mixed and homogenized at 20,000 psi for nine cycles. The resulting colloid was evaporated under reduced pressure then lyophilized. The NPs showed markedly greater water solubility than free drug (Kim et al., 2011). The nab technology platform has been successfully used to encapsulate other anticancer lipophilic drugs such as docetaxel, the mTOR inhibitor rapamycin, the heat-shock protein inhibitor Hsp90, the tubulin, polymerization/topoisomerase I dual inhibitor 5404, and a novel taxane CY196 (Elsadek & Kratz, 2012). The nab technology could be successfully extrapolated to proteins other than albumin. The mechanism of nab technology is generally considered to be associated with disulfide boundings by homogenization, which avoids many potential drawbacks of solvent-based formulations in the traditional forms. Lactoferrin (Lf), being a glycoprotein rich in cysteine (Cys), has a great potential to form NPs using the nab technology. Gambogic acid (GA)–Lf NPs were prepared by the nab technology to enhance oral absorption and anticancer activity of GA, hence reducing the related toxic effect (Zhang et al., 2013).

2.7 Self-assembly

Hydrophilic proteins such as albumin and GEL were hydrophobically modified through to impart amphiphilicity. In aqueous solution, the hydrophobically modified protein molecules were self-assembled into micelle-like nanospheres where the hydrophobic core of micelles acts as a microreservoir for poorly water-soluble anticancer drugs. Octyl-modified bovine serum albumin (OSA) has been synthesized based on the specific reaction between the aldehyde group of octaldehyde and primary amino group of albumin in order to improve the lipophilicity of albumin and thus form core–shell nanomicelles (Gong et al., 2009). Paclitaxel was successfully loaded into OSA micelles by a dialysis method with a high drug loading (33.1%) and entrapment efficiency (90.5%). Similarly, hexanoyl anhydride and α-tocopheryl succinate (TOS) were utilized as hydrophobic groups to chemically modify the hydrophilic GEL for loading of the lipophilic drugs camptothecin and 17-AAG (17-allylamino-17-demethoxygeldanamycin) into the hydrophobic core of the NPs (Li, Liu, & Chen, 2011; Li, Liu, Zhao, et al., 2011; Won, Yoon, Sonn, Lee, & Kim, 2011). On the other hand, self-assembled micelles synthesized by conjugating hydrophilic methoxy poly(ethylene glycol) to the hydrophobic plant protein, zein, significantly enhanced the aqueous solubility and stability of curcumin with higher cellular uptake than free curcumin in drug-resistant ovarian cancer cells (Podaralla, Averineni, Alqahtani, & Perumal, 2012).

2.8 Polyelectrolyte ionic complexation

Formation of protein NPs via electrostatic complexation represents an alternative to chemical cross-linking. Razmi et al. (2013) developed ionic nanocomplexes for the delivery of a novel platinum complex (bipyridine morpholine dithiocarbamate Pt(II) nitrate) based on the electrostatic interaction between negatively charged β-CN and positively charged chitosan (CS) molecules. The presence of hydrophilic CS molecules sterically stabilizes the NPs against aggregation. Other protein–polysaccharide electrostatic nanocomplexes including protamine–polyaspartate (Cheng et al., 2009) and lysozyme–carboxymethyl cellulose (CMC) (Zhu et al., 2013) were reported for the delivery of the anticancer drugs doxorubicin (Dox) and 5-fluorouracil, respectively.

2.9 Maillard conjugation

Maillard reaction is a naturally occurring reaction involving condensation of a reducing sugar or polysaccharide with protein or peptide, by linking the reducing end carbonyl groups in the former to the amino groups in the latter upon heating at 60 °C and relative humidity of 78.9%. Dox could be effectively loaded into the BSA–dextran Maillard conjugates by virtue of the electrostatic and hydrophobic interactions (Hao, Ma, Huang, He, & Yao, 2013). After heat treatment, the gelation of BSA results in fixation of Dox inside the gelated protein core whereas the dextran shell makes the NPs dispersible in solution.

2.10 Electrospray deposition technique

The electrospray deposition system which uses liquid atomization by means of electrical forces generates a plume of droplets by charging the liquid at a high voltage. The charged droplets were sprayed from the tip of the nozzle. The electrospray deposition system has several advantages over conventional methods, such as high drug loading efficiency and self-dispersion. Furthermore, NPs synthesized by electrospray deposition systems do not require any template or surfactant. Gulfam et al. (2012) used the electrospray deposition system to synthesize cyclophosphamide-loaded gliadin and gliadin–GEL composite NPs. A mixture of gliadin and drug in 70% ethanol was loaded in a 10-mL syringe and continuously infused using the syringe pump (flow rate of 0.5 mL/h) and a stainless steel nozzle with a diameter of 32 GA, which was connected to the high-voltage power supply. The high-voltage power supply generated a 12–14 kV potential difference between the nozzle and the ground stainless steel sheet.

2.11 Cross-linking of protein NPs

Cross-linking of protein nanocarriers is important to achieve sustained drug release and to increase the protein stability. Different types of cross-linkers were used to cross-link protein NPs including chemical (e.g., glutaraldehyde, genipin, and 1-Ethyl-3-(3-dimethylaminopropyl)carbodiimide/*N*-hydroxysuccinimide (EDC/NHS)), ionic (sodium tripolyphosphate, TPP), and enzymatic cross-linkers (e.g., microbial transglutaminase (MTG)) in addition to thermal cross-linking (gelation). Glutaraldehyde, the most commonly used cross-linker, induces poly- or bifunctional cross-links into the network structure of proteins by bridging free amino groups of lysine or hydroxylysine residues (Elzoghby et al., 2012b). However, the use of nontoxic cross-linking agents seems important for future applications of protein NPs. Genipin, a naturally derived cross-linking agent with a negligible cytotoxicity, was used to cross-link GNPs and CN NPs as a biodegradable drug delivery system with no toxic side effects. One genipin molecule cross-links two free amino groups of lysine residues on the protein macromolecular chains. Genipin-cross-linked FLT-loaded CN micelles showed slower drug release, greater stability, and more prolonged drug systemic circulation after i.v. administration into rats compared to uncross-linked one (Elzoghby, Helmy, Samy, & Elgindy, 2013a, 2013c). A mixture of water-soluble carbodiimide and *N*-hydroxysuccinimide (CDI/NHS) was successfully used as a nontoxic cross-linking system of protein NPs, e.g., albumin, GEL, and CN. HSA–methotrexate conjugate NPs were stabilized by the addition of EDC, which acts as the cross-linking agent between carboxylic acid groups and free amino groups of the protein (Taheri, Atyabi, et al., 2011; Taheri, Dinarvand, Ahadi, Khorramizadeh, & Atyabi, 2012; Taheri, Dinarvand, Atyabi, Ghahremani, & Ostad, 2012; Taheri, Dinarvand, et al., 2011). Ionic cross-linking was suggested as a potential alternative to the chemical cross-linkers, so as to avoid their possible toxicity. In our work, ionically cross-linked CN NPs were prepared using TPP as a nontoxic polyanionic cross-linker for controlled delivery of FLT (Elzoghby et al., 2013b; Elzoghby, Saad, et al., 2013). By decreasing the pH of CN solution below its isoelectric point (4.8), the amino groups of CN become positively charged and strongly attracted the negatively charged TPP phosphate groups. Enzymatic cross-linking is another attractive approach due to high specificity of the enzyme catalysis. Cisplatin-loaded CN–polyacrylic acid (PAA) NPs were successfully cross-linked with recombinant MTG (Zhen, Wang, Xie, Wu, & Jiang, 2013). TGase forms intra- and intermolecular isopeptide bonds in and between proteins by cross-linking the ε-amino groups of lysine to the side chain amide group of glutamine.

3. PROTEIN NPs IN CANCER THERAPY

3.1 Albumin

HSA is the most abundant plasma protein. HSA is a globular protein consisting of 585 amino acid residues with a relative molecular weight of 66,500 Da (Elzoghby et al., 2012a). It is not a standard protein since it is extremely robust toward pH (stable in the pH range of 4–9), temperature (can be heated at 60 °C for up to 10 h), and organic solvents. The use of albumin for preparing biodegradable NPs has gained great interest in therapeutics, especially in cancer treatments. Albumin contains amino and carboxylic groups, and these groups can be used for surface modifications. Moreover, HSA-based NPs were well tolerated without any serious side effects in clinical trial studies. As an alternative to blood-derived albumin, recombinant HSA (rHSA), a genetically engineered protein expressed in yeast cells, has shown comparable safety, tolerability, pharmacokinetics, and dynamics to native HSA (Elzoghby et al., 2012a).

From *in vitro* studies, it can be seen that albumin NPs as a carrier of anticancer drugs offer many advantages: (1) Albumin NPs exhibited extraordinary binding efficiency of many anticancer drugs due to the defined albumin primary structure and high content of charged amino acids (e.g., lysine). For water-soluble drugs, drugs can be loaded either by incubation with the just-formed and hardened albumin NPs or by incorporation of the drug into the solution of albumin prior to the formation and cross-linkage of the NPs. Whereas water-insoluble drugs, e.g., paclitaxel (Zhao et al., 2010) and HCPT (Zu et al., 2013), can be loaded by desolvation, emulsification in addition to nab technology, and self-assembly. (2) In most studies, the drug-loaded albumin NPs exerted a sustained drug release, which is beneficial for tumor drug delivery (Varshosaz, Hassanzadeh, Sadeghi, Ghelich Khan, & Rostami, 2013). (3) Albumin NPs demonstrated a small particle size that helps high tumor penetration capability. (4) The anticancer drug-loaded albumin NPs showed superior cytotoxicity against different cancer cell lines compared to the free drug by decreasing its IC_{50} value. However, it is possible that the slower release of the drug from the NPs reduces the cytotoxicity especially when cross-linked by higher glutaraldehyde concentrations (Dreis et al., 2007). (5) HSA NPs showed excellent stability with no changes in particle size when stored in aqueous suspension for up to 6 months at 4 °C.

Concerning *in vivo* studies, the anticancer drug-loaded albumin NPs showed superior anticancer therapeutic effects than free drug (Table 1). This potent antitumor activity of drug-loaded albumin NPs was attributed to (a) *enhanced water solubility of the insoluble drugs.* (b) *increased intratumoral accumulation* due to enhanced uptake of albumin-based NPs in solid tumors mediated by enhanced permeability and retention (EPR) and also due to increased transendothelial transport initiated by binding of native albumin to a cell surface, 60-kDa glycoprotein (gp60) receptor (albondin), as well as binding of albumin to SPARC, secreted protein acidic and rich in cysteine, an extracellular matrix glycoprotein that is overexpressed in a variety of cancers. Endothelial binding of paclitaxel increased 9.9-fold and transport of paclitaxel across the endothelial cell monolayer increased 4.2-fold with nab–paclitaxel compared with Cremophor EL–paclitaxel (Kratz, 2014). (c) *Improved the pharmacokinetic behavior of anticancer drugs*: albumin-bound form of Dox has a large area under the curve (AUC), a small volume of distribution, and low clearance compared to Dox. Radiopharmacokinetic studies using radiolabeled HSA have revealed enhanced uptake and degradation of albumin by solid tumors. As a consequence, the therapeutic index of Dox was significantly enhanced allowing high doses to be administered to tumor-bearing animals with a concomitant increase in antitumor activity compared to free Dox (Kouchakzadeh, Shojaosadati, Maghsoudi, & Farahani, 2010). (d) *HSA can be used for systemic tumor-targeted siRNA delivery*: both HSA and siRNA molecules were thiolated to improve the binding affinity for each other and then the thiolated HSA (tHSA) formed stable NPs with poly-siRNA (psi) through oxidative intermolecular disulfide self-cross-linking without using glutaraldehyde. The self-cross-linked psi–tHSA NPs provided efficient cellular uptake and gene-silencing activity in mouse melanoma cell lines. After intravenous injection, psi–tHSAs accumulated specifically at the tumor sites, leading to efficient gene silencing in the tumors in a sequential manner (Son et al., 2013).

Table 1 Albumin nanoparticles loaded with anticancer drugs (with unmodified surface)

Albumin nanocarrier	Drug	Preparation technique	Key outcomes	References
HSA NPs	Nab–paclitaxel (Abraxane)	Nab technology	Superior antitumor activity of nab–paclitaxel over paclitaxel in human tumor xenograft models (breast, lung, and pancreatic)	Elsadek and Kratzz (2012) and Maham et al. (2009)
Octyl-BSA micelles	Paclitaxel	Self-assembly	Small size (123.3–152.8 nm), high drug loading (33.1%), and safer than Tween 80 and Cremophor EL	Gong et al. (2009)
BSA NPs	10-Hydroxycamptothecin (HCPT)	Emulsification	Improving the stability and solubility of HCPT, 55.3% of the lactone HCPT accumulated in liver 24 h after injection showing the liver targeting potential of HCPT–BSA NPs	Yang et al. (2007)
HSA NPs	Doxorubicin	Desolvation	Size (150–500 nm), loading efficiency (70–95%), increased anticancer effects against neuroblastoma cells	Dreis et al. (2007)
BSA–dextran	Doxorubicin	Millard conjugation/ thermal gelation	Significant reduction of doxorubicin toxicity and increased survivability of hepatoma H22 tumor-bearing mice	Qi, Yao, He, Yu, and Huang (2010)
BSA NPs	Doxorubicin	Conjugation	Superior efficacy in murine tumor model, significantly reduced cardiotoxicity, and enhanced therapeutic index	Kratz (2014)
BSA NPs	5-Fluorouracil	Desolvation	Constant release of 5-fluorouracil for 20 h	Maghsoudi, Shojaosadati, and Farahani (2008)
BSA-alginate NP	5-Fluorouracil	Emulsion solidification	After oral administration, the NPs were mainly distributed in the liver, spleen, lungs, and kidneys and slightly in brain	Yi, Yang, and Pan (1999)

Continued

Table 1 Albumin nanoparticles loaded with anticancer drugs (with unmodified surface)—cont'd

Albumin nanocarrier	Drug	Preparation technique	Key outcomes	References
HSA	Methotrexate	Desolvation of HSA–MTX conjugates	MTX–HSA conjugate NPs (90–150 nm) were more cytotoxic on T47D cells compared to free methotrexate and showed promising antitumor efficacy in various animal models	Burger, Hartung, Stehle, Sinn, and Fiebig (2001)
BSA-SH and ALG–CYS NPs	Tamoxifen	Coacervation disulfide bond stabilization	The NPs showed enhanced antitumor activity of TMX in MCF-7 nude mice showing more accumulation of the drug into the tumor site	Martínez, Muñiz, Iglesias, Teijón, and Blanco (2012)
HSA NPs	Noscapine	pH coacervation	The NPs showed 35% cell viability against HER-2-positive SK-BR-3 breast cancer cells compared to 80% for free drug after 144 h	Sebak, Mirzaei, Malhotra, Kulamarva, and Prakash (2010)
HSA NPs	Curcumin	Nab technology	Enhanced water solubility, increased tumor accumulation, and an ability to traverse vascular endothelial cell	Kim et al. (2011)
HSA NPs	VEGF siRNA	psi–tHSA disulfide self-cross-linking	Specific tumor accumulation, efficient gene silencing, significant inhibition of angiogenesis, and retarding the growth in PC-3 mice	Son et al. (2013)
BSA lipoprotein-like NPs	Lapatinib	Modified desolvation	NPs (62.1 nm) induced a significant BT-474 cell arrest at G0/G1 phase and could target subcutaneous BT-474 xenografts and accumulate in tumor through EPR- and SPARC-mediated targeting	Zhang et al. (2014)

3.2 Gelatin

GEL is a denatured protein that is obtained by either partial acid or alkaline hydrolysis of animal collagen (Elzoghby, 2013). It is one of the most versatile natural biopolymers widely used in pharmaceutical industries due to its biocompatibility, biodegradability, low cost, readily available, and numerous available active groups for attaching targeting molecules (Karthikeyan, Prasad, Ganamani, & Balamurugan, 2013). GNPs have been extensively used for the delivery of both hydrophilic and hydrophobic anticancer drugs (Table 2). They bear multiple modification opportunities for coupling with cross-linkers and targeting-ligands, which may be especially useful in developing tumor-targeted drug delivery vehicles. The *in vitro* characteristics of GNPs make them good candidate for the delivery of anticancer drugs. (1) *Drug loading*: various mechanisms may be involved in drug loading into GNPs including physical entrapment, electrostatic attraction, or covalent conjugation. Furthermore, hydrogen bonding and hydrophobic interactions between drug and GEL were also reported. (2) *Drug release*: the drug release from GNPs may be due to three predominant mechanisms including desorption, diffusion, and biodegradation of GNPs (Bajpai & Choubey, 2006; Madan et al., 2011). Many factors were shown to influence the rate of drug release from GNPs including the cross-linking density and concentration of GEL. (3) *Particle size*: particle size in the 10–400 nm range can accumulate preferentially in tumor cells due to the EPR effect. The effect of various parameters such as temperature, pH, degree of cross-linking, and nature of GEL on the size of GNPs has been investigated by several groups (Wu et al., 2013).

GNPs could increase the antitumor efficacy of the encapsulated anticancer drugs through different mechanisms: (1) *Enhancing tumor penetration*: GNPs loaded with anticancer agents can be endocytosed by cells, resulting in internalization of the encapsulated drug by the cancer cell thus successfully increase the drug concentration in cancer tissues, and enhancing antitumor efficacy (Sun et al., 2014). Gaowa et al. (2014) found that EGFR2R-lytic peptide–GEL nanocomplex exhibited higher antitumor activity than that of peptide alone. The complex is relatively small in size (36 nm) in comparison with the peptide (454 nm), therefore, accumulates better in tumor tissue than in normal tissues. In the study of Ding et al. (2011), GEL–PAA core–shell NPs showed significantly superior anticancer efficacy in hepatic H22 tumor-bearing mice in comparison with free drug. Upon peritumoral implantation of GEL hydrogel (jelly) encapsulating cisplatin-loaded

Table 2 Gelatin nanoparticles loaded with anticancer drugs (with unmodified surface)

Gelatin nanocarrier	Drug	Preparation technique	Key outcomes	References
Gelatin type A	Camptothecin	Self-assembly	High drug loading efficiency (43.3–70.1%), prolonged release profile	Li, Liu, and Chen (2011) and Li, Liu, Zhao, et al. (2011)
Gelatin–PLA–DPPE	Doxorubicin	Double emulsion or nanoprecipitation	NPs showed faster release at pH 5.0 than at pH 7.4 with comparable anticancer efficacy with the free drug *in vitro* and *in vivo*	Han, Li, Liu, Gao, and Wu (2013)
Gelatin type A	Doxorubicin	Coacervation–phase separation	NPs improved the cytotoxicity allowing DXR to escape from the multidrug resistance mechanism	Leo, Vandelli, Cameroni, and Forni (1997)
Gelatin type B	Doxorubicin	Emulsification–solvent evaporation	Iron-oxide-incorporated GNPs increase the cumulative DOX release with increasing magnetic field strength	Choubey and Bajpai (2010)
Gelatin type B	Doxorubicin	Gelatin–DXR conjugation	pH-dependent DOX release; 50% drug release at the lysosomal pH of 4.8, but only 10% release at pH 7.4	Wu, Cammarata, Park, Rhodes, and Ofner (2013)
Gelatin type A	Paclitaxel	Coacervation–phase separation	NPs yielded 40% smaller AUC, longer retention, and 3.2-fold higher accumulation in organs especially in liver, small intestine, and kidney	Yeh, Lu, Wientjes, and Au (2005)

Gelatin type A	Paclitaxel	Desolvation	Rapid release from NPs in urine (90% after 2 h) increases the drug penetration into the bladder tissue after intravesical administration	Lu et al. (2004)
Gelatin type B	Methotrexate	Emulsification–solvent evaporation	Size of NPs 100–200 nm, low encapsulation efficiency (15.9%), and diffusion-controlled release mechanism	Cascone et al. (2002)
Gelatin type A	Cytarabine	Emulsification	Size of NPs was 100–300 nm. An optimum drug release was obtained at pH 1.8 while a lower release is obtained in pH 7.4	Bajpai and Choubeyy (2006)
Recombinant (rHG)	17-AAG	Self-assembly	rHG–TOS NPs passively targeted the tumor tissue via the EPR effect with long residence time and higher anticancer effect than free 17-AAG	Won et al. (2011)
Gelatin type B	Resveratrol (RSV)	Coacervation–phase separation	RSV–GNPs with size 294 nm exhibited very rapid cellular uptake and higher anticancer activity than free RSV to NCI-H160 cells	Karthikeyan et al. (2013)
Gelatin type A	Epigallocatechin gallate (EGCG)	Desolvation	The maximum EGCG release was reached at 8 h as compared with 15 min EGCG release from uncoated gelatin NPs	Shutava et al. (2009)

Continued

Table 2 Gelatin nanoparticles loaded with anticancer drugs (with unmodified surface)—cont'd

Gelatin nanocarrier	Drug	Preparation technique	Key outcomes	References
Gelatin	Dihydroartemisinin (DHA)	Electrostatic field system	Cells treated with GEL/DHA NPs exhibited a higher early apoptosis (25%) compared with pure DHA (11.5%)	Sun et al. (2014)
Gelatin type B–dextran	Tea polyphenol	Millard conjugation	The obtained micelles are of nanosize (86 nm in diameter) and stronger cytotoxicity against MCF-7 cells than free tea polyphenols	Zhou et al. (2012)
Gelatin type B	Ammonium heptamolybdate (AHM)	Electrostatic interaction	Gelatin/AHM NPs achieved a very high drug loading of around 70% with higher antitumor activity than AHM solution	Guo et al. (2011)
Gelatin	Cocoa procyanidins (CPCs)	Desolvation	CPCs loaded gelatin–chitosan NPs improved both stability and apoptotic activity of CPCs	Zou, Li, Percival, Bonard, and Gu (2012) and Zou, Percival, et al. (2012)
Gelatin type B	Cisplatin	Electrostatic interaction	The jelly implant containing cisplatin-loaded gelatin–PAA NPs showed much higher drug retention in tumor with superior anticancer effect compared to free drug	Ding et al. (2011)
Gelatin	Platinum (IV) prodrug	Carbodiimide conjugation	Gel/Pt(IV) NPs demonstrated high anticancer efficacy with lower IC_{50} than free Pt(IV) prodrug	Cheng et al. (2014)

GEL–PAA NPs in a murine hepatoma H22 cancer model, a significantly superior efficacy in impeding tumor growth and prolonging the lifetime of mice than those treated with i.v. injection of drug-loaded NPs was observed. (2) *Increasing the solubility and bioavailability of poorly soluble anticancer drugs*: the bioavailability of dihydroartemisinin (DHA) was increased by encapsulation within GNPs so that DHA GNPs had higher anticancer proliferation activities than DHA alone in A549 cells (Sun et al., 2014). (3) *Prolongation of systemic circulation*: GNPs have several advantages as a nonviral gene delivery vector in cancer therapy. They can be conjugated to moieties that stimulate receptor-mediated endocytosis, multiple plasmids can be encapsulated, and the bioactivity of the encapsulated DNA could be improved by preventing digestion by nucleases. In the study of Lee, Yhee, Kim, Kwon, and Kim (2013), psi–tGel NPs protected the formulated siRNA molecules from degradation by nuclease enzyme. psi–tGel NPs showed prolonged blood circulation without recognition by RES without poly(ethylene glycol) modification. The property of longer blood circulation without recognition by RES may be contributed to the accumulation of psi–tGel NPs in the target site. (4) *Gene/drug codelivery*: GNPs were used for gene/drug combination treatment, which significantly improved the therapeutic performance of the delivery system compared to the gene or drug alone treated groups. In the study of Xu, Singh, and Amiji (2014), the antitumor activity of the thiolated GEL-loaded wt-*p53* plasmid and gemcitabine was significantly increased due to synergistic effect. (5) *Localized pulmonary drug delivery*: GNPs may be promising vehicles of transporting drugs efficiently to the lung via inhalation, while avoiding unwanted mucociliary clearance. Tseng, Su, Yen, Yang, and Lin (2009) reported that inhalable GNPs did not cause lung inflammation as demonstrated by myeloperoxidase activity assay and are thus safe for use. Therefore, inhalative GNPs as carriers of cisplatin were designed for the treatment of lung cancer with anticipated improved therapeutic effect and reduced side effects. (6) *Protection of sensitive nutraceuticals with promising anticancer activity*: nanoencapsulation of cocoa procyanidins (CPCs) into GNPs significantly improved their storage stability and enhanced their apoptotic activity in human acute monocytic leukemia cells compared with the free CPCs (Zou, Li, et al., 2012; Zou, Percival, et al., 2012). The CPCs' content decreased drastically over time during the storage. The remaining percentages of CPCs in the NPs were significantly higher than those in the free CPCs, which suggested that encapsulation of CPCs in the NPs protected them from degradation and oxidation.

3.3 Milk proteins

Milk proteins are widely available, inexpensive, natural, and GRAS raw materials with high nutritional value, which make them highly suitable as vehicles for delivering various bioactives (Elgindy, Samy, & Elzoghby, 2014; Elzoghby, Abo El-Fotoh, & Elgindy, 2011; Livney, 2010). They can be classified into two categories according to their structure: flexible proteins for CNs and globular proteins for whey proteins.

3.3.1 Casein

Many of the structural and physicochemical properties of CNs, the major milk proteins, facilitate their functionality in drug delivery systems (Elzoghby et al., 2011; Livney, 2010). CN comprises about 94% protein and 6% low-molecular-weight compounds collectively called colloidal calcium phosphate. Mainly four CN phosphoproteins, αS1-, αS2-, β-, and κ-CN, exist in cow milk with molecular weights between 19 and 25 kDa and an average isoelectric point (p*I*) between 4.6 and 4.8. CNs are characterized by thermal stability due to their proline-rich open-structure with no disulfide bridges. Due to their amphiphilicity, with high levels of hydrophobic and hydrophilic amino acid residues, CN molecules self-assemble into aqueous solutions spherical micelles (50–500 nm in diameter) with a hydrophobic core and a hydrophilic-charged κ-CN layer (Elzoghby et al., 2011; Livney, 2010). Hence, hydrophobic anticancer drugs could be easily loaded into the hydrophobic core of self-assembled CN micelles via addition of the drug solution in a water miscible organic solvent (e.g., ethanol or DMSO) to aqueous CN micellar solution followed by solvent evaporation or dialysis to remove the residual solvent. Lyophilization or spray-drying was successfully used to obtain redispersible drug-loaded CN micelles. Celecoxib-loaded β-CN micelles were lyophilized without any cryoprotectants and were fully reconstituted by rehydration (Bachar et al., 2012). Based on the thermal stability of CN, FLT-loaded CN micelles were transformed into redispersible powder with improved long-term storage stability via spray-drying with no need for drying adjuvant (Elzoghby et al., 2013a, 2013c; Elzoghby, Samy, & Elgindy, 2013). CN micelles were recently used as carriers for anticancer drugs for various purposes (Table 3):

(1) *Solubilization of hydrophobic anticancer drugs*: β-CN NPs may serve as effective nanovehicles for solubilization of hydrophobic anticancer drugs including celecoxib (Bachar et al., 2012), mitoxantrone (Shapira, Assaraf, Epstein, et al., 2010; Shapira, Markman, Assaraf,

Table 3 Casein-based nanocarriers loaded with anticancer drugs

Casein (CN) nanocarrier	Drug	Preparation technique	Key outcomes	References
CN micelles	Celecoxib	Self-assembled micellization/ lyophilization	>100-fold higher encapsulation loads than other casein/drug formulations	Bachar et al. (2012)
TPP-cross-linked CN micelles	Flutamide	Emulsification-ionic gelation	NPs showed a higher antitumor efficacy in prostate cancer-bearing rats with a significantly higher percentage reduction in PSA serum level (75%) compared to free FLT (55%)	Elzoghby et al. (2013b) and Elzoghby, Saad, et al. (2013)
CN micelles	Flutamide	Self-assembled micellization/ spray-drying	Micelles demonstrated a marked decrease in relative weights of both prostate tumor and seminal vesicle (34.62% and 24.59%) compared to free FLT (11.86% and 17.74%), respectively	Elzoghby et al. (2013a, 2013c) and Elzoghby, Samy, et al. (2013)
CN micelles	Mitoxantrone	Self-assembled micellization/ dialysis	The gastric digestibility of β-CN suggests possible targeting to stomach tumors	Shapira, Assaraf, Epstein, and Livney (2010) and Shapira, Assaraf, and Livney (2010)
CN micelles	Paclitaxel, docetaxel, vinblastine, and irinotecan	Self-assembled micellization/ dialysis	β-CN micelles (below 100 nm) stabilized all studied drugs in aqueous solution by hydrophobic and electrostatic interactions	Shapira, Assaraf, and Livney (2010) and Shapira, Davidson, Avni, Assaraf, and Livney (2012)
CN micelles	Paclitaxel	Self-assembled micellization/ dialysis	Following simulated digestion with pepsin (2 h at pH 2), paclitaxel retained its cytotoxic activity to human N-87 gastric cancer cells	Shapira et al. (2012)

Continued

Table 3 Casein-based nanocarriers loaded with anticancer drugs—cont'd

Casein (CN) nanocarrier	Drug	Preparation technique	Key outcomes	References
CN–chitosan micelles	Platinum complex	Polyelectrolyte ionic complexation	The micelles enhanced the cytotoxicity and cellular uptake of platinum on colorectal carcinoma HCT116 cells	Razmi et al. (2013)
CN–PLGA NPs	Paclitaxel and EGCG	Emulsion–precipitation	NPs showed a sustained and sequential release of both drugs in plasma, indicating prolonged circulation of NPs	Narayanan et al. (2014)
CN–PAA NPs	Cisplatin	Noncovalent polymerization	NPs showed extraordinary capability to penetrate cell membranes, target tumor, and inhibit tumor growth 1.5-fold higher than free drug	(Zhen et al., 2013)
CN–dextran conjugates	EGCG	Maillard conjugation	Nanoconjugates could effectively protect EGCG from degradation in alkaline pH and displayed a sustained release in intestinal fluid	Xue, Tan, Zhang, Feng, and Xia (2014)
CN micelles	Curcumin	Self-assembled micellization	Increased aqueous solubility and enhanced cytotoxicity of curcumin to human leukemia cells	Esmaili et al. (2011)
Na caseinate micelles	Curcumin	Self-assembled micellization/spray-drying	Improved solubility, thermal stability, dispersibility, and antioxidant activity of curcumin	Pan, Zhong, and Baek (2013)

et al., 2010), docetaxel (Shapira, Assaraf, & Livney, 2010), and paclitaxel (Shapira et al., 2012), via incorporation into the hydrophobic core of CN micelles, thus, allowing them to be thermodynamically stable in aqueous solutions for oral-delivery applications. Moreover, i.v. nanoformulation of the poorly soluble anticancer drug, FLT, was enabled in our laboratory by virtue of the high solubilizing power of CN micelles (Elzoghby et al., 2013a; Elzoghby, Saad, et al., 2013).

(2) *Improved pharmacokinetics of anticancer drugs*: CN micelles could control the release of anticancer drugs and prolong their systemic circulation. After i.v. administration into rats, CN micelles exhibited a prolonged plasma circulation of FLT compared to drug solution (Elzoghby et al., 2013a; Elzoghby, Saad, et al., 2013). Uncross-linked and genipin-cross-linked CN micelles could extend the half-life of FLT from 0.88 to 10.60 h and 13.85 h, respectively. Similarly, polylactic-co-glycolic acid (PLGA)/CN NPs showed prolonged circulation and enhanced availability of the drugs paclitaxel and EGCG (Narayanan et al., 2014). The hydrophilic κ-CN "hairy" layer covering the surface of CN micelles was postulated to simulate the role of PEG hydrophilic shell that suppresses opsonization through generating a steric barrier.

(3) *Increased targetability to stomach cancer*: the gastric digestibility of β-CN was suggested as a possible targeting mechanism for stomach cancer. β-CN has high proline content (17% proline residues), leading to an open tertiary structure, which is easily accessible to gastric proteases, thus, providing an excellent target-activated release mechanism for unloading the drug in the stomach (Shapira et al., 2012). Following simulated digestion of paclitaxel-loaded β-CN nanomicelles with pepsin, paclitaxel retained its cytotoxic activity to human gastric carcinoma cells. Without prior simulated gastric digestion, the nanovehicles were noncytotoxic, suggesting that β-CN may protect the upper GIT regions including buccal and esophageal epithelia from untoward toxicity of paclitaxel.

(4) *Enhanced tumor penetration*: cisplatin-loaded CN hollow nanospheres demonstrated an extraordinary capability to penetrate the tumor after their extravasation through the vasculature and efficiently propagate throughout the entire tumor, thus, making cisplatin permeates the tumor more deeply (Zhen et al., 2013). These findings indicated that CN spheres can directly deliver cargoes into cytoplasm and thus offer a class of cell-penetrating and protein-based drug carriers. The cell-penetrating nature of CN spheres may be associated with the peptide sequence of CN macromolecule with a high level of proline, similar to the structure of some cell penetrating peptides (CPPs).

(5) *Enhanced antitumor efficacy*: cisplatin-loaded CN NPs showed superior antitumor efficacy on hepatic H22 tumor-bearing mice (1.5-fold higher) than free cisplatin (Zhen et al., 2013). In our work, the antitumor activity of FLT–CN micelles in prostate tumors-bearing rats was higher than free drug with lower hepatotoxicity. Treatment of rats with free FLT at a dose of 12 mg/kg body weight for 28 days resulted in 55.43% reduction in prostate specific antigen (PSA) level compared to 65.95% for FLT-loaded CN micelles (Elzoghby et al., 2013a; Elzoghby, Saad, et al., 2013). The significantly antitumor efficacy caused by FLT–CN micelles (74.6 nm) compared to free FLT could be attributed to the EPR effect of the nanosized FLT delivery system.

(6) *Sequential release of both hydrophobic and hydrophilic anticancer drugs*: a sequential release of two drugs from core–shell PLGA/CN NPs in PBS, pH 7.4, was revealed with paclitaxel releasing 28% and EGCG releasing 60% of the total entrapped drugs in 10 days. Paclitaxel-resistant tumors could be sensitized to paclitaxel therapy through the use of EGCG which additionally has chemotherapeutic and chemopreventive properties. The use of CN for encapsulating EGCG can be justified by the affinity of polyphenols such as EGCG toward proline, whereas the hydrophobic paclitaxel could be associated with PLGA (Narayanan et al., 2014).

(7) *Stabilization of sensitive chemopreventive nutraceuticals*: CN–dextran Maillard nanoconjugates could effectively protect EGCG from degradation in alkaline pH and displayed a sustained release in intestinal fluid, which may be attributed to the inhibiting effects of CN binding on the motion freedom of EGCG (Xue et al., 2014). Similar protective effect was observed for curcumin (Esmaili et al., 2011; Pan et al., 2013).

(8) *Hemo-, cyto-, and immune-compatibility of CN nanocarriers*: upon i.v. injection of both bare CN and PLGA–CN NPs in Sprague–Dawley rats, no hemolysis, platelet aggregation/activation, blood coagulation, or immuno- or cytotoxicity was observed over a concentration range of 50–1000 μg/mL demonstrating its suitability for i.v. use (Narayanan et al., 2014).

3.3.2 Whey proteins

3.3.2.1 β-Lactoglobulin

β-Lactoglobulin (βLG), the major whey protein in cow milk, accounts for about 10% of the total milk proteins. βLG is a small (18.3 kDa, 162 amino acids) globular protein with two disulfide bridges and a free thiol group. At room temperature and at neutral to slightly acidic pH, the protein is mostly present at a native dimeric state, whereas between pH 2 and 3 it dissociates into monomers due to electrostatic repulsion between the subunits. βLG is edible and easily solubilized with good emulsifying property due to its amphiphilic nature (Sneharani, Karakkat, Singh, & Rao, 2010). Its cost effectiveness, abundant availability, and acceptability make it an attractive alternative protein that can be used as a drug carrier. Moreover, the small molecular weight, highly unfolding, and less hydrophobicity of βLG make it a good candidate for preparing NPs. The main applications of βLG as a carrier for anticancer drugs can be summarized as follows:

1. Solubilization of hydrophobic anticancer drugs
   It has been reported that βLG has an observable affinity for many biochemically important hydrophobic compounds, e.g., retinol, cholesterol, vitamin D, and fatty acids. βLG has been identified as a member of the lipocalins superfamily of transporter molecules for small hydrophobic ligands (Sneharani et al., 2010). Members of this family share the common structural feature of a central β-barrel calyx, built from eight antiparallel β-sheets, arranged as an ideal site for hydrophobic ligands. This explains its ability as a natural carrier of many hydrophobic compounds. Curcumin could bind to the central calyx of βLG through hydrophobic interaction-forming βLG–curcumin NPs where the solubility of curcumin was significantly enhanced to 625 μM compared to its aqueous solubility (30 nM). The binding of curcumin to βLG decreased when the pH was decreased from 7.0 to 5.5. The central calyx of βLG is lined by hydrophobic amino acid residues that are gated by the protonation/deprotonation of Glu89 of the EF loop (Sneharani et al., 2010). At acidic pH (5.5), the EF loop (85–90) of βLG is in closed conformation, rendering the hydrophobic cavity inaccessible to the entry of ligands. With the increase in pH, the EF loop folds back, opening the gate for the access to the binding of ligands allowing curcumin to get into the hydrophobic pocket.
2. Targeting colon carcinoma
   Because it can maintain a stable globular conformation, βLG is known to be highly resistant to proteolytic degradation in the low pH of stomach. Thus, βLG was suggested to be a pH-sensitive hydrogel former, useful for controlled delivery of biologically active substances. Additionally, βLG has been successfully used for oral targeting of anticancer drugs (such as palladium complex, a new heavy metal complex designed for anticancer property, Ghalandari, Divsalar, Saboury, & Parivar, 2014,

and conjugated linoleic acid, (CLA), Jiang & Liu, 2010) against colon cancer. Oxali-palladium [2,2′-bipyridin *n*-hexyl dithiocarbamato Pd(II) nitrate complex] and CLA interacted to βLG by hydrophobic interaction-forming nanocomplexes. Low methoxyl pectin (LMP) was used for coating βLG/oxali-palladium nanocapsules by virtue of the electrostatic interaction between the anionic pectin and the cationic βLG (below its isoelectric point, p*I* 5.18). The release of oxali-palladium from βLG/LMP complex nanocapsules to its target place was pH sensitive. At pH 1.2 and 4.5, no dramatic change in oxali-palladium release because LMP tends to be converted from carboxylate anions to free carboxyl groups in acidic conditions, resulting in an increase in the number of hydrogen ions leading to highly stable complexes with very low erosion profile. On the other hand, at high pH values, LMP can be ionized and is susceptible to degradation so that the erosion contribution is at its highest so that the maximum oxali-palladium release occurs at pH 7.5 and 7, i.e., the oxali-palladium release starts from the terminal ileum and increases in the colon (Ghalandari et al., 2014). Similarly, the self-assembled βLG–CLA complex could resist peptic hydrolysis and remain stable and intact until reaching the upper portion of the small intestine. βLG–CLA complex significantly increased the level of CLA in Caco-2 cells (85.9 μM) compared with free CLA (45.8 μM) which may be explained by the ability of βLG to cross the Caco-2 cell monolayer membrane; or as a member of the lipocalin family, bovine βLG has specific membrane receptors for transcellular transport, such as cytoplasmic cellular retinol-binding protein and human STRA6 protein (Jiang & Liu, 2010).

3. Protection of unstable nutraceuticals with anticancer potential
The stability of curcumin bound to βLG was enhanced 6.7 ×, in comparison to curcumin alone. In the presence of βLG, the half-life of curcumin increased to 206 min compared to 30.8 min for free curcumin, indicating the protection of curcumin from hydrolytic degradation (Sneharani et al., 2010). Native βLG was also suggested as a carrier to improve the stability of resveratrol (Liang, Tajmir-Riahi, & Subirade, 2008) and ω-3 polyunsaturated fatty acid DHA (Girard, Turgeon, & Gauthier, 2002) against photo- and oxidative-degradation, respectively. In another study, thermally induced βLG–EGCG coassembled nanovehicles (smaller than 50 nm) conferred significant protection to the potent antioxidant and chemopreventive nutraceutical EGCG against oxidative degradation (Shpigelman, Israeli, & Livney, 2010). For the first 48 h, a 33-fold lower degradation rate was observed for the protected EGCG at room

temperature, and a 3.2-fold lower average degradation was obtained over 8 days. The following mechanisms of protection were proposed: (1) upon mild heating, the alpha helix slightly unfolds, improving its binding affinity to EGCG by a combination of hydrophobic interactions and H-bonds, thus, enhancing its antioxidative capacity. (2) βLG can form nanoaggregates upon heating at low ionic strength and pH not close to the p*I* (5.35). By entrapping EGCG in βLG nanocomplexes, it is immobilized and sterically shielded, which protects it from oxidizing agents. (3) The native protein also provided some protection, due to the antioxidative properties of available free thiols.

3.3.2.2 Lactoferrin

Bovine lactoferrin is a single-chain cationic iron-binding glycoprotein belonging to transferrin (Tf) family with a molecular weight around 80 kDa and approximately 700 amino acid residues. Its structure consists of two lobes with a reversible iron-binding capacity of two atoms per Lf molecule (one for each lobe). The relatively high p*I* of Lf (around 8.5) is explained by a unique basic region in the N-terminal region, which can bind acidic molecules (Elzoghby et al., 2012b; Livney, 2010). Lf possesses various biological properties such as antimicrobial, anti-inflammatory, antitumor, immunemodulatory, and enzymatic activities. Lf is suggested to mediate its multiple functions by means of specific receptors, Lf receptor (LfR), on various cells such as apical surface of enterocytes, platelets, megakaryocytes, dopaminergic neurons, and endothelial cells of mesencephalic microvessels. Especially in recent decade years, Lf has been widely reported as a ligand to modify the nanocarrier and cross the blood–brain barrier via LfR-mediated transcytosis. Furthermore, it has been observed that Lf is relatively resistant to proteolysis in the gastrointestinal tract.

Lf NPs were successfully developed for delivery of the anticancer drugs (Dox hydrochloride and gambogic acid, GA) via sol–oil chemistry (Zhang et al., 2013) and nab technology (Golla, Bhaskar, Ahmed, & Kondapi, 2013; Golla, Cherukuvada, Ahmed, & Kondapi, 2012), respectively. The developed Lf NPs exhibited a number of features suitable for the delivery of anticancer drugs: (1) *Small particle size* (around 68–150 nm) with positive zeta potential up to 32 mV and up to 79–92% drug encapsulation. (2) *Improved solubility and bioavailability of oral chemotherapeutics*: Lf as a potential NP carrier succeeded to enhance the solubility of hydrophobic drugs, e.g., gambogic acid (GA) without using solubilizing agents such as Cremophor EL (Zhang et al., 2013). The *in situ* intestinal perfusion test confirmed the

LfR role in the transcytosis of GA NPs via either active transport or facilitated diffusion. (3) *Increased drug stability*: incorporation of GA into NPs is suggested to improve the stability in harsh environment of GIT. In addition, Lf NPs might protect GA from drug-metabolizing enzymes in liver. (4) *Controlled release of anticancer drugs*: in SGF, 8.74% of GA was released from GA NPs after 2 h; on changing the medium to SIF, only 15.4% of GA was released after 24 h. These results suggested that upon oral administration the GA NPs might remain in their particular structure in the gastrointestinal tract (Zhang et al., 2013). (5) *Improved drug pharmacokinetics*: the plasma C_{max} and AUC were found to be increased by more than 50% for the nanoformulations, compared to Dox while free Dox showed a steep two- to fourfold increase in its level in heart, kidney, spleen, bone marrow, and lungs at 8 h postinjection, there was a concomitant decrease of Dox level in plasma and an increase in liver tumor when injected with Lf NPs (Golla et al., 2012). Since the liver is the site of action as well as the site of cancer induction, the presence of higher drug levels when administered through Lf NPs would help in enhancing the efficacy of treatment. (6) *Increasing tumor targetability*: since Lf receptors are highly expressed on the surface of metabolically active cancer cells, the specific and enhanced Dox accumulation in liver, i.e., at the site of the tumor, relative to the other tissues at 8 h post-injection may be attributed to the specific interaction of the NPs with the corresponding receptors that gets overexpressed (Golla et al., 2012, 2013). (7) *Enhanced antitumor efficacy*: GA NPs inhibited HepG2 cells viability in a concentration-dependent manner and displayed a stronger antitumor effect than the same dose of GA–Arg treated groups with inhibitory rate against S180 tumor mice after oral administration of GA NPs was up to 86.01%, which was 1.39-fold of intravenous injection of Arg–GA (Zhang et al., 2013). Similarly, the Dox–Lf nanoformulation showed significantly enhanced expression of the tumor suppressor genes *p53* and *p21*, down-regulation of elevated VEGFR expression and increased expression of the apoptotic marker, cleaved caspase-3 compared to free Dox. In rats treated with nanoformulation, the number of liver nodules was found reduced by >93%. (8) *Reduced toxicity of anticancer drugs*: cardiotoxicity and renal toxicity of Dox were significantly reduced with these new formulations as demonstrated by the increased glutathione peroxidase activity in heart tissue homogenate and the absence of lactate dehydrogenase (LDH) levels in serum of animals treated with nanoformulations (Golla et al., 2012, 2013). This suggests that Dox indeed caused damage to heart, which is prevented when treated with nanoformulations.

3.4 Zein

Zein is a plant protein isolated from corn with a molecular weight varies from 22 to 27 kDa and an isoelectric pH of 6.228. Zein belongs to a family of prolamines which are composed of high amounts (>50%) of hydrophobic amino acids, such as proline, glutamine, and asparagines (Elzoghby et al., 2012b). This water-insoluble but alcohol-soluble corn storage protein is one of the best understood biomacromolecules and classified as GRAS by the US Food and Drug Administration. It is soluble in binary solvents containing a lower aliphatic alcohol (ethanol, methanol, and isopropanol) and water at various concentrations of alcohols range from 55% to 90% (v/v). Zein is widely used in food and packaging industry as a coating for candy, nuts, fruits, pills, and other encapsulated foods due to its film-forming property and ability to provide a moisture barrier and has been studied as potential biomaterial for the development of colloidal delivery systems.

Due to its inherent hydrophobicity and biodegradability, zein NPs were successfully applied as a carrier for controlled release of hydrophobic drugs. Zein NPs loaded with anticancer drugs including curcumin (Patel et al., 2010; Podaralla et al., 2012), quercetin (Patel et al., 2012), daidzin (Zou & Gu, 2013), lutein (Hu, Lin, Liu, Li, & Zhao, 2012), and procyanidins (Zou, Li, et al., 2012; Zou, Percival, et al., 2012) were successfully prepared via different techniques such as phase separation procedure (nanoprecipitation), supercritical antisolvent technique, self-assembly, and electrospinning. Numerous merits were demonstrated by zein NPs as delivery vehicles of anticancer drugs. (1) *Improved drug oral bioavailability*: D-α-tocopheryl polyethylene glycol succinate (TPGS) 1000-emulsified zein NPs enhanced the oral absorption of daidzin by over twofold compared with that of daidzin solution after administration in mice, which is likely due to P-glycoprotein inhibition by TPGS and transcytosis of the NPs (Zou & Gu, 2013). TPGS consists of a lipophilic α-tocopherol and a hydrophilic polyethylene glycol chain. α-Tocopherol of the TPGS was partly inserted in the zein and the polyethylene glycol chains protrude to cover the surface of the zein. (2) *Improved drug stability*: in the study of Podaralla et al. (2012), the photostability of curcumin was enhanced approximately by sixfold after encapsulation in mPEG–zein micelles. The absorbance of free curcumin decreased rapidly to 50% within 10 min, while the absorbance of curcumin in mPEG–zein micelles decreased gradually in 6 h. (3) *Enhanced liver targeting*: after intravenous delivery of 5-fluorouracil-loaded zein NPs, the drug was the highest in the liver and was still observed in blood at 24 h. The optimized average size of NPs is 114.9 nm, therefore, the drug encapsulated in zein NPs could be efficiently targeted to liver's parenchymal cells via passive delivery (Lai & Guo, 2011). The relative uptake rate of zein NPs was in the order of liver > spleen > plasma > lung > heart > kidney. (4) *Improved drug pharmacokinetics*: after administration of TPGS 1000-emulsified zein NPs in mice, the plasma concentration of daidzein was significantly higher than those treated with daidzin solution. The C_{max} and AUC_{0-12h} of daidzein after oral administration was over 2- and 2.4-fold higher for daidzin loaded in zein NPs than that of daidzin solution, respectively (Zou & Gu, 2013). (5) *Mucoadhesive properties*: Zein and CN are known as natural polymers with mucoadhesive properties. Using zein–curcumin particles, an *in vitro* mucoadhesion study showed retention of more than 60% curcumin for 150 min (Patel et al., 2010). The mucoadhesion property was further confirmed from a mucin association study carried out on Caco-2 cells. The association was increased with time and seemed saturated at 90 min. However, both the kinetics (time required for association) and saturation (maximum affinity) were lower for pure curcumin as compared to the zein–curcumin composite colloidal particles, thus, confirming their mucoadhesive properties. (6) *Amphiphilic zein micelles*: biodegradable methoxy poly(ethylene glycol)–zein micelles were developed for effective delivery of curcumin with the hydrophobic zein forming the inner core, while the hydrophilic PEG is in the outer shell (Podaralla et al., 2012). The micelles could demonstrate the following advantages: (a) conjugation of zein with hydrophilic mPEG can prevent opsonization and also provide steric hindrance for enzymatic degradation of zein; whereas zein NPs can be rapidly taken up by macrophages because of its protein origin and hydrophobicity. (b) The micelles significantly enhanced the aqueous solubility (by 1000- to 2000-fold) and stability (by sixfold) of the hydrophobic anticancer drug curcumin. (c) Due to very small size (20–200 nm), the micelles can easily extravasate through the leaky vasculature and accumulate in the tumor tissue. This provides a unique advantage to passively target anticancer drugs to the tumor by the EPR effect. (d) The CMC for PEG–zein micelles is relatively low (5.5×10^{-2} g/L), indicating the thermodynamic stability of zein micelles especially on dilution. (e) PEG–zein was found to be nonimmunogenic and did not elicit any anti-zein antibodies when injected subcutaneously in mice (Podaralla et al., 2012). PEGylation of zein provides a hydrophilic surface and hence can prevent the macrophage uptake and the resultant immune response to zein.

3.5 Gliadin

Gliadin, the major component of wheat gluten, is made up of single-chain polypeptides with an average molecular weight of 25–100 kDa linked by intramolecular disulfide bonds. The term gliadin defines a group of proteins extracted from gluten by 70% ethanol. All fractions have remarkably low solubility in aqueous solution except at extreme pH (Elzoghby et al., 2012b). This low water solubility has been attributed to the presence of disulfide bonds and to the cooperative hydrophobic interactions, which cause the protein chains to assume a folded shape. The amino acid composition shows that gliadin has equal amounts of polar and neutral amino acids, mainly glutamine (about 40%) in addition to high proline content (14%). Gliadins do not present the common drawbacks of synthetic materials, related to the presence of monomer or initiator residues. As plant proteins, they are recognized as prion-free unlike animal proteins (Elzoghby et al., 2012b). Moreover, gliadin can adhere to the mucus layer of the stomach due to its mucoadhesive property, which is a result of various interactions (e.g., hydrogen bonding, van der Waals force, and mechanical penetration).

Gliadin NPs have previously been generated for drug delivery and controlled release applications. Although the gliadin particles can be synthesized using the desolvation method, it has some limitations such as low drug loading efficiency and inability to separate particles from the aqueous phase (Xie, Marijnissen, & Wang, 2006). Therefore, cyclophosphamide-loaded gliadin NPs were successfully prepared via the electrospray deposition technique. The lipophilic amino acid, one of the components in gliadin protein, bound to cyclophosphamide due to hydrophobic interactions. Higher percentage (72.02%) of drug loading in the gliadin NPs was attributed to the adhesive properties of gliadin polymers as compared to the drug loading (52.77–64.23%) of gliadin–GEL composite NPs (Gulfam et al., 2012). Cyclophosphamide was gradually released from the gliadin NPs for 48 h, while the gliadin–GEL composite NPs released cyclophosphamide in a rapid manner. This result suggests that the hydrophobicity of gliadin NPs could manipulate the drug release in a controlled manner. Furthermore, breast cancer cells cultured with cyclophosphamide-loaded gliadin NPs for 24 h became apoptotic, confirmed by downregulation of Bcl-2 protein.

3.6 Silk proteins

Silk proteins represent a unique family of natural fibrous proteins due to their unique structure. The molecular structure of many silks consists of large regions or domains of hydrophobic amino acids, segregated by relatively short and more hydrophilic regions. The p*I* of silk is around 4, and most charged amino acids are located in the hydrophilic regions as well as in the N- and C-termini (Altman et al., 2003; Numata & Kaplan, 2010). Silk proteins are promising materials for drug delivery and tissue engineering due to their biocompatibility, slow biodegradability, self-assembly, and controllable structure and morphology. Additionally, silk materials exhibit high encapsulation efficiency and controllable drug release kinetics due to the control of crystalline β-sheet formation. Compared to other natural degradable materials, silk proteins exhibit superior mechanical properties, tunable degradation rates ranging from weeks to months *in vivo* (Altman et al., 2003; Numata & Kaplan, 2010). The silkworm *Bombyx mori* produces silk to weave its cocoon and its major components are fibroin and sericin. Fibroin is a fibrous protein constituting the core of silk, while sericin is a glue-like protein that envelops fibroin fibers with successive sticky layers that help in the formation of a cocoon.

3.6.1 Silk fibroin

Silk fibroin (SF) protein obtained from mulberry silkworms of *B. mori* is a protein-based biomacromolecule composed of 5507 amino acid repeats consisting of repeated sequences of six residues of (Gly–Ala–Gly–Ala–Gly–Ser)$_n$ repeats (Altman et al., 2003). Biopolymer is a heterodimeric protein with a heavy chain (395 kDa) and a light chain (25 kDa), which are linked by a single disulfide bond. SF NPs (below 100 nm) were fabricated using a novel capillary-microdot technique for local and sustained curcumin delivery to cancer cells (Gupta, Aseh, Ríos, Aggarwal, & Mathur, 2009). In this technique, curcumin was suspended in pure SF or blended silk fibroin/chitosan CS (SFCS) solution. The suspension was dispensed on glass slides via a microcapillary then frozen and lyophilized. Dry dots containing SF–curcumin NPs were scraped off the slides and crystallized. One of the limitations of this study was the drug loss during the transfer of NPs from the glass slides. Pure SF–curcumin NPs showed the highest curcumin entrapment and controlled release of curcumin over 8 days as compared to all SFCS blends with a diffusion-based release mechanism (Gupta et al., 2009). This was explained by the hydrophilic character of CS, which resulted in reducing the entrapment efficiency of the hydrophobic curcumin. SF–curcumin showed significantly higher intracellular uptake and efficacy than SFCS–curcumin against breast cancer cells.

In another study, the potential of silk to act as a pH-dependent lysosomotropic anticancer nanomedicine was assessed (Seib, Jones,

Rnjak-Kovacina, Lin, & Kaplan, 2013). Dox-loaded silk NPs (98 nm diameter, polydispersity index 0.109) showed a high Dox-binding capacity at pH 7.4, by virtue of electrostatic interactions between Dox as a weak base and negatively charged silk. The NPs showed pH-dependent release (pH $4.5 \gg 6.0 > 7.4$) indicating that silk NPs could potentially serve as a lysosomotropic drug delivery platform. At low pH values, silk loses its overall acidic surface properties and negative net charge. Therefore, the low pH values encountered in lysosomes were predicted to increase drug release by changing the electrostatic interaction between the silk and Dox (Seib et al., 2013). The drug-loaded silk NPs were significantly more cytotoxic than the free drug to Dox-resistant human breast cancer cells (MCF-7/DOX) indicating their ability to serve as a lysosomotropic drug delivery platform that can overcome drug resistance.

3.6.2 Silk sericin

Silk sericin is a hydrophilic natural polymer produced mainly by Lepidopteron insects of the family Bombycidae and Saturniidae. Silk sericin protein contains predominantly amino acid groups like serine (40%), glycine (16%), glutamic acid, aspartic acid, threonine, and tyrosine. It consists of polar side chains made up of hydroxyl, carboxyl, and amino groups that enable easy cross-linking, copolymerization, and blending with other polymers to form improved biodegradable materials (Numata & Kaplan, 2010). Recent reports suggest immunogenic inertness of silk sericin in both *in vitro* and *in vivo* studies.

Mandal and Kundu (2009) successfully prepared self-assembled silk sericin micellar nanostructures using pluronic stabilizer, capable of carrying both hydrophilic (FITC-inulin having potential antitumor, immunostimulating, and prebiotic effects) and hydrophobic (paclitaxel) drugs. The hydrophobic drug, paclitaxel, was effectively encapsulated within the hydrophobic core of the sericin–pluronic micellar structure, while the hydrophilic drug inulin may get entrapped within the hydrophilic silk sericin in the corona region of the pluronic micellar structure Mandal and Kundu (2009). Sericin NPs were efficiently taken up and showed significant cytotoxicity and apoptosis against MCF-7 breast cancer cells comparable to free paclitaxel.

3.7 Elastin

Elastin is an essential component of animal tissues such as skin, lung, artery, ligament, and cartilage. Native elastin is composed of elastomeric domains and cross-linking domains. The cross-linking domains are rich in alanines and lysines, which are involved in the cross-linking reactions. Elastin molecules form covalently cross-linked network which can stretch and recoil like a rubber band. These molecules provide elasticity to skin, blood vessels, lung, etc. Elastin is produced from a water-soluble precursor, tropoelastin, which spontaneously aggregates into a covalently cross-linked fibrillar polymeric structure (MacEwan & Chilkoti, 2014).

Elastin-like polypeptides (ELPs) are biopolymers inspired by the repetitive hydrophobic domains of human tropoelastin. ELPs are composed of short repeating peptide motifs of which the most common is the pentapeptide VPGXG where X is a guest residue that is any amino acid except proline. ELPs offer several advantages for anticancer drug delivery to solid tumors. First, ELPs are biocompatible and are therefore suitable for local and systemic administration, as they induce minimal inflammatory and immune effects in animal models (MacEwan & Chilkoti, 2014; McDaniel, Callahan, & Chilkoti, 2010). Second, ELP can be conveniently synthesized as a monodisperse macromolecule by recombinant DNA synthesis resulting in high yield, high purity, and low cost without using deleterious solvents. Additionally, the genetically encoded design of ELPs permits exact control over the sequence of the ELP, which can be exploited to precisely specify the location at which a biological drug, peptide or protein, is fused to an ELP. Moreover, the MW and size of ELPs can be precisely tailored so they can exploit the EPR effect, resulting in an increased local concentration of an ELP–drug conjugate in the tumor compartment. Third, ELPs are thermally responsive biopolymers that undergo an inverse phase transition, also called a lower critical solution temperature transition, in response to an increase in temperature (MacEwan & Chilkoti, 2014; McDaniel et al., 2010). The ELP is soluble at temperatures below the transition temperature (Tt) but aggregates at temperatures greater than the Tt; moreover, this process is fully reversible. Due to these attributes, ELPs have been used extensively for drug delivery to tumors in a variety of different embodiments.

3.7.1 ELP–drug conjugates

The recombinant design of ELPs allows the site-specific incorporation of reactive residues such as lysine and Cys that can be used for covalent conjugation of small hydrophobic chemotherapeutics, thus, imparting the amphiphilicity needed to mediate the self-assembly of NPs (Dreher et al., 2003; Furgeson, Dreher, & Chilkoti, 2006). Thus, chimeric polypeptides (CPs) were developed from two components: (1) the N-terminal segment is a hydrophilic ELP and (2) the C-terminal segment contains a variable number of periodically spaced Cys residues arranged in a Cys–(Gly–Gly–Cys)$_n$ motif to which small molecule can be attached. In these systems, release of free drug is enabled by the incorporation of a cleavable linker between the ELP and the drug molecule, such as an acid-labile hydrazone bond or a cathepsin enzyme-cleavable Gly–Phe–Leu–Gly (GFLG) peptide linker. These linkers release the free drug intracellularly after endocytic uptake and accumulation in the acidic, enzyme-rich environment of endosomes, and lysosomes. In the study of MacKay et al. (2009), multiple copies of Dox were attached to Cys residues at the C-terminus of a recombinant CP resulting in the spontaneous self-assembly of CP–Dox conjugates into near monodisperse NPs (40 nm diameter), such that Dox is sequestered in the NP core while the hydrophilic CP forms the hydrated NP corona, thus, limiting the drug toxicity to healthy tissues while targeting the drug to tumor via the EPR effect. The NPs showed a 3.5-fold increase in tumor concentration after 24 h over free drug and a 2.6-fold decrease in heart concentration after 24 h. Mice with subcutaneous C26 colon carcinomas were intravenously injected with CP–Dox at 20 mg/kg Dox significantly reduced the tumor volume and increased the mouse survival compared to free Dox-treated mice.

3.7.2 Thermal targeting of thermoresponsive ELPs by focused, local hyperthermia of solid tumors

An ELP with a Tt between 37 °C and 42 °C (Th) was designed so that the ELP remained soluble upon i.v. injection but underwent its phase transition and formed insoluble aggregates in a tumor that was heated locally to 42 °C, thus, resulting in increased tumor accumulation beyond the passive accumulation of soluble ELP through the EPR effect. A thermoresponsive, genetically engineered ELP containing a C-terminal Cys residue was synthesized by Furgeson et al. (2006) and conjugated to Dox molecules through four different pH-sensitive hydrazone linkers. The highest release of the ELP–Dox conjugate by cleavage of the hydrazone bond at pH 4 was nearly 80% over 72 h and was exhibited by the conjugate with the shortest linker. It was shown that conjugation of Dox to the C-terminal Cys residue through short aliphatic spacers minimally perturbs the Tt and kinetics of the native ELP, while bulky and longer spacers significantly perturb the transition kinetics. In the study of McDaniel et al. (2014), the NPs aggregate in the vasculature of tumors heated to 42 °C and upon cessation of tumor heating, the ELP resolubilization creates a steep concentration gradient, such that the ELP was then pumped out by diffusion from regions of high concentration in the vasculature to regions of low concentration in the extravascular space. Thermally cycling the tumors following injection of the thermally responsive CP–Dox NPs resulted in significantly enhanced drug accumulation (a 2.6-fold greater) over nonthermally responsive NPs.

To control the size of ELP aggregates, ELP-diblock copolymers (ELPBCs) composed of hydrophobic and hydrophilic ELP blocks were designed by Dreher et al. (2003). Above the Tt of the hydrophilic ELP block, but below the Tt of the hydrophobic ELP block, the ELPBC protein self-assembled to form spherical micelles ranging about 30 nm in size. In another study, a diblock ELP was designed composed of poly glutamic/aspartic acids for facile drug conjugation and an ELP block for thermo-targeted chemotherapy of hyperthermic tumor margins was developed by Chen, Youn, and Furgeson (2011) as a Ringsdorf macromolecular solubilizer for geldanamycin (GDA), a potent Hsp90 inhibitor. GDA could be covalently conjugated in nearly 50–67% capacity while retaining thermal-responsive properties to target hyperthermic hepatocellular carcinoma (HCC) tumor margins.

Callahan et al. (2012) designed ELPBCs with histidine-rich hydrophobic blocks to create triple stimulus-responsive NPs that dissociate at the low extracellular pH of solid tumors. Such ELPBC self-assembles at 37 °C into spherical micelles that are stabilized by Zn^{2+} and are disrupted as the pH drops from 7.4 to 6.4. These systems self-assemble in response to the three orthogonal triggers of temperature, pH, and cosolutes; temperature can be used to self-assemble the ELPBCs into micelles below physiological temperature, a drop in pH that corresponds to tumor pH leads to micelle disassembly, and addition of physiological concentrations of Zn^{2+} can further stabilize these micelles or alternatively can be used to trigger their self-assembly at lower temperatures with no adverse impact on their pH sensitivity. After injected i.v. into HCT-15 tumor-bearing nude mice, pH-sensitive ELPBCs achieved a more homogeneous spatial distribution and a greater penetration within the tumor compared with the pH-insensitive control.

3.7.3 Intratumoral administration of ELP–drug conjugate (ELP coacervate depots)

Local administration of thermally sensitive ELP, designed to have a Tt below 37 °C, forms a depot by coacervation of the ELP at body temperature. These ELPs are injected as soluble unimers into a tumor and form a depot *in situ* because of their temperature-triggered coacervation (Liu et al., 2010, 2012). Thermal coacervation of ^{125}I-labeled ELP following intratumoral injection successfully prolonged the residence of radionuclide in murine tumors and decreased the release of active radiotherapeutics into the systemic circulation which translated to a delay in tumor growth relative to controls. The design of an ELP depot for local tumor radiation therapy was further optimized where the ELP composition was changed by appending the ELP C-terminus with increasing number of tyrosine residues, which served as both reactive residues for iodine conjugation and as hydrophobic moieties to depress the ELP Tt. The incorporation of seven tyrosine residues was found to maximize retention, as an increasing number of these hydrophobic residues depressed the ELP Tt (Liu et al., 2010, 2012). An optimal ELP with a length of 120 pentapeptides (50 kDa), functionalized with seven tyrosine residues, which was delivered at a concentration of 1000 μmol/L was labeled with ^{131}I and demonstrated 100% response and over 67% complete regression over 60 days in two xenografts—FaDu squamous cell carcinoma and PC-3 prostate cancer—implanted subcutaneously on the mouse hind limb.

A series of silk–elastin-like protein polymers (SELPs) have been produced, which combine repeating sequences derived from silk and elastin to thus exhibit the mechanical and biological properties of both proteins. Compared with elastin-based materials, the silk blocks in SELPs are able to crystallize into β-sheets via hydrogen bonding enabling robust materials formation (Gustafson & Ghandehari, 2010). In addition, SELPs form materials with unique mechanical properties, avoiding the need for chemical cross-linking. Xia, Wang, Lin, Xu, and Kaplan (2014) demonstrated that Dox can efficiently trigger the self-assembly of SE8Y (SELPs with silk to elastin ratio of 1:8) into uniform micellar-like NPs with silk blocks buried in the core, thus, the hydrophobic compounds likely bind to the silk blocks. The drug can be loaded in the SE8Y NPs with efficiency around 6.5% (65 ng Dox/μg SE8Y). The protein polymers were not cytotoxic to HeLa cell lines ($IC_{50} > 200$ μg/mL), while the Dox-loaded SE8Y NPs showed a 1.8-fold higher cytotoxicity than the free drug.

3.8 Protamine

Protamine sulfate (PS), mainly composed of poly(L-arginine), is biodegradable and has been widely used for biomedical and biotechnological applications. In the study of Cheng et al. (2009), polyionic complexes (PICs) were prepared based on the ionic interaction between the polycation PS and the polyanion poly(L-aspartic acid) (PASP) for combined delivery of the anticancer drug, Dox, and plasmid DNA delivery. Dox was covalently conjugated to PASP via CDI coupling reaction whereas PS forms ionic complex with DNA. Then, DOX–PASP solution was then added into PS/DNA solution to form PICs (lower than 210 nm; Cheng et al., 2009). The PICs efficiently associated DNA with above 80% viability of HEK293T cells and HeLa cells against PICs showing the much lower cytotoxicity of PICs when compared with that of 25 kDa polyethyleneimine (PEI). While the cell viability decreased dramatically from 80% to 20% in the presence of DOX–PICs.

3.9 Lysozyme

Lysozyme (Lys) is a globular protein extracted from egg white with a molecular weight of 14.3 kDa and an isoelectric point value of 10.7 (Zhu et al., 2013). Besides, it consists of 129 amino acid residues and has four disulfide bridges and an ellipsoidal shape. Like most of the globular food proteins, Lys has a gelation property where heating Lys solution leads to the formation of transparent solution, turbid suspension, transparent gel, translucent gel, or opaque gel depending on the concentration of the protein, pH, and ionic strength of the solution. A novel nanogel fabricated via electrostatic complexation between positively charged Lys and highly anionic sodium CMC followed by simple heating for 60 min at 80 °C to achieve the denaturation temperature of Lys (Zhu et al., 2013). After heating process, the Lys denatured and aggregated to force the CMC chains to form spherical aggregates. For drug loading, the carboxyl groups of CMC form hydrogen bonds with the 5-Fu amide groups, as well as the physical entrapment effect when the molecules change from chains to nanogels during heating, resulting in high drug loading (Zhu et al., 2013). Besides, the nanogels obviously have sustained release effect, and the release of 5-Fu in SGF was more slowly compared with SIF (33.9% and 58.6% in a period of 16 h), which could protect the 5-Fu in stomach and ensure most of them released in intestine.

4. TUMOR-TARGETED PROTEIN NPs

Protein-based NPs offer various possibilities for surface modification due to the presence of functional groups (i.e., carboxylic and amino groups) on the surface of the NPs (Elzoghby et al., 2012a, 2012b). Surface modifications of protein NPs with the aim of enhancing tumor-targeting properties comprise two potential strategies: passive targeting and active targeting.

4.1 Passive-targeted protein NPs

In passive targeting, the drug carriers are designed to exhibit prolonged blood circulation with reduced accumulation in normal organs. Chemical coupling of PEG to proteins or particles (PEGylation) is known to prolong their circulation half-life by greater than 50-fold, reduce their immunogenicity, and also promote their accumulation in tumors due to the EPR effect (Lee, Park, Nam, Kim, & Byun, 2006). PEGylation increases the hydrodynamic size of the particles decreasing their clearance through the kidney and prolonged circulatory time. It could form a dense hydrophilic shell of long chains and protect the core from nonspecific hydrophobic interaction with serum proteins (e.g., opsonins) resulting in reduced uptake by RES (Elzoghby et al., 2012a, 2012b). IV administration of noscapine, the tubulin-binding anticancer agent, resulted in rapid elimination of the drug with a half-life of 0.39 h. A 1.43-fold increase in the AUC for noscapine-loaded PEGylated GNPs over unmodified GNPs and a 13.09-fold increase over free noscapine were noted with significantly enhanced cytotoxicity (Madan et al., 2011). Other PEGylated protein NPs for the delivery of anticancer drugs are illustrated in Table 4.

4.2 Active-targeted protein NPs

In active targeting, different types of targeting ligands were successfully attached to the surface of protein NPs to enhance their targetability resulting in specific accumulation within tumors (Table 5). These ligands can be categorized into peptides (e.g., RGD, EGF, NGR, CREKA, AP1, Hph1, LyP-1, Antp, Tat, and MTS), antibodies (e.g., trastuzumab, DI17E6, hKC4, and anti-CD3 antibodies), vitamins (e.g., folic acid and biotin), small molecules (e.g., retinoic acid and glycyrrhizic acid), carbohydrates (e.g., heparin and galactose), hormones (e.g., luteinizing hormone-releasing hormone), enzymes (e.g., collagenase), and polymers (e.g., Lf, Tf, PEI, and the thermoresponsive polymer poly(*N*-isopropylacrylamide-co-acrylamide)-block-polyallylamine). These ligands are commonly attached covalently to the functional groups on the surface of NPs via chemical conjugation reactions, e.g., CDI coupling and maleimide–thiol coupling (Elzoghby, 2013). On the other hand, cationic Lf as a targeting ligand to glioma was attached to negatively charged BSA NPs via electrostatic attraction (Su et al., 2014). Alternatively, fusion with peptide ligands that recognize receptors upregulated in tumor cells via recombinant functionalization of ELP unimers results in the formation of protein-based NPs that exhibit

Table 4 Passively tumor-targeted anticancer drug-loaded protein nanoparticles

Protein	Drug	Surface ligand	Surface modification method	Key outcomes	References
HSA	Doxorubicin	PEG	Conjugation of HSA NH_2 to mPEG-SPA	Sucrose and trehalose were superior to mannitol as cryoprotectants for usage in freeze-drying	Anhorn, Mahler, and Langer (2008)
BSA	5-Fluorouracil	PEG	Conjugation of HSA NH_2 to mPEG-SPA	PEGylation prolonged circulation half-life (50-fold greater), reduced immunogenicity, and increased tumor accumulation	Kouchakzadeh et al. (2010)
Gelatin type B	Doxorubicin	PEG	Conjugation of PEG-epoxide to gelatin NH_2	PEGylation significantly inhibited tumor growth by 82% compared to 62% for free DOX	Lee et al. (2006)
Gelatin type B	Noscapine	PEG	Conjugation of PEG-epoxide to gelatin NH_2	PEGylation showed 2.5-fold greater uptake in Caco-2 cells and increased AUC in comparison with non-PEGylated NPs and free noscapine	Madan et al. (2011)
Gelatin type B	pDNA-encoding EGFP-N1	PEG-thiolated gelatin	Conjugation of mPEG-succinimidyl glutarate to gelatin NH_2	Long-circulating NPs respond to intracellular glutathione concentrations, enhance DNA transfection, and resist proteolytic enzymes	Kommareddy and Amiji (2007)
Gelatin type B	pDNA-encoding β galactosidase	PEG	Conjugation of PEG-epoxide to gelatin NH_2	Preserved supercoiled structure of the plasmid DNA at the perinuclear region and efficient transfection	Kaul and Amiji (2005)
HSA	Doxorubicin	Polysorbate 80	Surface coating	Reduced hematological, cardiac, and testicular toxicities of doxorubicin	Pereverzeva et al. (2007)
Gelatin type B	Polymerized siRNA psi (RFP)	Thiolated gelatin	Conjugation of gelatin COOH to cystamine	Protected siRNA molecules from enzymatic degradation and $2.8\times$ higher tumor accumulation than naked poly-siRNA	Lee et al. (2013)
Gelatin type B	pDNA EGFP-N1	Thiolated gelatin	Conjugation of gelatin NH_2 to iminothiolane	The release from thiolated gelatin NPs was enhanced in the presence of GSH	Kommareddy and Amiji (2005)

Table 5 Actively tumor-targeted anticancer drug-loaded protein nanoparticles

Protein	Drug	Surface ligand	Surface modification method	Key outcomes	References
BSA–dextran	Doxorubicin	Folate	Carbodiimide coupling between folate COOH and dextran OH	Folate-targeted NPs achieved 88.9% inhibition of murine ascites hepatoma H22 tumor and increased the average life time	Hao et al. (2013)
BSA	Paclitaxel	Folate	Carbodiimide coupling between folate COOH and BSA NH_2	Folate-decorated BSA NPs could effectively targeted human prostate cancer cell line	Zhao et al. (2010)
HSA	Micronized 10-HCPT	Folate	Carbodiimide coupling between folate COOH and HSA NH_2	FA-HSA NPs showed sustained release for 120 h, and higher internalization in SGC7901 cells in comparison with HSA NPs	Li, Liu, and Chen (2011) and Li, Liu, Zhao, et al. (2011)
HSA-magnetic NPs	Cisplatin	Folate	Carbodiimide coupling between folate COOH and HSA NH_2	FA NPs prolonged the half release time ($t_{1/2}$) of cisplatin to 24 h compared to 65 min for drug solution	Chen et al. (2010)
BSA	Vinblastine sulfate	Folate	Carbodiimide coupling between folate COOH and BSA NH_2	Improved both drug loading and entrapment efficiency with increased particle size	Zu et al. (2009)
HSA	Doxorubicin	DI17E6 (mAb)	Conjugation to HSA via maleimide–thiol coupling	NPs specifically targeted $\alpha v\beta 3$ integrin-positive melanoma cells showing increased cytotoxic activity than free drug	Wagner et al. (2010)
HSA	Methotrexate	Trastuzumab (TMAB)	Conjugation to HSA via maleimide–thiol coupling	Higher uptake and cytotoxicity of TMAB–MTX–HSA NPs on HER-2-positive cells than that for nontargeted NPs and free MTX	Taheri, Dinarvand, Atyabi, Ghahremani, et al. (2012)

Continued

Table 5 Actively tumor-targeted anticancer drug-loaded protein nanoparticles—cont'd

Protein	Drug	Surface ligand	Surface modification method	Key outcomes	References
HSA	ASOs P12 (targeted against Plk1)	Trastuzumab	Conjugation of sulfhydryl-reactive NPs with thiolated trastuzumab	Significant reduction of Plk1 mRNA and protein expression in BT-474 cells	Steinhauser, Langer, Strebhardt, and Spänkuch (2008)
BSA lyophilisomes	None	Humanized KC4 (hKC4) antibody	Conjugation of thiolated hKC4 to maleimide-activated lyophilisomes	hKC4-conjugated lyophilisomes achieved specific targeting to MUC1-overexpressing tumor cells (HeLa, OVCAR-3, and SKOV-3) compared to MUC1-negative cells (LS174T)	van Bracht et al. (2014)
HSA	Methotrexate	Biotin	Carbodiimide coupling of biotin onto HSA	Biotin-targeted NPs showed stronger *in vivo* antitumor activity in mice bearing 4T1 breast carcinoma than nontargeted NPs	Taheri, Dinarvand, et al. (2011)
Gelatin type A	Cisplatin	Epidermal growth factor (EGF)	Binding of biotinylated EGF to NeutrAvidin-conjugated NPs	EGF NPs enhanced cisplatin accumulation in cancerous lungs via inhalation and reduced its nephrotoxicity	Tseng et al. (2009)
Type B gelatin	pDNA-encoding wt-*p53*	EGF	Covalent conjugation through a PEG spacer	Improved tumor growth suppression in combination therapy (77.3%) compared to wt-*p53* (50.1%) or gemcitabine (61.7%) alone	Xu et al. (2014)
Type B gelatin	EGFR2R-lytic peptide	EGF	Chemical conjugation	Better accumulation in tumor due to small size (36 nm) compared with the peptide (454 nm)	Gaowa et al. (2014)

HSA	Doxorubicin	TRAIL and transferrin	Conjugation of HSA with thiolated TRAIL or transferrin	TRAIL/transferrin/doxorubicin HSA NPs had remarkable cytotoxic and apoptotic activities in all cancer cells examined	Bae et al. (2012)
BSA	Doxorubicin	Lactoferrin	Electrostatic adsorption of cationic Lf onto anionic BSA	Lf NPs showed longer circulation time, high uptake and cytotoxicity in glioma cells, and improved accumulation of DOX in brain tissue	Su et al. (2014)
HSA	Methotrexate	LHRH	Carbodiimide coupling of LHRH COOH to HSA NH_2	LHRH-targeted NPs showed the greatest tumor growth inhibition due to the active targeting to the 4T1 breast tumors using LHRH molecules	Taheri, Dinarvand, Ahadi, Khorramizadeh, et al. (2012)
OVA–chitosan	Doxorubicin	Retinoic acid	Carbodiimide coupling of retinoic acid COOH to chitosan NH_2	Reduced IC_{50} of drug to half and one-third compared to nontargeted NPs and free drug, respectively, reduced drug cardiotoxicity	Varshosaz et al. (2013)
BSA	10-HCPT	Glycyrrhizic acid (GL)	Coupling of GL to BSA in the presence of $NaIO_4$	GL–BSA NPs showed higher inhibition against hepatoma cells (SMMC7721) suggesting hepatocellular carcinoma-targeting ability	Zu et al. (2013)
Gelatin type A	Cisplatin	Heparin	Carbodiimide coupling of heparin COOH to gelatin NH_2	H-GNPs have 4× greater accumulation in tumor-bearing mice compared to GNPs via binding with FGF2 receptor	Jain et al. (2012)

Continued

Table 5 Actively tumor-targeted anticancer drug-loaded protein nanoparticles—cont'd

Protein	Drug	Surface ligand	Surface modification method	Key outcomes	References
Gelatin type B	Doxorubicin	Galactose	Carbodiimide coupling of lactobionic acid COOH to gelatin NH_2	Galactosylated NPs exhibited enhanced uptake and cytotoxicity (IC_{50}; 0.35 μg/mL) by HepG2 cells compared with GN-DOX (0.75 μg/mL)	Garga et al. (2012)
Type B gelatin	siRNA	Polyethyleneimine (PEI)	Carbodiimide coupling of gelatin COOH and PEI NH_2	The gelatin–PEI nanogels were 4 × less toxic than the native PEI and were able to effectively deliver the siRNA into HeLa cells	Mimi, Ho, Siu, Wu, and Li (2012)
BSA	Doxorubicin	PNIPAM-AAm-AA	Carbodiimide coupling	Selective accumulation into tumors above the cloud-point temperature of PAN via local hyperthermia	Shen et al. (2008)
HSA	Curcumin and riluzole	Collagenase	Incubation	Collagenase-modified NPs exhibited one- to twofold greater tumor penetration into spheroid mass	Cui et al. (2013)
BSA	5-Fluorouracil	RGD/RAD peptide	Conjugation of RGD peptides with MPB and PEG–MAL NPs	RGD–SN NPs exhibited eightfold higher binding to human umbilical vein endothelial cells	Dubey, Singodia, Verma, and Vyas (2010)
HSA	Abraxane (nab–paclitaxel)	Tumor-homing peptides (CREKA and LyP-1)	Coupling peptides to abraxane through their cysteine sulfhydryl using sulfo-SMCC cross-linker	CREKA was used to target abraxane to tumors while LyP-1-abraxane produced a statistically highly significant inhibition of tumor growth compared with untargeted abraxane	Karmali et al. (2009)

HSA	Paclitaxel	Cell-penetrating peptide (Hph1)	Hph1 was conjugated to HSA NPs using a MAL–PEG–NHS cross-linker	Hph1 facilitated intracellular delivery of paclitaxel via a Hph1-mediated transcytosis and efficiently induced cancer cell apoptosis	Lee, Bae, et al. (2011) and Lee, Heng, et al. (2011)
ELP	*p21* peptide drug	TAT, MTS, penetratin (Antp) CPPs	Recombinant genetic fusion to the end of ELP hydrophilic blocks	Antp-ELP was most efficient in intracellular delivery of a *p21*-derived peptide, a kinase inhibitor drug which enhanced apoptosis to SKOV-3 and HeLa cells	Massodi, Bidwell, and Raucher (2005)
ELP	Doxorubicin	TAT CPP	Recombinant genetic fusion to the end of ELP hydrophilic blocks	TAT-ELP enhanced cellular uptake and accumulation of ELP–Dox in MES-SA/Dx5 cells. Both cytotoxicity and apoptosis were enhanced by hyperthermia	Bidwell, Davis, Fokt, Priebe, and Raucher (2007)
ELPBC	Rapamycin	FK506-binding protein 12 (FKBP)	Recombinant genetic fusion to the end of ELP hydrophilic blocks	ELP_{BC} NPs encapsulated rapamycin in the hydrophobic core and on the FKBP-functionalized corona and inhibited tumor growth of breast cancer xenografts in mice	Shi et al. (2013)
ELPBC	None	RGD and NGR peptides	Recombinant genetic fusion to the end of ELP hydrophilic blocks	Greater internalization of NGR-functionalized ELP_{BC} micelles in HT-1080 cells expressing the NGR receptor, aminopeptidase N (CD13)	Dreher et al. (2008)

an active targeting ability. These ligands were fused on the genetic level to the end of the ELP hydrophilic blocks, to be easily encoded into the biopolymer sequence without the need for a separate conjugation step resulting in the multivalent display of the peptide ligands on the exterior of the NP (Massodi et al., 2005).

5. CONCLUSIONS

Proteins can be posed as the natural counterpart to synthetic polymers for the development of NPs for tumor-targeted drug delivery. From the studies reviewed, it can be seen that protein nanocarriers possess various applications in the field of anticancer drug delivery including solubilization of poorly soluble anticancer drugs, improved drug pharmacokinetics and tissue distribution, enhanced tumor targetability, improved antitumor efficacy, and reduced toxicity of anticancer drugs. Some proteins such as albumin and Lf can enhance tumor targeting of anticancer drugs via interaction with specific receptors overexpressed on tumor cells. Moreover, some other proteins such as CN demonstrated extraordinary cancer cell penetration ability similar to cell-penetrating peptides. Thus, the biocompatible and non-immunogenic milk protein CN can be proposed as a viable alternative to proteins such as albumin and GEL for developing tumor-targeted nanocarriers. On the other hand, some proteins such as zein and gliadin can replace the commonly used synthetic polymers, e.g., PLGA by virtue of their hydrophobicity. Their NPs may not need any further chemical or physical treatment to harden them, thus, avoiding the toxic chemical cross-linkers. Finally, studies will continue to further improve tumor-targeted delivery of anticancer drugs using protein nanocarriers. Special emphasis seems to be given for using recombinant protein-based NPs for anticancer drug delivery such as rHSA, recombinant GEL, and ELP NPs as alternatives to native ones.

REFERENCES

Altman, G. H., Diaz, F., Jakuba, C., Calabro, T., Horan, R. L., Chen, J., et al. (2003). Silk-based biomaterials. *Biomaterials*, *24*, 401–416.

Anhorn, M., Mahler, H., & Langer, K. (2008). Freeze-drying of human serum albumin (HSA) nanoparticles with different excipients. *International Journal of Pharmaceutics*, *363*(1–2), 162–169.

Bachar, M., Mandelbaum, A., Portnaya, I., Perlstein, H., Even-Chen, S., & Barenholz, Y. (2012). Development and characterization of a novel drug nanocarrier for oral delivery, based on self-assembled β-casein micelles. *Journal of Controlled Release*, *160*, 164–171.

Bae, S., Ma, K., Kim, T. H., Lee, E. S., Oh, K. T., Park, E. S., et al. (2012). Doxorubicin-loaded human serum albumin nanoparticles surface-modified with TNF-related apoptosis-inducing ligand and transferrin for targeting multiple tumor types. *Biomaterials*, *33*(5), 1536–1546.

Bajpai, A. K., & Choubey, J. (2006). In vitro release dynamics of an anticancer drug from swellable gelatin nanoparticles. *Journal of Applied Polymer Science*, *101*, 2320–2332.

Bertrand, N., Wu, J., Xu, X., Kamaly, N., & Farokhzad, O. C. (2014). Cancer nanotechnology: The impact of passive and active targeting in the era of modern cancer biology. *Advanced Drug Delivery Reviews*, *66*, 2–25.

Bidwell, G. L., III, Davis, A. N., Fokt, I., Priebe, W., & Raucher, D. (2007). A thermally targeted elastin-like polypeptide–doxorubicin conjugate overcomes drug resistance. *Investigational New Drugs*, *25*, 313–326.

Burger, A., Hartung, G., Stehle, G., Sinn, H., & Fiebig, H. (2001). Pre-clinical evaluation of a methotrexate–albumin conjugate (MTX–HSA) in human tumor xenografts in vivo. *International Journal of Cancer*, *92*(5), 718–724.

Callahan, D. J., Liu, W., Li, X., Dreher, M. R., Hassouneh, W., Kim, M., et al. (2012). Triple stimulus-responsive polypeptide nanoparticles that enhance intratumoral spatial distribution. *Nano Letters*, *12*(4), 2165–2170.

Cascone, M. G., Lazzeri, L., Carmignani, C., & Zhu, Z. (2002). Gelatin nanoparticles prepared by a simple w/o emulsion as delivery system for methotrexate. *Journal of Materials Science. Materials in Medicine*, *13*(5), 523–526.

Chen, D., Tang, Q., Xue, W., Xiang, J., Zhang, L., & Wang, X. (2010). The preparation and characterization of folate-conjugated human serum albumin magnetic cisplatin nanoparticles. *Journal of Biomedical Research*, *24*(1), 26–32.

Chen, Y., Youn, P., & Furgeson, D. Y. (2011). Thermo-targeted drug delivery of geldanamycin to hyperthermic tumor margins with diblock elastin-based biopolymers. *Journal of Controlled Release*, *155*(2), 175–183.

Cheng, Z., Dai, Y., Kang, X., Li, C., Huang, S., Lian, H., et al. (2014). Gelatin-encapsulated iron oxide nanoparticles for platinum (IV) prodrug delivery, enzyme-stimulated release and MRI. *Biomaterials*, *35*, 6359–6368.

Cheng, H., Li, Y. Y., Zeng, X., Sun, Y. X., Zhang, X. Z., & Zhuo, R. X. (2009). Protamine sulfate/poly(L-aspartic acid) polyionic complexes self-assembled via electrostatic attractions for combined delivery of drug and gene. *Biomaterials*, *30*(6), 1246–1253.

Choubey, J., & Bajpai, A. K. (2010). Investigation on magnetically controlled delivery of doxorubicin from superparamagnetic nanocarriers of gelatin crosslinked with genipin. *Journal of Materials Science. Materials in Medicine*, *21*, 1573–1586.

Coester, C. J., Langer, K., van Briesen, H., & Kreuter, J. (2000). Gelatin nanoparticles by two-step desolvation—A new preparation method, surface modification and cell uptake. *Journal of Microencapsulation*, *17*, 187–193.

Cui, M., Naczynski, D., Zevon, M., Griffith, C., Sheihet, L., Poventud-Fuentes, I., et al. (2013). Multifunctional albumin nanoparticles as combination drug carriers for intratumoral chemotherapy. *Advanced Healthcare Materials*, *2*(9), 1236–1245.

Desai, N. (2007). Nanoparticle albumin bound (nab) technology: Targeting tumors through the endothelial gp60 receptor and SPARC. *Nanomedicine*, *3*, 337–346.

Ding, D., Zhu, Z., Li, R., Li, X., Wu, W., Jiang, X., et al. (2011). Nanospheres-incorporated implantable hydrogel as a trans-tissue drug delivery system. *ACS Nano*, *5*, 2520–2534.

Dreher, M. R., Raucher, D., Balu, N., Michael, C. O., Ludeman, S. M., & Chilkoti, A. (2003). Evaluation of an elastin-like polypeptide–doxorubicin conjugate for cancer therapy. *Journal of Controlled Release*, *91*(1–2), 31–43.

Dreher, M. R., Simnick, A. J., Fischer, K., Smith, R. J., Patel, A., Schmidt, M., et al. (2008). Temperature triggered self-assembly of polypeptides into multivalent spherical micelles. *Journal of the American Chemical Society*, *130*, 687–694.

Dreis, S., Rothweiler, F., Michaelis, M., Cinatl, J., Kreuter, J., & Langer, K. (2007). Preparation, characterisation and maintenance of drug efficacy of doxorubicin-loaded human serum albumin (HSA) nanoparticles. *International Journal of Pharmaceutics*, *341*(1–2), 207–214.
Dubey, P., Singodia, D., Verma, R., & Vyas, S. (2010). RGD modified albumin nanospheres for tumour vasculature targeting. *The Journal of Pharmacy and Pharmacology*, *63*(1), 33–40.
Egusquiaguirre, S. P., Igartua, M., Hernández, R. M., & Pedraz, J. L. (2012). Nanoparticle delivery systems for cancer therapy: Advances in clinical and preclinical research. *Clinical & Translational Oncology*, *14*(2), 83–93.
Elgindy, N., Elkhodairy, K., Molokhia, A., & Elzoghby, A. (2010). Lyophilization monophase solution technique for improvement of the physicochemical properties of an anticancer drug, flutamide. *European Journal of Pharmaceutics and Biopharmaceutics*, *74*, 397–405.
Elgindy, N., Elkhodairy, K., Molokhia, A., & Elzoghby, A. (2011a). Biopolymeric microparticles combined with lyophilized monophase dispersions for controlled flutamide release. *International Journal of Pharmaceutics*, *411*, 113–120.
Elgindy, N., Elkhodairy, K., Molokhia, A., & Elzoghby, A. (2011b). Biopolymeric nanoparticles for oral protein delivery: Design and in vitro evaluation. *Journal of Nanomedicine & Nanotechnology*, *2*, 1–8.
Elgindy, N., Elkhodairy, K., Molokhia, A., & Elzoghby, A. (2011c). Lyophilization monophase solution technique for preparation of amorphous flutamide dispersions. *Drug Development and Industrial Pharmacy*, *37*, 754–764.
Elgindy, N., Elkhodairy, K., Molokhia, A., & Elzoghby, A. (2011d). Lyophilized flutamide dispersions with polyols and amino acids: Preparation and in vitro evaluation. *Drug Development and Industrial Pharmacy*, *37*, 446–455.
Elgindy, N. A., Samy, W. A., & Elzoghby, A. O. (2014). Casein-based micelles: A novel vector for delivery of the poorly soluble anticancer drug, flutamide? *Therapeutic Delivery*, *5*, 7–9.
Elsadek, B., & Kratz, F. (2012). Impact of albumin on drug delivery—New applications on the horizon. *Journal of Controlled Release*, *157*(1), 4–28.
Elzoghby, A. O. (2013). Gelatin-based nanoparticles as drug and gene delivery systems: Reviewing three decades of research. *Journal of Controlled Release*, *172*, 1075–1091.
Elzoghby, A. O., Abo El-Fotoh, W. S., & Elgindy, N. A. (2011). Casein-based formulations as promising controlled release drug delivery systems. *Journal of Controlled Release*, *153*, 206–216.
Elzoghby, A. O., Helmy, M. W., Samy, W. M., & Elgindy, N. A. (2013a). Micellar delivery of flutamide via milk protein nanovehicles enhances its anti-tumor efficacy in androgen-dependent prostate cancer rat model. *Pharmaceutical Research*, *30*, 2654–2663.
Elzoghby, A. O., Helmy, M. W., Samy, W. M., & Elgindy, N. A. (2013b). Novel ionically crosslinked casein nanoparticles for flutamide delivery: Formulation, characterization, and in vivo pharmacokinetics. *International Journal of Nanomedicine*, *8*, 1721.
Elzoghby, A. O., Helmy, M. W., Samy, W. M., & Elgindy, N. A. (2013c). Spray-dried casein-based micelles as a vehicle for solubilization and controlled delivery of flutamide: Formulation, characterization, and in vivo pharmacokinetics. *European Journal of Pharmaceutics and Biopharmaceutics*, *84*, 487–496.
Elzoghby, A. O., Saad, N. I., Helmy, M. W., Samy, W. M., & Elgindy, N. A. (2013). Ionically-crosslinked milk protein nanoparticles as flutamide carriers for effective anticancer activity in prostate cancer-bearing rats. *European Journal of Pharmaceutics and Biopharmaceutics*, *85*, 444–451.
Elzoghby, A. O., Samy, W. M., & Elgindy, N. A. (2012a). Albumin-based nanoparticles as potential controlled release drug delivery systems. *Journal of Controlled Release*, *157*, 168–182.

Elzoghby, A. O., Samy, W. M., & Elgindy, N. A. (2012b). Protein-based nanocarriers as promising drug and gene delivery systems. *Journal of Controlled Release, 161*, 38–49.

Elzoghby, A. O., Samy, W. M., & Elgindy, N. A. (2013). Novel spray-dried genipin-crosslinked casein nanoparticles for prolonged release of alfuzosin hydrochloride. *Pharmaceutical Research, 30*, 512–522.

Esmaili, M., Ghaffari, S. M., Moosavi-Movahedi, Z., Atri, M. S., Sharifizadeh, A., & Farhadi, M. (2011). Beta casein-micelle as a nanovehicle for solubility enhancement of curcumin; food industry application. *LWT—Food Science and Technology, 44*, 2166–2172.

Furgeson, D. Y., Dreher, M. R., & Chilkoti, A. (2006). Structural optimization of a "smart" doxorubicin–polypeptide conjugate for thermally targeted delivery to solid tumors. *Journal of Controlled Release, 110*(2), 362–369.

Gaowa, A., Horibe, T., Kohno, M., Sato, K., Harada, H., Hiraoka, M., et al. (2014). Combination of hybrid peptide with biodegradable gelatin hydrogel for controlled release and enhancement of anti-tumor activity in vivo. *Journal of Controlled Release, 176*, 1–7.

Garga, M., Madan, J., Pandey, R. S., Sardana, S., Katyal, A., & Chandra, R. (2012). Galactosylated gelatin nanovectors of doxorubicin inhibit cell proliferation and induce apoptosis in hepatocarcinoma cells. *Anti-Cancer Drugs, 23*, 836–845.

Ghalandari, B., Divsalar, A., Saboury, A. A., & Parivar, K. (2014). The new insight into oral drug delivery system based on metal drugs in colon cancer therapy through b-lactoglobulin/oxali-palladium nanocapsules. *Journal of Photochemistry and Photobiology B: Biology, 140*, 255–265.

Girard, M., Turgeon, S. L., & Gauthier, S. F. (2002). Interbiopolymer complexing between β-lactoglobulin and low- and high-methoxylated pectin measured by potentiometric titration and ultrafiltration. *Food Hydrocolloids, 16*, 585–591.

Golla, K., Bhaskar, C., Ahmed, F., & Kondapi, A. K. (2013). A target-specific oral formulation of doxorubicin–protein nanoparticles: Efficacy and safety in hepatocellular cancer. *Journal of Cancer, 4*, 644–652.

Golla, K., Cherukuvada, B., Ahmed, F., & Kondapi, A. K. (2012). Efficacy, safety and anticancer activity of protein nanoparticle-based delivery of doxorubicin through intravenous administration in rats. *PLoS One*, 7, e51960.

Gong, J., Huo, M., Zhou, J., Zhang, Y., Peng, X., Yu, D., et al. (2009). Synthesis, characterization, drug-loading capacity and safety of novel octyl modified serum albumin micelles. *International Journal of Pharmaceutics, 376*(1–2), 161–168.

Gulfam, M., Kim, J. E., Lee, J. M., Ku, B., Chung, B. H., & Chung, B. G. (2012). Anticancer drug-loaded gliadin nanoparticles induce apoptosis in breast cancer cells. *Langmuir, 28*, 8216–8223.

Guo, R., Cheng, Y., Ding, D., Li, X., Zhang, L., Jiang, X., et al. (2011). Synthesis and antitumoral activity of gelatin/polyoxometalate hybrid nanoparticles. *Macromolecular Bioscience, 11*, 839–847.

Gupta, V., Aseh, A., Ríos, C. N., Aggarwal, B. B., & Mathur, A. B. (2009). Fabrication and characterization of silk fibroin-derived curcumin nanoparticles for cancer therapy. *International Journal of Nanomedicine, 4*, 115–122.

Gustafson, J. A., & Ghandehari, H. (2010). Silk-elastin like protein polymers for matrix-mediated cancer gene therapy. *Advanced Drug Delivery Reviews, 62*(15), 1509–1523.

Han, S., Li, M., Liu, X., Gao, H., & Wu, Y. (2013). Construction of amphiphilic copolymer nanoparticles based on gelatin as drug carriers for doxorubicin delivery. *Colloids and Surfaces B: Biointerfaces, 102*, 833–841.

Hao, H., Ma, Q., Huang, C., He, F., & Yao, P. (2013). Preparation, characterization, and in vivo evaluation of doxorubicin loaded BSA nanoparticles with folic acid modified dextran surface. *International Journal of Pharmaceutics, 444*(1–2), 77–84.

Hawkins, M. J., Soon-Shiong, P., & Desai, N. (2008). Protein nanoparticles as drug carriers in clinical medicine. *Advanced Drug Delivery Reviews*, *60*(8), 876–885.

Hu, D., Lin, C., Liu, L., Li, S., & Zhao, Y. (2012). Preparation, characterization, and in vitro release investigation of lutein/zein nanoparticles via solution enhanced dispersion by supercritical fluids. *Journal of Food Engineering*, *109*, 545–552.

Jain, A., Gulbake, A., Jain, A., Shilpi, S., Hurkat, P., Jain, A., et al. (2012). Development of surface-functionalised nanoparticles for FGF2 receptor-based solid tumor targeting. *Journal of Microencapsulation*, *29*, 95–102.

Jiang, H. R., & Liu, N. (2010). Self-assembled β-lactoglobulin-conjugated linoleic acid complex for colon cancer-targeted substance. *Journal of Dairy Science*, *93*, 3931–3939.

Karmali, P., Kotamraju, V., Kastantin, M., Black, M., Missirlis, D., Tirrell, M., et al. (2009). Targeting of albumin-embedded paclitaxel nanoparticles to tumors. *Journal of Nanomedicine & Nanotechnology*, *5*(1), 73–82.

Karthikeyan, S., Prasad, N. R., Ganamani, A., & Balamurugan, E. (2013). Anticancer activity of resveratrol-loaded gelatin nanoparticles on NCI-H460 non-small cell lung cancer cells. *Biomedicine and Preventive Nutrition*, *3*, 64–73.

Kaul, G., & Amiji, M. (2005). Tumor-targeted gene delivery using poly(ethylene glycol)-modified gelatin nanoparticles: In vitro and in vivo studies. *Pharmaceutical Research*, *22*, 951–961.

Kim, T. H., Jiang, H. H., Youn, Y. S., Park, C. W., Tak, K. K., Lee, S., et al. (2011). Preparation and characterization of water-soluble albumin-bound curcumin nanoparticles with improved antitumor activity. *International Journal of Pharmaceutics*, *403*, 285–291.

Kommareddy, S., & Amiji, M. (2005). Preparation and evaluation of thiol-modified gelatin nanoparticles for intracellular DNA delivery in response to glutathione. *Bioconjugate Chemistry*, *16*, 1423–1432.

Kommareddy, S., & Amiji, M. (2007). Poly (ethylene glycol)-modified thiolated gelatin nanoparticles for glutathione-responsive intracellular DNA delivery. *Nanomedicine*, *3*, 32–42.

Kouchakzadeh, H., Shojaosadati, S., Maghsoudi, A., & Farahani, E. (2010). Optimization of PEGylation conditions for BSA nanoparticles using response surface methodology. *AAPS PharmSciTech*, *11*(3), 1206–1211.

Kratz, F. (2014). A clinical update of using albumin as a drug vehicle—A commentary. *Journal of Controlled Release*, *190*, 331–336.

Lai, L. F., & Guo, H. X. (2011). Preparation of new 5-fluorouracil-loaded zein nanoparticles for liver targeting. *International Journal of Pharmaceutics*, *404*, 317–323.

Langer, K., Balthasar, S., Vogel, V., Dinauer, N., von Briesen, H., & Schubert, D. (2003). Optimization of the preparation process for human serum albumin (HSA) nanoparticles. *International Journal of Pharmaceutics*, *257*, 169–180.

Lee, J., Bae, K., Kim, J., Nam, Y., & Park, T. (2011). Intracellular delivery of paclitaxel using oil-free, shell cross-linked HSA—Multi-armed PEG nanocapsules. *Biomaterials*, *32*(33), 8635–8644.

Lee, S. H., Heng, D., Ng, W. K., Chan, H. K., & Tan, R. B. (2011). Nano spray drying: A novel method for preparing protein nanoparticles for protein therapy. *International Journal of Pharmaceutics*, *403*, 192–200.

Lee, E. J., Khan, S. A., Park, J. K., & Lim, K. H. (2012). Studies on the characteristics of drug-loaded gelatin nanoparticles prepared by nanoprecipitation. *Bioprocess and Biosystems Engineering*, *35*, 297–307.

Lee, G. Y., Park, K., Nam, J. H., Kim, S. Y., & Byun, Y. (2006). Anti-tumor and anti-metastatic effects of gelatin–doxorubicin and PEGylated gelatin–doxorubicin nanoparticles in SCC7 bearing mice. *Journal of Drug Targeting*, *14*, 707–716.

Lee, S. J., Yhee, J. Y., Kim, S. H., Kwon, I. C., & Kim, K. (2013). Biocompatible gelatin nanoparticles for tumor-targeted delivery of polymerized siRNA in tumor-bearing mice. *Journal of Controlled Release*, *172*(1), 358–366.

Leo, E., Vandelli, M. A., Cameroni, R., & Forni, F. (1997). Doxorubicin-loaded gelatin nanoparticles stabilized by glutaraldehyde involvement of the drug in the cross-linking process. *International Journal of Pharmaceutics*, *155*, 75–82.

Li, W., Liu, D., & Chen, S. (2011). Amphiphilically-modified gelatin nanoparticles: Self-assembly behavior, controlled biodegradability, and rapid cellular uptake for intracellular drug delivery. *Journal of Materials Chemistry*, *21*, 12381–12388.

Li, Q., Liu, C., Zhao, X., Zu, Y., Wang, Y., Zhang, B., et al. (2011). Preparation, characterization and targeting of micronized 10-hydroxycamptothecin-loaded folate-conjugated human serum albumin nanoparticles to cancer cells. *International Journal of Nanomedicine*, *6*, 397–405.

Liang, L., Tajmir-Riahi, H. A., & Subirade, M. (2008). Interaction of beta-lactoglobulin with resveratrol and its biological implications. *Biomacromolecules*, *9*, 50–56.

Liu, W., MacKay, J. A., Dreher, M. R., Chen, M., McDaniel, J. R., Simnick, A. J., et al. (2010). Injectable intratumoral depot of thermally responsive polypeptide–radionuclide conjugates delays tumor progression in a mouse model. *Journal of Controlled Release*, *144*(1), 2–9.

Liu, W., McDaniel, J., Li, X., Asai, D., Quiroz, F. G., Schaal, J., et al. (2012). Brachytherapy using injectable seeds that are self-assembled from genetically encoded polypeptides in situ. *Cancer Research*, *72*(22), 5956–5965.

Livney, Y. D. (2010). Milk proteins as vehicles for bioactives. *Current Opinion in Colloid & Interface Science*, *15*, 73–83.

Lu, Z., Yeh, T. K., Tsai, M., Au, J. L., & Wientjes, M. G. (2004). Paclitaxel-loaded gelatin nanoparticles for intravesical bladder cancer therapy. *Clinical Cancer Research*, *10*(22), 7677–7684.

MacEwan, S. R., & Chilkoti, A. (2014). Applications of elastin-like polypeptides in drug delivery. *Journal of Controlled Release*, *190*, 314–330.

MacKay, J. A., Chen, M., McDaniel, J. R., Liu, W., Simnick, A. J., & Chilkoti, A. (2009). Self-assembling chimeric polypeptide–doxorubicin conjugate nanoparticles that abolish tumors after a single injection. *Nature Materials*, *8*(12), 993–999.

Madan, J., Dhiman, N., Sardana, S., Aneja, R., Chandra, R., & Katyal, A. (2011). Long-circulating poly (ethylene glycol)-grafted gelatin nanoparticles customized for intracellular delivery of noscapine preparation, in-vitro characterization, structure elucidation, pharmacokinetics, and cytotoxicity analyses. *Anti-Cancer Drugs*, *22*, 543–555.

Maghsoudi, A., Shojaosadati, S., & Farahani, E. (2008). 5-Fluorouracil-loaded BSA nanoparticles: Formulation optimization and in vitro release study. *AAPS PharmSciTech*, *9*(4), 1092–1096.

Maham, A., Tang, Z., Wu, H., Wang, J., & Lin, Y. (2009). Protein-based nanomedicine platforms for drug delivery. *Small*, *5*(15), 1706–1721.

Mandal, B. B., & Kundu, S. C. (2009). Self-assembled silk sericin/poloxamer nanoparticles as nanocarriers of hydrophobic and hydrophilic drugs for targeted delivery. *Nanotechnology*, *20*(35), 355101.

Martínez, A., Muñiz, E., Iglesias, I., Teijón, J., & Blanco, M. (2012). Enhanced preclinical efficacy of tamoxifen developed as alginate–cysteine/disulfide bond reduced albumin nanoparticles. *International Journal of Pharmaceutics*, *436*(1–2), 574–581.

Massodi, I., Bidwell, G. L., III, & Raucher, D. (2005). Evaluation of cell penetrating peptides fused to elastin-like polypeptide for drug delivery. *Journal of Controlled Release*, *108*, 396–408.

McDaniel, J. R., Callahan, D. J., & Chilkoti, A. (2010). Drug delivery to solid tumors by elastin-like polypeptides. *Advanced Drug Delivery Reviews*, *62*, 1456–1467.

McDaniel, J. R., MacEwan, S. R., Li, X., Radford, D. C., Landon, C. D., Dewhirst, M., et al. (2014). Rational design of "heat seeking" drug loaded polypeptide nanoparticles that thermally target solid tumors. *Nano Letters*, *14*(5), 2890–2895.

Mimi, H., Ho, K. M., Siu, Y. S., Wu, A., & Li, P. (2012). Polyethyleneimine-based core–shell nanogels: A promising siRNA carrier for argininosuccinate synthetase mRNA knockdown in Hela cells. *Journal of Controlled Release, 158*, 123–130.

Narayanan, S., Pavithran, M., Viswanath, A., Narayanan, D., Mohan, C. C., Manzoor, K., et al. (2014). Sequentially releasing dual-drug-loaded PLGA–casein core/shell nanomedicine: Design, synthesis, biocompatibility and pharmacokinetics. *Acta Biomaterialia, 10*, 2112–2124.

Numata, K., & Kaplan, D. L. (2010). Silk-based delivery systems of bioactive molecules. *Advanced Drug Delivery Reviews, 62*, 1497–1508.

Pan, K., Zhong, Q., & Baek, S. J. (2013). Enhanced dispersibility and bioactivity of curcumin by encapsulation in casein nanocapsules. *Journal of Agricultural and Food Chemistry, 61*, 6036–6043.

Patel, A. R., Heussen, P. C. M., Hazekamp, J., Drost, E., & Velikov, K. P. (2012). Quercetin loaded biopolymeric colloidal particles prepared by simultaneous precipitation of quercetin with hydrophobic protein in aqueous medium. *Food Chemistry, 133*, 423–429.

Patel, A., Hu, Y., Tiwari, J. K., & Velikov, K. P. (2010). Synthesis and characterization of zein–curcumin colloidal particles. *Soft Matter, 6*, 6192–6199.

Pereverzeva, E., Treschalin, I., Bodyagin, D., Maksimenko, O., Langer, K., Dreis, S., et al. (2007). Influence of the formulation on the tolerance profile of nanoparticle-bound doxorubicin in healthy rats: Focus on cardio- and testicular toxicity. *International Journal of Pharmaceutics, 337*(1–2), 346–356.

Podaralla, S., Averineni, R., Alqahtani, M., & Perumal, O. (2012). Synthesis of novel biodegradable methoxypoly(ethylene glycol)–zein micelles for effective delivery of curcumin. *Molecular Pharmaceutics, 9*, 2778–2786.

Qi, J., Yao, P., He, F., Yu, C., & Huang, C. (2010). Nanoparticles with dextran/chitosan shell and BSA/chitosan core—Doxorubicin loading and delivery. *International Journal of Pharmaceutics, 393*(1–2), 176–184.

Razmi, M., Divsalar, A., Saboury, A. A., Izadi, Z., Haertlé, T., & Mansuri-Torshizi, H. (2013). Beta-casein and its complexes with chitosan as nanovehicles for delivery of a platinum anticancer drug. *Colloids and Surfaces B: Biointerfaces, 112*, 362–367.

Sebak, S., Mirzaei, M., Malhotra, M., Kulamarva, A., & Prakash, S. (2010). Human serum albumin nanoparticles as an efficient noscapine drug delivery system for potential use in breast cancer: Preparation and in vitro analysis. *International Journal of Nanomedicine, 5*, 525–532.

Seib, F. P., Jones, G. T., Rnjak-Kovacina, J., Lin, Y., & Kaplan, D. L. (2013). pH-dependent anticancer drug release from silk nanoparticles. *Advanced Healthcare Materials, 2*(12), 1606–1611.

Shapira, A., Assaraf, Y. G., Epstein, D., & Livney, Y. D. (2010). Beta-casein nanoparticles as an oral delivery system for chemotherapeutic drugs: Impact of drug structure and properties on co-assembly. *Pharmaceutical Research, 27*, 2175–2186.

Shapira, A., Assaraf, Y. G., & Livney, Y. D. (2010). Beta-casein nanovehicles for oral delivery of chemotherapeutic drugs. *Nanomedicine, 6*, 119–126.

Shapira, A., Davidson, I., Avni, N., Assaraf, Y. G., & Livney, Y. D. (2012). β-Casein nanoparticle-based oral drug delivery system for potential treatment of gastric carcinoma: Stability, target-activated release and cytotoxicity. *European Journal of Pharmaceutics and Biopharmaceutics, 80*, 298–305.

Shapira, A., Markman, G., Assaraf, Y. G., & Livney, Y. D. (2010). β-Casein-based nanovehicles for oral delivery of chemotherapeutic drugs: Drug–protein interactions and mitoxantrone loading capacity. *Nanomedicine, 6*, 547–555.

Shen, Z., Wei, W., Zhao, Y., Ma, G., Dobashi, T., Maki, Y., et al. (2008). Thermosensitive polymer-conjugated albumin nanospheres as thermal targeting anti-cancer drug carrier. *European Journal of Pharmaceutical Sciences, 35*(4), 271–282.

Shi, P., Aluri, S., Lin, Y. A., Shah, M., Edman, M., Dhandhukia, J., et al. (2013). Elastin-based protein polymer nanoparticles carrying drug at both corona and core suppress tumor growth in vivo. *Journal of Controlled Release*, *171*, 330–338.

Shpigelman, A., Israeli, G., & Livney, Y. D. (2010). Thermally-induced protein polyphenol co-assemblies: Betalactoglobulin-based nanocomplexes as protective nanovehicles for EGCG. *Food Hydrocolloids*, *24*, 735–743.

Shutava, T. G., Balkundi, S. S., Vangala, P., Steffan, J. J., Bigelow, R. L., Cardelli, J. A., et al. (2009). Layer-by-layer-coated gelatin nanoparticles as a vehicle for delivery of natural polyphenols. *ACS Nano*, *3*, 1877–1885.

Sneharani, A. H., Karakkat, J. V., Singh, S. A., & Rao, A. G. A. (2010). Interaction of curcumin with β-lactoglobulin; stability, spectroscopic analysis, and molecular modeling of the complex. *Journal of Agricultural and Food Chemistry*, *58*, 11130–11139.

Son, S., Song, S., Lee, S. J., Min, S., Kim, S. A., Yhee, J. Y., et al. (2013). Self-crosslinked human serum albumin nanocarriers for systemic delivery of polymerized siRNA to tumors. *Biomaterials*, *34*(37), 9475–9485.

Steinhauser, I., Langer, K., Strebhardt, K., & Spänkuch, B. (2008). Effect of trastuzumab-modified antisense oligonucleotide-loaded human serum albumin nanoparticles prepared by heat denaturation. *Biomaterials*, *29*(29), 4022–4028.

Su, Z., Xing, L., Chen, Y., Xu, Y., Yang, F., Zhang, C., et al. (2014). Lactoferrin-modified poly(ethylene glycol)-grafted BSA nanoparticles as a dual-targeting carrier for treating brain gliomas. *Molecular Pharmaceutics*, *11*(6), 1823–1834.

Sun, Q., Teong, B., Chen, I., Chang, S. J., Gao, J., & Kuo, S. (2014). Enhanced apoptotic effects of dihydroartemisinin-aggregated gelatin and hyaluronan nanoparticles on human lung cancer cells. *Journal of Biomedical Materials Research Part B: Applied Biomaterials*, *102B*, 455–462.

Taheri, A., Atyabi, F., Nouri, F. S., Ahadi, F., Derakhshan, M., Amini, M., et al. (2011). Nanoparticles of conjugated methotrexate–human serum albumin: Preparation and cytotoxicity evaluations. *Journal of Nanomaterials*, *2011*, 1–7.

Taheri, A., Dinarvand, R., Ahadi, F., Khorramizadeh, M., & Atyabi, F. (2012). The in vivo antitumor activity of LHRH targeted methotrexate–human serum albumin nanoparticles in 4T1 tumor-bearing Balb/c mice. *International Journal of Pharmaceutics*, *431*(1–2), 183–189.

Taheri, A., Dinarvand, R., Atyabi, F., Ghahremani, M., & Ostad, S. (2012). Trastuzumab decorated methotrexate–human serum albumin conjugated nanoparticles for targeted delivery to HER2 positive tumor cells. *European Journal of Pharmaceutical Sciences*, *47*(2), 331–340.

Taheri, A., Dinarvand, R., Nouri, F. S., Khorramizadeh, M. R., Borougeni, A. T., Mansoori, P., et al. (2011). Use of biotin targeted methotrexate–human serum albumin conjugated nanoparticles to enhance methotrexate antitumor efficacy. *International Journal of Nanomedicine*, *6*, 1863–1874.

Tseng, C., Su, W., Yen, K., Yang, K., & Lin, F. (2009). The use of biotinylated-EGF-modified gelatin nanoparticle carrier to enhance cisplatin accumulation in cancerous lungs via inhalation. *Biomaterials*, *30*, 3476–3485.

van Bracht, E., Stolle, S., Hafmans, T. G., Boerman, O. C., Oosterwijk, E., van Kuppevelt, T. H., et al. (2014). Specific targeting of tumor cells by lyophilisomes functionalized with antibodies. *European Journal of Pharmaceutics and Biopharmaceutics*, *87*(1), 80–89.

Varshosaz, J., Hassanzadeh, F., Sadeghi, H., Ghelich Khan, Z., & Rostami, M. (2013). Retinoic acid decorated albumin–chitosan nanoparticles for targeted delivery of doxorubicin hydrochloride in hepatocellular carcinoma. *Journal of Nanomaterials*, *2013*, 1–12.

Wagner, S., Rothweiler, F., Anhorn, M., Sauer, D., Riemann, I., Weiss, E., et al. (2010). Enhanced drug targeting by attachment of an anti αv integrin antibody to doxorubicin loaded human serum albumin nanoparticles. *Biomaterials*, *31*(8), 2388–2398.

Won, Y., Yoon, S., Sonn, C. H., Lee, K., & Kim, Y. (2011). Nano self-assembly of recombinant human gelatin conjugated with α-tocopheryl succinate for Hsp90 inhibitor, 17-AAG, delivery. *ACS Nano*, *5*, 3839–3848.

Wu, D. C., Cammarata, C. R., Park, H. J., Rhodes, B. T., & Ofner, C. M. (2013). Preparation, drug release, and cell growth inhibition of a gelatin–doxorubicin conjugate. *Pharmaceutical Research*, *30*, 2087–2096.

Xia, X. X., Wang, M., Lin, Y., Xu, Q., & Kaplan, D. L. (2014). Hydrophobic drug-triggered self-assembly of nanoparticles from silk-elastin-like protein polymers for drug delivery. *Biomacromolecules*, *15*(3), 908–914.

Xie, J., Marijnissen, J. C., & Wang, C. H. (2006). Microparticles developed by electrohydrodynamic atomization for the local delivery of anticancer drug to treat C6 glioma in vitro. *Biomaterials*, *27*, 3321–3332.

Xu, J., Singh, A., & Amiji, M. M. (2014). Redox-responsive targeted gelatin nanoparticles for delivery of combination wt-p53 expressing plasmid DNA and gemcitabine in the treatment of pancreatic cancer. *BMC Cancer*, *14*(75), 1–12.

Xue, J., Tan, C., Zhang, X., Feng, B., & Xia, S. (2014). Fabrication of epigallocatechin-3-gallate nanocarrier based on glycosylated casein: Stability and interaction mechanism. *Journal of Agricultural and Food Chemistry*, *62*, 4677–4684.

Yang, L., Cui, F., Cun, D. M., Tao, A., Shi, K., & Lin, W. H. (2007). Preparation, characterization and biodistribution of the lactone form of 10-hydroxycamptothecin (HCPT)-loaded bovine serum albumin (BSA) nanoparticles. *International Journal of Pharmaceutics*, *340*, 163–172.

Yeh, T. K., Lu, Z., Wientjes, M. G., & Au, J. L. S. (2005). Formulating paclitaxel in nanoparticles alters its disposition. *Pharmaceutical Research*, *22*(6), 867–874.

Yi, Y. M., Yang, T. Y., & Pan, W. M. (1999). Preparation and distribution of 5-fluorouracil (125)I sodium alginate–bovine serum albumin nanoparticles. *World Journal of Gastroenterology*, *5*(1), 57–60.

Zhang, Z. H., Wang, X. P., Ayman, W. Y., Munyendo, W. L., Lv, H. X., & Zhou, J. P. (2013). Studies on lactoferrin nanoparticles of gambogic acid for oral delivery. *Drug Delivery*, *20*, 86–93.

Zhang, L., Zhang, S., Ruan, S. B., Zhang, Q. Y., He, Q., & Gao, H. L. (2014). Lapatinib-incorporated lipoprotein-like nanoparticles: Preparation and a proposed breast cancer-targeting mechanism. *Acta Pharmacologica Sinica*, *35*(6), 846–852.

Zhao, D., Zhao, X., Zu, Y., Li, J., Zahng, Y., Jiang, R., et al. (2010). Preparation, characterization, and in vitro targeted delivery of folate-decorated paclitaxel-loaded bovine serum albumin nanoparticles. *International Journal of Nanomedicine*, *5*, 1–9.

Zhen, X., Wang, X., Xie, C., Wu, W., & Jiang, X. (2013). Cellular uptake, antitumor response and tumor penetration of cisplatin-loaded milk protein nanoparticles. *Biomaterials*, *34*(4), 1372–1382.

Zhou, H., Sun, X., Zhang, L., Zhang, P., Li, J., & Liu, Y. (2012). Fabrication of biopolymeric complex coacervation core micelles for efficient tea polyphenol delivery via a green process. *Langmuir*, *28*, 14553–14561.

Zhu, K., Ye, T., Liu, J., Peng, Z., Xu, S., Lei, J., et al. (2013). Nanogels fabricated by lysozyme and sodium carboxymethyl cellulose for 5-fluorouracil controlled release. *International Journal of Pharmaceutics*, *41*(1–2), 721–727.

Zou, T., & Gu, L. (2013). TPGS emulsified zein nanoparticles enhanced oral bioavailability of daidzin: In vitro characteristics and in vivo performance. *Molecular Pharmaceutics*, *10*, 2062–2070.

Zou, T., Li, Z., Percival, S. S., Bonard, S., & Gu, L. (2012). Fabrication, characterization, and cytotoxicity evaluation of cranberry procyanidins-zein nanoparticles. *Food Hydrocolloids*, *27*, 293–300.

Zou, T., Percival, S. S., Cheng, Q., Li, Z., Rowe, C. A., & Gu, L. (2012). Preparation, characterization, and induction of cell apoptosis of cocoa procyanidins–gelatin–chitosan nanoparticles. *European Journal of Pharmaceutics and Biopharmaceutics*, *82*, 36–42.
Zu, Y., Meng, L., Zhao, X., Ge, Y., Yu, X., Zhang, Y., et al. (2013). Preparation of 10-hydroxycamptothecin-loaded glycyrrhizic acid-conjugated bovine serum albumin nanoparticles for hepatocellular carcinoma-targeted drug delivery. *International Journal of Nanomedicine*, *8*, 1207–1222.
Zu, Y., Zhang, Y., Zhao, X., Zhang, Q., Liu, Y., & Jiang, R. (2009). Optimization of the preparation process of vinblastine sulfate (VBLS)-loaded folate-conjugated bovine serum albumin (BSA) nanoparticles for tumor-targeted drug delivery using response surface methodology (RSM). *International Journal of Nanomedicine*, *4*, 321–333.

CHAPTER SEVEN

Polysaccharide Nanoparticles for Protein and Peptide Delivery: Exploring Less-Known Materials

Marlise A. dos Santos*, Ana Grenha[†,1]

*Joan-Vernikos Aerospace Pharmacy Laboratory, School of Pharmacy, Microgravity Centre, Pontifical Catholic University of Rio Grande do Sul PUCRS, Porto Alegre, Brazil

[†]CBME—Centre for Molecular and Structural Biomedicine/IBB—Institute for Biotechnology and Bioengineering, Faculty of Sciences and Technology, University of Algarve, Campus de Gambelas, Faro, Portugal

[1]Corresponding author: e-mail address: amgrenha@ualg.pt

Contents

Advances in Protein Chemistry and Structural Biology, Volume 98
ISSN 1876-1623
http://dx.doi.org/10.1016/bs.apcsb.2014.11.003

Abstract

Finding adequate carriers for protein and peptide delivery has become an urgent need, owing to the growing number of macromolecules identified as having therapeutic potential. Nanoparticles have emerged in the field as very promising vehicles and much work has been directed to testing the capacity of different materials to compose the matrix of these carriers. Natural materials and, specifically, polysaccharides have been taking the forefront of the challenge, because of several favoring properties that include the higher propensity to exhibit biodegradability and biocompatibility, and also the high structural flexibility. The majority of works found in the literature regarding polysaccharide nanoparticles uses very popular materials like chitosan or hyaluronic acid. This review is aimed at describing and exploring the potential of polysaccharides that are not so well known or that are less explored. For those, the main properties will be described, together with an overview of the reported applications as nanoparticle matrix materials.

1. INTRODUCTION

The strong evolution of the biotechnology industry has resulted in a great number of macromolecules identified to have therapeutic potential, many being protein based. Alongside with the new therapeutic possibilities, questions arose on the feasibility of their administration, because of the inherent physicochemical and biopharmaceutical properties that make their delivery a challenging issue. The therapeutic effect of proteins is not only limited by the potential degradation in biological environments, but is also compromised by their low ability to reach the therapeutic site of action (Antosova, Mackova, Kral, & Macek, 2009; Casettari & Illum, 2014; Janes, Calvo, & Alonso, 2001). These limitations are related with the presence of a high number of functional groups susceptible of chemical degradation and, also, with the high hydrophilic character of the proteins (de la Fuente, Csaba, Garcia-Fuentes, & Alonso, 2008; Li et al., 2013). Soon it was, thus, assumed that specific carriers were needed to provide an efficient delivery. The efforts of many scientists have turned to the task, resulting in the development of a wide variety of drug delivery systems. Nanoparticles, which are solid submicron-sized carriers (Pinto Reis, Neufeld, Ribeiro, & Veiga, 2006), are one of the most promising and popular systems. This is due to several interesting features that include the high surface-to-volume ratio and the surface functionality, thus offering potential for the association of macromolecules (Csaba, Garcia-Fuentes, & Alonso, 2006). Additionally, nanosized carriers provide increased contact with epithelial surfaces (Desai, Labhasetwar, Amidon, & Levy, 1996) and, regarding protein delivery strategies, an improvement of proteins' stability, bioavailability, targeting, uptake, and biological activity has been shown (Dombu & Betbeder, 2013). Finally, biological transport processes, anatomically and down to the cellular and subcellular levels, are considered to be affected by the physical attributes of the nanocarriers (Elsabahy & Wooley, 2012).

On selecting the appropriate material for drug delivery approaches, several requirements must be met: biodegradability, biocompatibility, versatility, and low overall costs of production (Chiellini, Piras, Errico, & Chiellini, 2008; Karewicz, Bielska, & Nowakowska, 2012). By definition, nanoparticles can be produced with either synthetic or natural materials, but the latter are known to comply more easily with the mentioned requisites (Beneke, Viljoen, & Hamman, 2009; Malafaya, Silva, & Reis, 2007). Actually, it has been claimed that one of the ways of avoiding the potential toxicity of nanocarriers, an issue believed to be one of the most relevant on preventing diverse clinical applications so far, may be using natural materials (Anwunobi & Emeje, 2011). Moreover, these have some remarkable merits over synthetic ones, namely improved capacity for cell adhesion and mechanical properties similar to natural tissues (Pollard et al., 2008). Additionally, they are among the most versatile building units of drug delivery systems, as their properties can be easily tailored to meet specific requirements.

Polysaccharides are a class of natural materials, along with proteins and lipids. Many are obtained from plants or marine organisms, therefore being less probable to induce adverse immunological reactions, as compared with proteins (Karewicz et al., 2012). Importantly, they are economical, readily available, usually biodegradable and with few exceptions, also biocompatible (Beneke et al., 2009; Copetti, Grassi, Lapasin, & Pricl, 1997). These are the reasons justifying that polysaccharides assume a relevant role as matrix materials for drug delivery systems and, namely, nanoparticles.

In this review, polysaccharide nanoparticles developed for protein and peptide delivery will be described. The emphasis will be placed on materials that were less explored or that are not so well-known regarding the development of nanoparticulate systems, thus excluding polymers like chitosan, alginate, cyclodextrins, and hyaluronic acid. For those polysaccharides selected to be described, a report on their main properties will be presented.

Additionally, an overview of their applications on the production of nanoparticles aimed at protein delivery will be provided.

2. POLYSACCHARIDES AS NANOPARTICLE MATRIX MATERIALS

Nature provides an impressive array of polymeric materials that have been finding very interesting applications in the biomedical field and, namely, in drug delivery. Polysaccharides are complex carbohydrates, composed of monosaccharides joined together by glycosidic bonds (Karewicz et al., 2012; Mizrahy & Peer, 2012). As mentioned above, they are one of the material classes applied more frequently in the production of nanoparticles. The most common basic unit of polysaccharides is the monosaccharide D-glucose, although D-fructose, D-galactose, L-galactose, D-mannose, L-arabinose, and D-xylose are also frequently present. Some polysaccharides comprise simple sugar acids (glucuronic and iduronic acids) and also monosaccharide derivatives in their structure, like the amino sugars D-glucosamine and D-galactosamine, as well as their derivatives *N*-acetylneuraminic acid and *N*-acetylmuramic acid (d'Ayala, Malinconico, & Laurienzo, 2008). The presence of several of these sugar units on the side chain of carbohydrate polymers makes them good candidates for targeted delivery by carbohydrate recognizing receptors found on the surface of several cells (Zhang, Ma, & Sun, 2010).

As evidenced in Table 1, polysaccharides have several possible origins, namely algal, plant, microbial, and animal.

Notwithstanding the variety of origins, they benefit from the general properties of natural polymers, including the propensity for biocompatibility, the low cost and hydrophilicity (Anwunobi & Emeje, 2011; Liu, Jiao, Wang, Zhou, & Zhang, 2008; Mizrahy & Peer, 2012). Table 2 presents the pros and cons of polysaccharides as nanoparticle matrix materials.

The general propensity for biocompatibility and biodegradability is perhaps the major advantage of these materials. Actually, carriers based on polysaccharides are expected to be easily eliminated from the organism, as these will be processed to result in small sugar units that integrate conventional metabolic processes, thereby permitting elimination or reabsorption (Anwunobi & Emeje, 2011). Polysaccharides present diverse physicochemical properties, deriving from multiple chemical structures that also translate into different molecular weights (MW) and intrinsic characteristics. Ionic nature, for instance, is one of the greatest items of variation, as all cationic,

Table 1 Main origin and sources of polysaccharides

Polysaccharide	Origin	Examples of sources
Cellulose	Plant	Cotton, wheat, potato tubers, sugar beet pulp, soybean stock
Pectin	Plant	Citric fruits, apples, pears
Starch	Plant	Potato, wheat, rice, maize
Mannan	Plant	Coffee bean, ivory nut
Glucomannan	Plant	Scotch pine, orchid tubers, *Amorphophallus konjac*[a]
Arabic gum	Plant	Species of *Acacia* (*Acacia senegal*)[a]
Tragacanth gum	Plant	Species of *Astragalus* (*Astragalus gummifer*)[a]
Carrageenan	Plant, algae	Red seaweed
Fucoidan	Plant, algae	Brown seaweed (*Fucus vesiculosus*)[a]
Pullulan	Fungus	*Aureobasidium pullulans*
Chondroitin sulfate	Animal	Cartilage
Heparin	Animal	Mucosal tissues

[a]Main source.

Table 2 Pros and cons of polysaccharides as nanoparticle matrix materials

Advantages	Limitations
Structural flexibility, stability	Inter-batch variability
Low cost	Limited availability (if widely used)
Bioadhesion capacity	Complex and varied composition
Hydrophilicity	Difficult to process
Biocompatibility, biodegradability	Immunogenicity

anionic, and neutral polysaccharides can be found (Beneke et al., 2009; Mizrahy & Peer, 2012). The property of hydrophilicity is of particular importance, as it permits the production of carriers by methods avoiding organic solvents. Additionally, polysaccharides also benefit from a great structural flexibility, forming either linear or branched structures and easily permitting chemical modifications (Karewicz et al., 2012; Malafaya et al., 2007). This characteristic might be beneficial for several nanoparticle production methods or the development of targeting strategies. The more sophisticated applications, which include controlled or triggered release, or even targeted delivery, demand chemical modifications of polysaccharides. These usually encompass the introduction of ionic or hydrophobic groups, degradable bonds, spacers, or targeting moieties (Karewicz et al., 2012). The ability to adhere to biological surfaces includes a relevant advantage in drug delivery applications (Mizrahy & Peer, 2012), as these frequently imply an interaction with cell surfaces. In this regard, the reactive functional groups of polysaccharides allow the formation of noncovalent bonds with cell surfaces, affording enhanced residence time and, consequently, increased drug absorption (Anwunobi & Emeje, 2011; Liu et al., 2008).

Notwithstanding the relevance of the mentioned advantages, some drawbacks should also be taken into account, such as the possibility of generating immunogenic responses and the polymer variability related to origin and supplier (Pollard et al., 2008). Regarding the latter, plant-derived polysaccharides pose potential challenges, as structural differences might occur according to the location and plant collection season (Beneke et al., 2009).

Traditionally, polysaccharides were included in formulations as inert materials, but in modern pharmaceutical design they increasingly assume relevant roles, endowing the dosage forms with multifunctional abilities, such as controlled release, stabilization, emulsification, or bioadhesiveness, among others (Beneke et al., 2009; Mizrahy & Peer, 2012). Nevertheless, the set of polysaccharide properties confers a relevant versatility that adds to a wide range of biological abilities, thus contributing to an increased application in drug carrier production.

Below, a description is given for each polysaccharide focused in this chapter as matrix material of nanoparticles aimed at protein delivery. The distribution of polysaccharides is presented according to their main origin, as described in Table 1, and the chemical structure of the polysaccharides is described in Fig. 1. The structure of arabic and tragacanth gums is not shown, as these are composed of polysaccharide mixtures.

2.1. Cellulose

Cellulose is the primary structural component of plant cell walls and it can also be extracted from marine invertebrate animals (tunicate), algae, fungi,

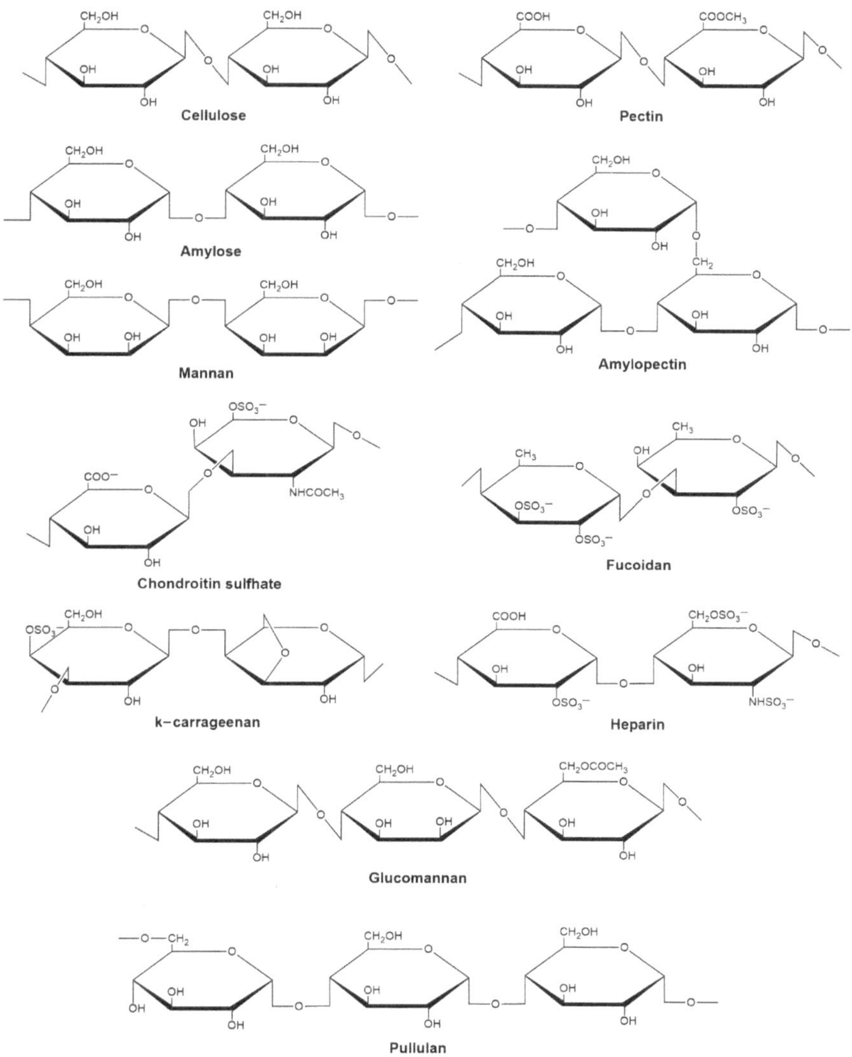

Figure 1 Chemical structure of polysaccharides.

and bacteria (Lavoine, Desloges, Dufresne, & Bras, 2012; Mangalam, Simonsen, & Benight, 2009). It is the most abundant polysaccharide available in the world, being thus readily accessible and relatively cheap (Ding et al., 2012; Padalkar et al., 2010; Song, Zhou, & Chen, 2012). The basic structure of cellulose consists of a linear chain of β-(1,4)-linked D-glucose units which are joined to form cellobiose repeating units (Ding et al.,

2012; Moura, Dias, Carvalho, & de Sousa, 2013). Cellulose glucan chains are parallel and are packed side-by-side, forming microfibrils, stabilizing the structure but reducing its flexibility. The degrees of polymerization and the molecular organization of the chains are the main characteristics affecting chemical and physical properties. Cellulose-based materials are considered biocompatible due to their reduced inflammatory response. Moreover, resorption of cellulose in tissues does not occur since cells are not able to synthesize cellulases (Moura et al., 2013). Due to its hydroxyl groups, cellulose can undergo chemical modifications producing, for instance, cellulose ether derivatives such as methyl cellulose (MC), hydroxypropyl cellulose (HPC), hydroxyethyl cellulose (HEC), hydroxypropyl methyl cellulose (HPMC), and carboxymethyl cellulose (CMC) that can be used to produce controlled delivery systems (Moura et al., 2013; Song et al., 2012). The MW can be varied, making aqueous viscosity and gelation properties tunable, and thus allowing different pharmaceutical and biomedical uses (Moura et al., 2013). MC, HPMC, and CMC are Generally Recognised as Safe (GRAS) excipients by the American Food and Drug Administration (FDA) since 1973 (FDA, 2014).

Microbial cellulose has attracted attention as a novel biomedical material due to its biocompatibility, good strength, and high swelling capacity. Bacterial cellulose is a good candidate for the fabrication of cellulose-based hydrogels, since it is produced as pure cellulose (Mangalam et al., 2009; Müller et al., 2013; Song et al., 2012; Valo et al., 2013). In other words, it does not contain compounds like lignin, pectin, and hemicellulose (Moon, Martini, Nairn, Simonsen, & Youngblood, 2011).

2.2. Pectin

Pectin is a complex polysaccharide also present in plant cell walls. It is an essential component in the initial growth of fruit and is often a waste material from food industry, with consequent high availability (Beneke et al., 2009). Its basic structure is a linear chain of α-(1,4)-linked D-galacturonic acid units, presenting highly branched regions. Like all other galacturonic acid-based polysaccharides, pectin is rich in neutral sugars such as L-rhamnose, L-arabinose, D-galactose, D-xylose, and D-glucose (Beneke et al., 2009; Ridley, O'Neill, & Mohnen, 2001). Nevertheless, the uronic acids have carboxyl groups, providing a negative charge, some of which naturally presented as methyl esters (Rowe, Sheskey, & Owen, 2009). The degree of esterification (percentage of the esterified carboxyl groups), is the basis of the classification of pectins (Sungthongjeen, Sriamornsak, Pitaksuteepong, Somsiri, & Puttipipatkhachorn, 2004). Pectin composition is, however, known to vary according to the botanical source, resulting for instance in different amounts of neutral sugars (Beneke et al., 2009).

Pectin has an average MW of about 30–100 kDa (Rowe et al., 2009) and was granted the GRAS statute in 1977 (FDA, 2014). It mainly finds applications as gelling and thickening agent in food industry (Opanasopit, Apirakaramwong, Ngawhirunpat, Rojanarata, & Ruktanonchai, 2008), but applications as pharmaceutical excipient have also been described, as adsorbent, bulk-forming agent, and emulsion stabilizer (Rowe et al., 2009). This polysaccharide shows ability for gel formation, lower degrees of esterification ($<50\%$) leading to the formation of rigid gels by the action of calcium or multivalent cations, which cross-link the galacturonic acid chains (Sungthongjeen et al., 2004). It has been shown particularly suitable for colonic delivery of drugs due to its selective digestion by colonic microflora, alongside with high resistance to proteases and amylases of the upper gastrointestinal tract (Alvarez-Lorenzo, Blanco-Fernandez, Puga, & Concheiro, 2013). A review on the structure and biosynthesis of pectin is available at Ridley et al. (2001).

2.3. Starch

Starch is a plant-derived storage carbohydrate composed of glucose monomers, in the form of both amylose (20–30%) and amylopectin (70–80%; Beneke et al., 2009; Malafaya et al., 2007). While amylose is a nonbranching helical polymer of α-(1,4)-linked D-glucose monomers, amylopectin consists mainly of α-(1–4)-linked D-glucose residues but it is a branched molecule with α-(1–6) branch points every 24–30 glucose residues on average. Most common cereal starches contain 20–30% amylose, as indicated above, but those genetically modified might have up to 70–90% (Beneke et al., 2009). Because starch composition can be considerably variable, depending for instance on the used source (potato, corn, rice, etc.), its MW is also very variable.

Starch is a GRAS excipient since 1979 (FDA, 2014) and is widely used in food and pharmaceutical industries, as thickener, binder, or diluent (Jobling, 2004; Rowe et al., 2009). It is also considered very interesting for biomedical applications, given its inherent biodegradability (it is degraded by amylases), overwhelming abundance and renewability. Additionally, its degradation products are oligosaccharides that can be readily metabolized to produce energy, thus not raising particular issues regarding host reaction to degradation products (Malafaya et al., 2007).

Native starch may not be suitable to produce controlled release drug carriers due to substantial swelling and rapid enzymatic degradation, which leads to very rapid drug release. This has led to the use of derivatives that are more resistant to enzymatic degradation, as well as cross-linking and formation of copolymers (Beneke et al., 2009).

2.4. Mannan

Mannan consists of D-mannose molecules linked together by β-(1–4) unions to form a polymer, usually with MW lower than 30 kDa. Mannan derivatives, so called heteromannans, are based on variations of β-mannan backbone, which might be interrupted with D-glucose to form glucomannan and/or branched with α-(1,6)-linked D-galactose to form galactomannan (Chauhan, Puri, Sharma, & Gupta, 2012; Pauly et al., 2013).

Linear chains of β-(1–4) mannan are found in the seed endosperms of certain plant species, like ivory nut and coffee bean (Chauhan et al., 2012; Hägglund et al., 2001; Wolfrom, Laver, & Patin, 1961). The biodegradability of the polymer is ensured by β-mannanase and β-mannosidase, thus providing its elimination (Chen et al., 1995; Dionísio & Grenha, 2012). Mannan is used mainly in food industry as a stabilizer and gelling agent (Pauly et al., 2013) and is being proposed in drug delivery, mainly to compose drug carriers aimed at targeted delivery (Sihorkar & Vyas, 2001; Yu, Liu, Liu, Zhang, & Xu, 2010).

2.5. Glucomannan

Glucomannan is a plant-derived polysaccharide belonging to the mannan family, with frequent application in food industry as thickener and emulsifier. It has a variety of sources, but the most common glucomannan is konjac glucomannan extracted from *Amorphophallus konjac,* known as konjac plant (Xiao, Gao, & Zhang, 2000). Irrespective of its origin, it is composed of β-(1–4)-linked D-mannose and D-glucose monomers, although different ratios of mannose/glucose can be found according to the source. Konjac glucomannan has a mannose/glucose molar ratio around 1.6:1 (Buchala, Franz, & Meier, 1974). The diverse types of glucomannan also vary in the acetylation degree, but the polymer typically presents 5–10% acetylation (Gao & Nishinari, 2004).

The polymer is water soluble, but its high MW, reported to be up to 1.9×10^6 (Davé & McCarthy, 1997), leads to very viscous solutions. Glucomannan presents gelling properties and is also able to synergistically interact with other polymers, such as carrageenan and xanthan gum, an advantageous characteristic for drug delivery applications (Alonso-Sande, Teijeiro-Osorio, Remuñán-López, & Alonso, 2009).

A comprehensive review on glucomannan potential in drug delivery is available in Alonso-Sande et al. (2009).

2.6. Arabic gum

Arabic gum is also known as Acacia gum and is obtained from the exudates of *Acacia* trees (Beneke et al., 2009), being a GRAS excipient since 1973 (FDA, 2014). Its chemical composition is complex, consisting in a branched molecule with a backbone of (1,3)-linked β-D-galactopyranosyl units, with side chains composed of two to five (1,3)-linked β-D-galactopyranosyl units, joined to the main chain by (1,6)-linkages (Ali, Ziada, & Blunden, 2009). Both the main and the side chains contain other carbohydrates, such as L-arabinose, L-rhamnose, and glucuronic acid (Ali et al., 2009; Beneke et al., 2009). This combination of sugars results in a MW of about 240–580 kDa (Rowe et al., 2009).

As happens with many plant-derived polysaccharides, the chemical composition of arabic gum is reported to vary with its source, the age of the trees from which it is obtained, climatic conditions, and soil environment (Ali et al., 2009).

Arabic gum has wide industrial uses as a stabilizer, thickening agent, and emulsifier, mainly in the food industry (Ali et al., 2009), but it also has an application in pharmaceutical industry as emulsifying and suspending agent, as well as tablet binder (Rowe et al., 2009). A review on the properties and applications of this material is available at Verbeken, Dierckx, and Dewettinckk (2003).

2.7. Tragacanth gum

Tragacanth gum is a plant-derived molecule obtained from the stems and branches of Asiatic species of *Astragalus*, having a MW of about 840 kDa (Rowe et al., 2009; Verbeken et al., 2003). It is a branched and anionic carbohydrate composed of a complex mixture of polysaccharides, being thus very heterogeneous in its structure. It was affirmed as GRAS excipient within the USA in 1972 (FDA, 2014). The composition of the gum shows considerable variation according to the *Astragalus* species used as source. Additionally, even when a specific species is used, some variations might be observed deriving from different seasons of collection and also geographical issues (Verbeken et al., 2003), which is typical for natural polymers. Nevertheless, tragacanth gum is composed of two major fractions: one water soluble (tragacanthin) and another that is not soluble but instead water-swellable (bassorin; Ghorbani Gorji, Ghorbani Gorji, Mohammadifar, & Zargaraan, 2014; Rowe et al., 2009). Tragacanthin is composed of (1) tragacanthic acid, which contains residues of D-galacturonic acid, D-xylose, L-fucose, and D-galactose, and (2) an arabinogalactan, containing residues of L-arabinose, D-galactose, and D-galacturonic acid (Aspinall & Baillie, 1963; Fattahi, Sadrjavadi, et al., 2013). Different species of *Astragalus* have been reported to display different ratios of the two fractions, resulting in diverse physicochemical and rheological properties and, consequently, in specific applications and functionalities (Ghorbani Gorji et al., 2014). Its most frequent pharmaceutical application occurs as suspending and emulsifying agent (Rowe et al., 2009). A comprehensive review is available at Verbeken et al. (2003).

2.8. Carrageenan

Carrageenan is an anionic, high MW (well above 100 kDa) sulfated GRAS polysaccharide, obtained from certain species of red seaweed (Beneke et al., 2009; Necas & Bartosikova, 2013).

It is composed of alternate units of β-D-galactose and 3,6-anhydro-α-D-galactose, linked by α-(1,3) and β-(1,4) glycosidic unions. Depending on the method and the algae from which carrageenan is extracted, three main types of carrageenans can be obtained, kappa (κ), iota (ι), and lambda (λ), although several other types are also reported (Alvarez-Lorenzo et al., 2013; Necas & Bartosikova, 2013; Sankalia, Mashru, Sankalia, & Sutariya, 2006). The main difference among subtypes is structural and concerns the degree of substitution (DS) of the sulfate group, the position of the ester sulfate groups, and the presence or absence of anhydrogalactose units (Rowe et al., 2009; Sankalia et al., 2006). The former is one of the features most influencing the properties of carrageenan type (Necas & Bartosikova, 2013). All the described subtypes of carrageenan have between 22% and 35% sulfate groups and it is known that higher levels of ester sulfate result in lower solubility temperature and lower gel strength (Necas & Bartosikova, 2013).

κ-Carrageenan has an ester sulfate content of 25–30% and an amount of anhydrogalactose units corresponding to 28–35%. ι–Carrageenan has similar ester sulfate content (28–30%) and about 25–30% anhydrogalactose units. In turn, λ-carrageenan has higher sulfate substitution (32–39%) and no anhydrogalactose content (Barbeyron, Michel, Potin, Henrissat, & Kloareg, 2000). From these three types, only κ- and ι-carrageenans evidence gel-forming ability, κ-carrageenan gels being more firm than those obtained with ι-carrageenan, which are more elastic and soft (Beneke et al., 2009; Bixler, 1994).

Carrageenan is susceptible to depolymerization through acid-catalyzed hydrolysis, as glycosidic linkages are hydrolyzed at pH below 3 (Necas & Bartosikova, 2013). Additionally, sulfatases are also involved on its metabolization (Ghosh, 2007). This polysaccharide is widely used in the food industry and also has applications in the pharmaceutical industry, as gel base, emulsifying and stabilizing agent (Rowe et al., 2009).

2.9. Fucoidan

Fucoidan is an anionic polysaccharide containing substantial amounts of L-fucose and sulfate ester groups. It is mainly extracted from brown seaweed, but different algae species can be used as source, such as *Fucus vesiculosus*, *Ascophyllum nodosum*, *Laminaria japonica*, and *Macrocystis pyrifera*, among others, resulting in different polymeric compositions (Berteau & Mulloy, 2003; Li, Lu, Wei, & Zhao, 2008). *F. vesiculosus* is the most common of the species being used and the one providing the simplest polymer of the whole group, with only L-fucose and sulfate units (Li et al., 2008). If other source is used, the structure might comprise units of D-mannose, D-galactose, D-xylose, L-rhamnose, and uronic acid, among others (Anastyuk, Shevchenko, Nazarenko, Dmitrenok, & Zvyagintseva, 2009; Huang & Lam, 2011; Li et al., 2008).

Structurally, the polysaccharide has a backbone of α-(1–3)-linked fucose units or is composed of repeating disaccharide units of α-(1–3)- and α-(1–4)-linked fucose residues with O-2 branches. Depending on the structure of the main chain, fucoidans may be sulfonated at O-4, O-2, or at both positions of fucose units. Moreover, some fucoidans are both sulfated and acetylated (Anastyuk et al., 2009). A wide range of MW has been reported, varying from 5–7 to several hundred kDa. Importantly, low MW fractions are suggested to be more biocompatible (Balboa, Conde, Moure, Falqué, & Domínguez, 2013).

Fucoidan has been identified to exhibit several interesting pharmacological effects, such as antithrombotic, antitumoral, antiviral, and antiinflamatory activity (Balboa et al., 2013).

2.10. Pullulan

Pullulan is a neutral polymer composed of α-(1,6)-linked maltotriose residues, which in turn are composed of three glucose molecules connected to each other by an α-(1,4) glycosidic bond (Chiellini et al., 2008; Mizrahy & Peer, 2012; Namazi, Fathi, & Heydari, 2011). It is produced from starch by strains of the fungus *Aureobasidium pullulans*, which cultivation parameters strongly affect the resulting MW, usually ranging from 4.5×10^4 to 6×10^5 Da (Cheng, Demirci, & Catchmark, 2011).

Pullulan evidences interesting features regarding drug delivery applications, including water solubility, the production of relatively low viscous aqueous solutions, and adhesiveness (Mizrahy & Peer, 2012; Rekha & Chandra, 2007). Regarding the latter property, it is believed to have a role on the adhesion of the referred fungus to biological surfaces, such as leafs (Pouliot, Walton, Nolen-Parkhouse, Abu-Lail, & Camesano, 2005), indicating potential bioadhesive properties that might be relevant for drug delivery purposes. In fact, pullulan-based nanoparticles were reported to adhere to the nasal epithelium in a study of nasal vaccination (Nochi et al., 2010). More recently, pullulan was described to have a role on hepatocyte targeting, because of its ability to interact with asialoglycoprotein receptors present on these cells, leading to cell internalization (Li et al., 2014; Wang et al., 2013). This ability has increased the interest in the application of pullulan to compose drug carriers specifically targeted to the liver, which is a particularly interesting approach in hepatic cancer.

The polymer is also expected to be biodegradable, being exposed to the hydrolysis of glycoside bonds and the subsequent metabolism of glucose (Teramoto & Shibata, 2006).

2.11. Chondroitin sulfate

Chondroitin sulfate is one of the most physiologically important glycosaminoglycans, appearing in the human organism covalently linked to a protein core and, thereby, forming a proteoglycan (Malafaya et al., 2007) or as a receptor on cell or basement membranes (Alvarez-Lorenzo et al., 2013). It is a component of connective tissues (cartilage, for instance), being therefore of animal origin. The polysaccharide has a linear structure that consists of repeating disaccharide units of (1–3)-β-*N*-acetyl-D-galactosamine and

(1–4)-β-glucuronic acid, which present sulfate, hydroxyl, and carboxylic acid functionalities, exhibiting a MW of approximately 463 Da. Each monosaccharide might have between zero and two sulfate groups (Alvarez-Lorenzo et al., 2013; Kaneiwa, Mizumoto, Sugahara, & Yamada, 2010), thus providing an overall negative charge. Given the human origin, no biocompatibility issues are expected and biodegradability is also ensured by either bacteria (Chourasia & Jain, 2004; Jain, Gupta, & Jain, 2007) or specific enzymes like hyaluronidases (Kaneiwa et al., 2010).

Considering its nature, the most frequent application of chondroitin sulfate has been in the production of tissue engineering scaffolds and associated materials (Malafaya et al., 2007).

2.12. Heparin

Heparin is an animal polysaccharide exhibiting a linear structure consisting of (1–4)-linked disaccharide repeating units of uronic acid (either D-glucuronic acid or L-iduronic acid) and D-glucosamine residues (Liang & Kiick, 2014; Mizrahy & Peer, 2012). The uronic acid typically comprises an iduronic acid/glucuronic acid proportion of 90–10% (Mizrahy & Peer, 2012). It has a highly anionic charge due to the prevalence of sulfate and carboxylate groups and is characterized by a heterogeneous sulfation pattern (Liang & Kiick, 2014; Mizrahy & Peer, 2012).

With an average MW of 12 kDa (Linhardt, 2003), heparin is produced exclusively by mast cells, where it has a function in storage of granular components such as histamine (Carlsson & Kjellén, 2012), and plays a key role in the regulation of varied biological activities, such as coagulation (Mizrahy & Peer, 2012). As referred above for chondroitin sulfate, considering the human origin of heparin, no biocompatibility or biodegradability issues are expected, heparanases ensuring its decomposition (Rönnberg, Melo, & Pejler, 2012). Heparin has been used as anticoagulant since 1935 (Liang & Kiick, 2014) and has recently been proposed for other applications in drug delivery such as, for instance, the treatment of asthma (Oyarzun-Ampuero, Brea, Loza, Torres, & Alonso, 2009).

3. POLYSACCHARIDE-BASED NANOPARTICLES

3.1. Cellulose

Positively charged cellulose derivatives have been reported as matrix of nanoparticles, mainly aimed at gene delivery. Two types of hydroxyethylcellulose, polyquaternium-4 (PQ-4) and polyquaternium-10

(PQ-10), were complexed with pDNA, forming nanoparticles of approximately 200 nm. The interactions between pDNA and these cationic celluloses were shown to be stronger than those of pDNA/polyethyleneimine (PEI; Fayazpour et al., 2006). In this context, another work reported the higher ability of strongly quaternized polymers to bind DNA, in comparison with polymers exhibiting lower quaternization degree (Ding et al., 2012). Notwithstanding the strong interaction, the produced complexes exhibited much lower ability for cell transfection compared with PEI complexes, which was attributed to the neutral surface charge in the case of PQ-4/pDNA nanoparticles, as it certainly hinders cell interaction. As for PQ-10/pDNA nanoparticles, in spite of the positive charge, the strong binding of pDNA to PQ-10 possibly impedes DNA release in cytosol. According to the authors, tailoring the nature and extent of cationic side chains may further enhance DNA delivery ability (Fayazpour et al., 2006).

Quaternized cellulose (QC) derivatives with different MW and DS (Song, Sun, Zhang, Zhou, & Zhang, 2008) were also tested for gene delivery. Eight different QC derivatives were used to associate pGL-3 and pEGFP-C1 pDNA and the resulting QC/DNA complexes exhibited effective transfection in comparison to naked DNA. QCs with MW of about 8×10^4 g/mol and DS of about 0.6 displayed relatively higher transfection efficiencies, attributed to the intermediate stability. Furthermore, both QCs and QC/DNA complexes presented low cytotoxicity in human embryonic kidney transformed (293T) cells, wherein it increased slightly with MW and DS increase (Song et al., 2010). The authors further reported cellulose-based polyelectrolyte complex nanoparticles upon mixing negatively charged CMC and positively charged QC. CMC/QC nanoparticles displayed high bovine serum albumin (BSA) encapsulation efficiency, which released steadily from nanoparticles within the first 8 h in PBS pH 7.4. Nanoparticle size and zeta potential could be modulated by adjusting polymer concentrations. CMC/QC nanoparticles exhibited both low cytotoxicity in Caco-2 cells (nanoparticle concentration of 0.8 mg/ml), and significant BSA cellular uptake (Song et al., 2012).

3.2. Pectin

One sole work is available using pectin as nanoparticle matrix material for protein delivery. Pectin was applied to produce a gene carrier by ionic gelation with different cations like calcium, magnesium, and manganese. The association of DNA took place by means of electrostatic interaction as well.

Only calcium and magnesium permitted the formation of spherical nanoparticles. Moreover, magnesium provided the smaller sizes, from 300 nm to approximately 1 μm, while calcium induced the formation of particles with size between 900 nm and 1 μm. Nevertheless, it was observed that calcium-based particles have a higher ability to associate DNA, possibly due to a better interaction between both the cation and the macromolecule. pDNA encoding the green fluorescent protein (GFP) was incubated with Huh7 cells (a human hepatocellular carcinoma cell line) and it was observed that both calcium- and magnesium-based pectin nanoparticles induced low levels of transfection, as well as low toxicity (Opanasopit et al., 2008). Nevertheless, magnesium nanoparticles permitted higher transfection, possibly due to the lower size and the positive charge of the nanoparticles. The low transfection obtained in general was possibly derived from the high size of nanoparticles, well above 100 nm, which resulted in low ability to reach the intracellular compartment, the first limitation to reach the nuclei.

3.3. Starch

Several starch derivatives were proposed as nanoparticle matrix with the aim of protein delivery. Starch acetate was proposed in the ambit of oral insulin delivery. The acetylation of starch provided a hydrophobic polymer, which was further pegylated. The resultant polymer self-aggregated to form micellar nanoparticles in presence of an aqueous solution of insulin. The nanoparticles have a size around 30–50 nm and negative zeta potential (−4 to −12 mV). The hydrophobic core of starch acetate was suggested to protect insulin from the gastric environment and a release of less than 20% was observed upon 2 h in simulated gastric fluid, as shown in Fig. 2. The nanoparticles were demonstrated to be strongly mucoadhesive as compared with chitosan, which is well known for its mucoadhesion characteristics. Furthermore, depending on the PEG MW, negligible cytotoxicity levels can be achieved in Caco-2 cells (Minimol et al., 2013).

Cationic starch was used to prepare nanoparticles using an emulsification technique. The nanoparticles were loaded with either BSA or bone morphogenetic protein (BMP)-4. Nanoparticle size was found to vary according to the used oil/water ratio, the amount of Span 80 (emulsifier) and starch, and also the amount of phosphoric trichloride ($POCl_3$), used as cross-linker. These variations permitted obtaining nanoparticles with sizes varying within 70 and 500 nm, approximately. Drug loading of both proteins varied within 17% and 27%, the lower value being registered for BMP-4. This protein

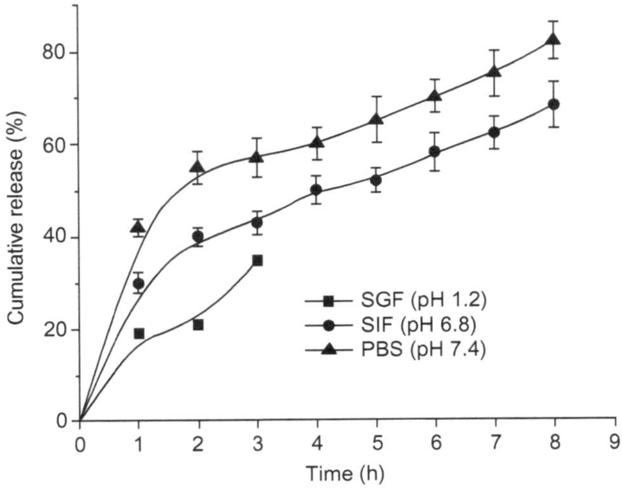

Figure 2 *In vitro* release profile of insulin from nanoparticles of SA-PEG1900 in SGF (pH 1.2), SIF (pH 6.8), and phosphate-buffered saline (pH 7.4). *Reprinted with permission from Minimol, Paul, and Sharma (2013).*

further evidenced a faster release in PBS pH 7.4 (80% in 4 h, in contrast with 58–67% of BSA-loaded formulations), which was justified by a predominant diffusional release of BMP-4, while that of BSA is mostly governed by electrostatic forces. Cytotoxicity was negligible in HeLa cells when these were exposed 24 h to a concentration of up to 0.3 mg nanoparticles/ml (Huang et al., 2013).

In a strategy of gene delivery, the copolymer hydroxyethyl starch (HES)-PEI was used to associate pDNA and form polyplexes. The idea supporting the approach was that HES could provide the controlled shielding/deshielding of polyplexes, due to its biodegradability. Deshielding was catalyzed enzymatically by amylase, which presence in a mixture of polyplexes and erythrocytes led to erythrocyte aggregation, an effect not observed in the absence of enzyme. The aggregation indicates deshielding, which exposes the positive charges of polyplexes and enables the electrostatic interaction with erythrocytes that results in aggregation. Using two distinct cell lines, Neuro2A (murine neuroblastoma) and Huh7 cells, it was demonstrated the ability of HES polyplexes to increase the transfection efficiency by 2 or 3 orders of magnitude when amylase was added to the culture medium (Noga et al., 2012).

In a different approach, nanoparticles comprised of modified starch (no specifications on the derivative were provided) were produced by a spray-drying technique using the Nano Spray Dryer from Buchi. This work was one of the first reporting the use of this equipment after its commercialization and was developed to show its potential to produce nanoparticles using different materials. The polysaccharide was sprayed at three different concentrations (0.1%, 1%, and 10%, w/v), resulting in the production of nanoparticles with sizes between 450 and 620 nm, all exhibiting a spherical shape (Li, Anton, Arpagaus, Belleteix, & Vandamme, 2010).

While in all the studies described above starch was used as nanoparticle matrix material, one study reports its application as coating of iron oxide magnetic nanoparticles. These were previously pegylated and functionalized with heparin, and further loaded with protamine. The interest in using starch as coating material relies on its biocompatibility and on the presence of several hydroxyl functional groups, which enable additional conjugations. The resulting nanoparticles (160 nm), demonstrated strong binding affinity and high protein loading, showing promise as carriers of cationic proteins and, thus, as a delivery platform (Zhang et al., 2013).

3.4. Mannan

Mannan residues are known to provide a preferential targeting to certain types of cells, including dendritic cells, macrophages, and epithelial M cells (Gupta, Gupta, & Gupta, 2009). Therefore, this material has been used to decorate drug delivery systems aimed at a specific targeting. Mannan-modified poly(ε-caprolactone)-polyethyleneglycol–poly(ε-caprolactone) nanoparticles encapsulating human basic fibroblast growth factor (bFGF) were demonstrated to induce higher immunization as compared with similar nanoparticles not modified with mannan. As observed in Fig. 3, upon subcutaneous administration to mice, IgG, IgG1, and IgG2a titers were higher, demonstrating the ability of mannan to mediate the specific targeting of dendritic cells, improving humoral immunity (Gou et al., 2008).

A modification with mannan was also used to potentiate the cell targeting ability of solid lipid nanoparticles (SLN), the system being applied in a gene delivery strategy. The nanoparticles were prepared at first and complexed with DNA. The mannan ligand consisted in a graft with phosphatidylethanolamine (PE). SLN–DNA complexes were then modified by adsorption of PE-mannan ligands. Mannan-modified complexes showed agglutination upon incubation with concanavalin A, thus demonstrating the existence of α-mannose residues at the surface of nanoparticles. No cytotoxicity was

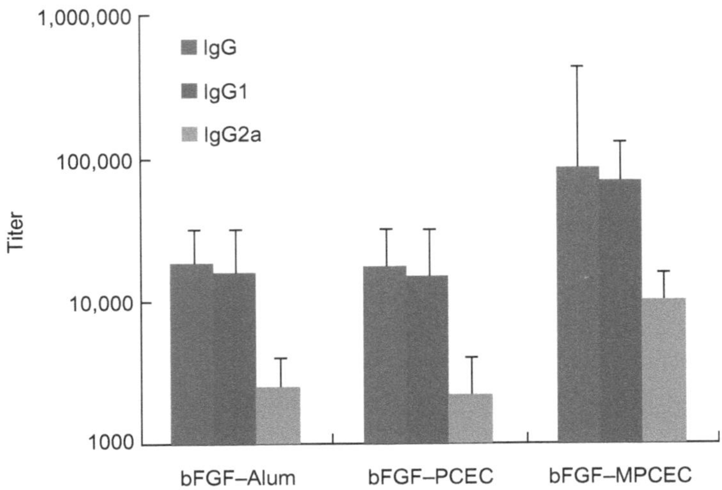

Figure 3 Serum total bFGF-special IgG, IgG1, and IgG2a titer created by bFGF–nanoparticles complex. Mice were immunized subcutaneously on week 0, 1, 2, and 3. The serum was collected on week 4. *Reprinted with permission from Gou et al. (2008).*

observed in RAW 264.7 cells (mouse leukemic monocyte macrophage cell line) and the mannan-modified nanoparticles produced higher gene expression than controls, including lipofectamine, especially *in vivo* in rats (Yu et al., 2010).

More recently, another work reported the preparation of dual-ligand-SLN, one of the ligands being PE-mannan and the other one transferrin-PEG-PE. This composition was aimed at providing carriers with the ability of targeting both cancer cells and macrophages in the liver, in a strategy of liver targeted gene delivery upon systemic administration. Evidencing absence of *in vitro* toxicity in HepG2 and Kupffer cells, the dual-modified vectors complexed with the pGFP (198 nm; gene loading of 89%) were tested in HepG2-induced tumor bearing mice. The transfection efficiency was determined to be approximately 50% higher than that obtained by unmodified SLN (Jing, Li, Liu, Wang, & Sui, 2013).

A different approach regarding gene delivery used mannan as emulsifying wax-nanoparticle coating for macrophage targeting. To demonstrate the system efficiency, nanoparticles (100 nm, positive surface charge) were incubated with J774E cells, a mannose receptor positive mouse macrophage cell line. When the incubation was performed at 37 °C (1 h), the uptake of mannan-coated nanoparticles was 50% higher than that of uncoated nanoparticles. An incubation at 4 °C for the same period resulted in twofold higher uptake for coated carriers. It was also shown that this increased uptake of coated nanoparticles was abolished by preincubating the cells with free mannan, confirming that uptake is receptor mediated (Cui, Hsu, & Mumper, 2003).

3.5. Glucomannan

Nanoparticles of carboxymethyl konjac glucomannan–chitosan were prepared by polyelectrolyte complexation. Sizes varied between 50 and 1200 nm, depending on the polymeric ratio, increasing when the concentration of polymers increases. Zeta potential was positive between +15 and +45 mV, being higher for the higher concentrations of chitosan. These physicochemical parameters were modified by changing the pH of the formulation and by adding NaCl to the formulation. The addition of either dilute HCl or NaOH results in size increase and zeta potential decrease. Similarly, adding NaCl to the formulation also leads to an increase in nanoparticle size. This is due to the destabilization caused by increased ionic strength due to electrostatic shielding. On the contrary, zeta potential decreases due to a phenomenon of double layer compression. BSA was successfully associated to these nanoparticles with efficiencies of 30–45%, depending on the specific ratios of chitosan/glucomannan (Du, Sun, et al., 2005). BSA registered a prolonged release up to 24 h in PBS pH 7.4 (80–90% BSA released). Higher concentration of chitosan resulted in slower release, which is explained by the formation of more entanglements between both polymers, resulting in stronger polymeric nets that retard BSA release. The release is much more influenced by the presence of chitosan than by the nanoparticle inner structure, and it was also observed a dependence on ionic strength (pH) and salt concentration (NaCl; Du, Zhang, et al., 2005).

Another work reported the preparation of glucomannan/chitosan nanoparticles, but this time a phosphorylated glucomannan was used as coating of preformed chitosan/tripolyphosphate nanoparticles. Nanoparticles evidenced sizes in the range of 170–300 nm and the zeta potential shifted from positive to negative upon glucomannan coating. The presence of glucomannan contributed to the stabilization of nanoparticles, preventing their aggregation in buffered medium (PBS pH 7.4). Nanoparticles only remained stable along time (up to 120 min) for the highest tested amount of glucomannan (chitosan/tripolyphosphate/glucomannan = 3/1/6.7 (w/w) in comparison with 6/1/23 (w/w)). Nanoparticles further demonstrated ability to associate the protein complex P1 (encapsulation efficiency around 25%), a mixture of polypeptides that upregulates natural and specific cell-mediated immune mechanisms (Cuña et al., 2006).

3.6. Arabic gum

Arabic gum–chitosan nanoparticles were proposed as insulin carriers for oral delivery. Taking benefit from its negative charge, arabic gum was complexed with the positively charged chitosan, resulting in nanoparticles of 150–200 nm and strong positive zeta potential, which facilitated the association of insulin (association efficiency of 25–35%). The concentration of the polymers was the parameter most affecting insulin association efficiency. The release was very rapid in pH 1.2 (approximately 90% insulin released in 1 h) and exhibited a more controlled pattern in pH 6.5 or 7.2 (30–60% in 4 h). The authors further suggested that the release resulted from several concomitant mechanisms, such as diffusion, dissolution, and relaxation of polymer chains (Avadi et al., 2010). In a subsequent work, the authors demonstrated a greater *ex vivo* absorption of insulin in rats when loaded in the nanoparticles as compared with the free protein (Avadi et al., 2011).

As described previously for starch, arabic gum nanoparticles were also produced with the Nano Spray Dryer. In this case, the polysaccharide was also sprayed at three different concentrations (0.1%, 1%, and 10%, w/v), resulting in spherical nanoparticles with a size of 350–600 nm (Li et al., 2010).

3.7. Tragacanth gum

The use of tragacanth gum in drug delivery is scarce (Kaffashi, Zandieh, & Khadiv-Parsi, 2006; Siahi, Barzegar-Jalali, Monajjemzadeh, Ghaffari, & Azarm, 2005) and its application as matrix material of nanoparticles aimed at protein delivery was reported only in one occasion, regarding the preparation of a gene delivery carrier. In that work, a water-soluble derivative of tragacanth gum (Fattahi, Petrini, et al., 2013) was complexed with an oppositely charged chitosan oligosaccharide to obtain nanoparticles. These were used to condense DNA, thus acting as gene carriers. After a screening of production conditions, a positively charged formulation (+30 mV) of 130 nm was selected. These chitosan/tragacanth gum nanoparticles evidenced better transfection efficiency in HeLa and HepG2 cells as compared with chitosan nanoparticles formulated in absence of tragacanth derivative; the difference being attributed to a weaker complexation of DNA, which facilitated its release and migration (Fattahi, Sadrjavadi, et al., 2013).

3.8. Carrageenan

Our group has recently presented several works on the preparation of nanoparticles based on chitosan and carrageenan (Grenha et al., 2010; Rodrigues, Cordeiro, Seijo, Remuñán-López, & Grenha, 2014; Rodrigues, Rosa da Costa, & Grenha, 2012). The polyelectrolyte complexation between polymers resulted in nanoparticles with a size around 600 nm and ability to associate ovalbumin. As seen in Fig. 4, the protein released according to an approximate zero order kinetics for 3 weeks in PBS pH 7.4 (Grenha et al., 2010).

Nevertheless, as nanoparticle size was out of the optimal interval for transmucosal delivery, tripolyphosphate was added to the formulation to act as cross-linker and decrease dimensions. The strategy was successful, nanoparticles evidencing sizes around 200 nm and strong positive zeta potential (around +50 mV), and the addition of tripolyphosphate was further demonstrated to endow the nanoparticles with increased storage stability (Rodrigues et al., 2012). Recently we have shown the ability of these nanoparticles to associate proteins with different characteristics (insulin and BSA) (Susana Rodrigues and Ana Grenha, unpublished results). An absence of toxicity was demonstrated in two cell lines representative of the respiratory epithelia (Calu-3 and A549 cells). This conclusion was drawn based on the results of a cytotoxicity test (MTT), the monitorization of transepithelial

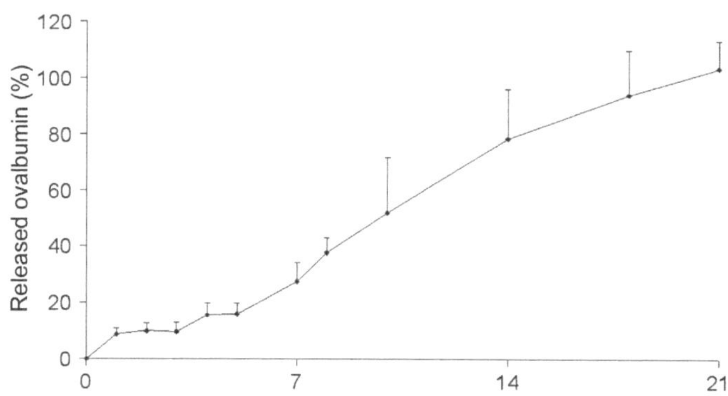

Figure 4 Ovalbumin release profile from chitosan–carrageenan (CS/CRG = 3.5/1) nanoparticles in PBS pH 7.4 at 37 °C (ovalbumin = 30% (w/w) based on chitosan; mean ± SD; $n \geq 3$). *Reprinted with permission from Grenha et al. (2010).*

electrical resistance and the study of the generated inflammatory response (Rodrigues et al., 2014). Based on these results, the nanoparticles were proposed as candidates for nasal and pulmonary transmucosal delivery of proteins.

Chitosan–carrageenan nanoparticles obtained by the same method were applied to encapsulate recombinant human erythropoietin. Depending on the process conditions (concentration of polymers, polymer mass ratio, pH of reaction medium), nanoparticles with size from 200 up to 1000 nm and zeta potential of +40–55 mV were obtained. Response surface modeling was applied to optimize nanoparticle characteristics and an encapsulation efficiency of approximately 50% was achieved. The protein released according to a controlled pattern in PBS pH 7.4, releasing 50% in 2 weeks (Bulmer, Margaritis, & Xenocostas, 2012).

Carrageenan was also used to encapsulate previously formed pDNA/PEI polyplexes. Upon encapsulation, the new carriers were called ternary complexes. The positive zeta potential of the polyplexes (+50 mV) enabled an electrostatic interaction with the sulfate groups of carrageenan, resulting in a shift of the zeta potential to −36 mV and a final size of 92 nm. Possibly because of the newly exhibited negative zeta potential, the ternary complexes provided a decreased in the cytotoxicity effect on B16-F10 cells (skin melanoma cell line). They further demonstrated to prevent the agglutination with erythrocytes that was observed for the original polyplexes. However, carrageenan-based ternary complexes have shown very little capacity for cell uptake and transfection, thus limiting the interest in the use of these carriers for gene delivery purposes (Kurosaki et al., 2009).

3.9. Fucoidan

The first reported application of fucoidan as nanoparticle matrix material regarding the delivery of protein-based drugs concerned the design of a gene carrier. As described above for carrageenan, pDNA/PEI polyplexes were also encapsulated using fucoidan to form ternary complexes. The positive zeta potential of polyplexes (+50 mV) enabled the subsequent complexation with fucoidan by electrostatic interaction, mediated by the sulfate groups. This process did not alter nanoparticle size (72 nm) but shifted the zeta potential to a strongly negative value (−27 mV). As reported for carrageenan, ternary complexes provided a decrease in the cytotoxicity effect on B16-F10 cells, further demonstrating to prevent the agglutination with erythrocytes that was observed for the original polyplexes. However, the ternary complexes have shown very little capacity for cell uptake and transfection, limiting their interest in gene delivery applications (Kurosaki et al., 2009).

Another group studied the use of fucoidan–chitosan nanoparticles in tissue engineering applications, acting as carriers of stromal cell-derived factor 1 (SDF-1), which is an important chemokine in stem cell mobilization, further playing a critical role in the biological and physiological functions of mesenchymal stem cells (MSC). Chitosan/fucoidan, chitosan/tripolyphosphate, and chitosan/tripolyphosphate/fucoidan nanoparticles were prepared. All the formulations evidenced tunable physicochemical properties (size and zeta potential) and efficiently associated the protein (encapsulation efficiency between 60% and 100%). Although chitosan/tripolyphosphate/fucoidan nanoparticles were those associating the lower amount of protein (60–68%), they evidenced the most favorable release profile in PBS pH 7.4. A burst release of approximately 20% was observed for the first 12 h, while a controlled release was then obtained up to 7 days with an accumulated release of 30%. In turn, the other formulations enabled the release of significantly lower amounts of protein (less than 10%) in the same period. SDF-1 released from these nanoparticles also evidenced higher mitogenic activity as compared with free SDF-1. In fact, MSC showed a much more intense growth upon 1 or 3 days of incubation with SDF-1-loaded nanoparticles. Moreover, the migration of MSC in response to the controlled release of SDF-1 closely follows the kinetics of SDF-1 release from nanoparticles. The carriers further protected SDF-1 against proteolysis and heat treatment, providing the means to overcome some of the major drawbacks of this protein, which are the short half-life and easy enzyme degradation (Huang & Liu, 2012). bFGF presents these same limitations and, therefore, the authors tested the fucoidan/chitosan nanoparticles for the delivery of this protein, as it has a role in nerve tissue engineering, namely having a positive effect in angiogenesis and neuronal cell survival. Nanoparticles with size around 200 nm can be modulated for zeta potential depending on the used polymeric ratios. In any case, bFGF encapsulation efficiencies were around 80–90%. As described above for SDF-1, the nanoparticles protect bFGF from heat and enzymatic degradation. Furthermore, the nanoparticles provide a controlled release of the protein (4% in 24 h), in a rate that is sufficient for neurite extension, as evidenced upon incubation with PC-12 cells (derived from a pheochromocytoma of the rat adrenal medulla). This permits using a low amount of bFGF for neurite extension and also a low concentration of nanoparticles. In fact, a cytotoxicity test in this cell line evidenced the absence of toxicity of all the nanoparticle formulations (either negatively or positively charged) up to 72 h, which is possibly due to the low concentration of nanoparticles used in the study (125 ng/ml; Huang & Yang, 2013). Fucoidan/chitosan nanoparticles thus showed potential for both stem cell mobilization and nerve tissue engineering.

Finally, the most recent report involves the preparation of fucoidan/chitosan nanoparticles by means of electrostatic interaction, as reported in the other cases, but assisted by ultrasonic energy. From a total of nine tested conditions, only in one case submicron particles were obtained, displaying a size of 920 nm. BSA was easily encapsulated in the nanoparticles (encapsulation efficiency ranging between 46% and 89%) and released according to a controlled profile, reaching 80% cumulative release in 96 h (Liu et al., 2014). Although the association of the protein and the release profile is interesting, these nanoparticles present the great disadvantage of requiring the use of aggressive conditions that are not applied in a simple polyelectrolyte complexation. Additionally, notwithstanding the higher energy that was used, the nanoparticles are almost out of the size range to be classified as such, with the consequent loss of favorable properties, such as for instance the ability to have an intimate interaction with mucosal epithelia or tissues.

3.10. Pullulan

Pullulan has been applied as nanocarrier matrix material regarding varied objectives, including immunization, gene delivery, and also the design of imaging tools. Pullulan exhibits high flexibility in aqueous solutions and its chemical structure can be modified by hydrophobic molecules, such as cholesterol (Rekha & Sharma, 2011). Cholesterol-bearing pullulan self-aggregates in nanoparticles (Akiyoshi et al., 1998; Lee & Akiyoshi, 2004) which have been reported as protein carriers. Insulin was tested in this regard, producing stable complexes (1 week, room temperature) which provide protection to the protein, preventing enzymatic (α-chymotrypsin) and thermal (90 °C) denaturation. Although no significant difference was found on glycemic levels upon IV administration to rats, in comparison with free insulin, it was demonstrated that insulin retained better its biological activity when administered in encapsulated form (Akiyoshi et al., 1998). The referred nanoparticles, however, evidence low intracellular uptake due to the neutral charge of pullulan. To improve this ability, cholesterol–pullulan was functionalized with ethylenediamine, endowing the final nanoparticles with positively charged surface (nanoparticle size was <50 nm). Different amounts of amino groups were tested and fluorescently labeled BSA (FITC-BSA) was used as model protein. Upon incubation of protein-loaded nanoparticles with HeLa cells, the uptake of FITC-BSA increased over 6 h, in a direct proportion of the content of amino groups. Importantly, no cytotoxic effect was observed at final nanoparticle concentration of 100 nM for up to 24 h (Ayame, Morimoto, & Akiyoshi, 2008). The same nanoparticles were used to associate quantum dots (QDs) to provide the release of those as cell imaging tools. Effective intracellular delivery systems of QDs are needed to enable cell application without interference with cellular functions. The association of a protein to QDs and the encapsulation of this QD–protein complex in ethylenediamine-functionalized cholesterol–pullulan nanoparticles permitted cellular internalization of QDs. In fact, nanoparticles remained detectable inside rabbit MSC for at least 2 weeks, with little effect on their *in vitro* chondrogenic ability. These systems were thus proposed as candidate cell tracers in tissue engineering (Toita et al., 2008).

The cholesterol–pullulan nanoparticles were also proposed for nasal vaccination, using *Clostridium botulinum* type-A neurotoxin (BoHc/A) as antigen. Nanoparticles prepared with conventional cholesterol–pullulan demonstrated no improved *in vivo* results (mice) as compared with naked antigen and, thus, the authors synthesized a cationic derivative containing 15 amino groups per 100 glucose units. Upon successful *in vitro* (HeLa cells) internalization of nanoparticles prepared with this derivative, an *in vivo* imaging study showed the effective intranasal delivery of nanoparticles carrying T^{18}FU-labeled BoHc/A which remained continuously retained by the nasal mucosa, while in the absence of nanoparticles, the antigen disappeared from the nasal cavity within 6 h. IgA titers were only detected upon treatment with antigen-loaded nanoparticles and not with non-encapsulated antigen. Additionally, it was shown that no transition into the olfactory bulbs or brain occurs over a 2-day period post-administration, indicating the safety of the carrier regarding dissemination to the central nervous system (Nochi et al., 2010).

Cholesterol–pullulan-based nanoparticles were also proposed in the ambit of cancer immunization. The overexpression of HER2 oncoprotein has been related with several adenocarcinomas (breast, ovarian, lung, gastric) and HER2-based peptides are known to induce CD8+ cytotoxic T lymphocytes (CTLs). These cells are the major *in vivo* effector cells for the rejection of HER2+ tumors. As evidenced in Fig. 5, HER2-loaded cholesterol–pullulan nanoparticles demonstrated *in vivo* ability to induce

CD8+ CTLs against HER2+ tumors and the complete rejection of tumors was observed when nanoparticles were administered early after tumor implantation in mice (up to day 3; Shiku et al., 2000).

Pullulan–PEI nanoparticles were proposed as gene carriers (size < 100 nm), benefiting from the positive charge attributed by PEI. Different degrees of PEI conjugation were tested, but only the lowest degree permitted acceptable HepG2 cell viability (above 80%). The conjugation of pullulan with PEI did not hinder the plasmid nuclear localization of the latter. Additionally, using pGFP and pGL3, the transfection efficiency provided by pullulan–PEI nanoparticles was similar to that obtained for PEI, but the former evidenced improved hemocompatibility (Rekha & Sharma, 2011).

Our group has also proposed pullulan nanoparticles as protein carriers for nasal and pulmonary transmucosal delivery. Upon synthesizing charged derivatives of pullulan, the aminated derivative was complexed with carrageenan and the sulfated derivative with chitosan. Positively charged nanoparticles (180–270 nm) were obtained which evidenced ability to associate BSA (34–44%). The protein showed a burst release of 30% in PBS pH 7.4, which maintains for up to 24 h. No cytotoxic effect was observed for nanoparticles upon exposure to Calu-3 cells (cell viability around 80–90%), a respiratory cell line representative of both pulmonary and nasal epithelia (Dionísio et al., 2013).

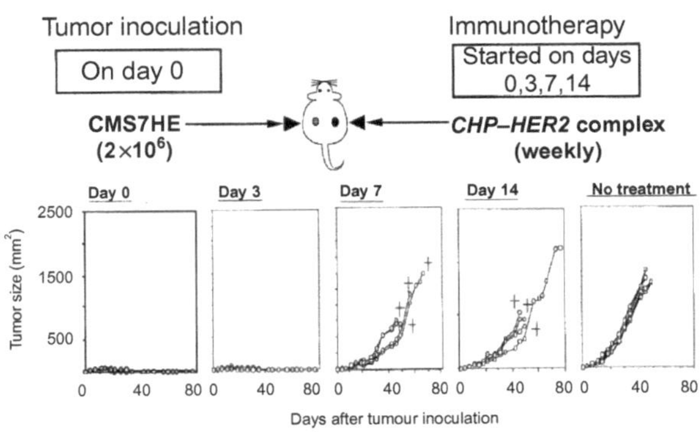

Figure 5 Therapeutic effects of CHP–HER2 vaccine. BALB/c mice were challenged with CMS7HE 2×10^6 sc and weekly received CHP–HER2 complex containing protein 20 μg starting on the day of challenge, or 3, 7, or 14 days after. Each group consisted of four mice. Each line represents a single mouse. *Reprinted with permission from Shiku et al. (2000).*

3.11. Chondroitin sulfate

The pDNA/PEI polyplexes referred above in the sections addressing fucoidan and carrageenan were also tested with chondroitin sulfate to obtain different ternary complexes. The polyplexes interacted with the sulfate groups of the polymer, inducing a change of the zeta potential to −29 mV. Nanoparticle size is 77 nm. The high agglutination activity with erythrocytes and the cytotoxicity observed for pDNA/PEI complexes in B16-F10 cells were reversed upon formation of this ternary complex. Actually, from all the tested materials, which include not only carrageenan and fucoidan reported herein, but also alginate and hyaluronic acid, only the ternary complex formed with chondroitin showed comparable transgene efficiency to the pDNA/PEI complex. The uptake and gene expression of this ternary complex were significantly inhibited by hypothermia and the addition of chondroitin sulfate, suggesting that the carrier was taken by chondroitin sulfate-specific receptor-mediated energy-dependent process (Kurosaki et al., 2009).

The first report to chitosan/chondroitin sulfate nanoparticles prepared by a simple polyelectrolyte complexation appeared in 2011, proposing these nanoparticles as protein carriers. FITC-BSA was the model protein (association efficiency up to 90%) and the nanoparticles exhibited sizes between 150 and 500 nm, depending on the polymer ratios and the amount of protein initially added. The protein released progressively in PBS pH 7.4, reaching approximately 70% release in 90 min. Uptake studies performed in Caco-2 and HEK293 cells indicated the occurrence of nanoparticle endocytosis while, as shown in Fig. 6, cytotoxicity tests in Caco-2 cells evidenced absence of toxicity after 24 h incubation, up to a concentration of 0.1 mg/ml (Yeh et al., 2011). However, this might be considered a low concentration to test.

A similar work also used FITC-BSA as model protein and demonstrated that varying polymer mass ratios provided different sizes and zeta potentials, the latter suffering a complete shift from positive to negative when chondroitin sulfate was present in higher amount. Nanoparticles of 250 nm and around −20 mV were selected for subsequent assays. A cytotoxicity assay performed on Caco-2 cells led to viabilities around 95% after 72 h incubation with FITC-BSA-loaded nanoparticles, but again the maximum concentration tested was only 0.1 mg/ml, which is considerably lower than usual concentrations reported in other studies available on the literature. Transport studies performed in the same cell line revealed the transport of

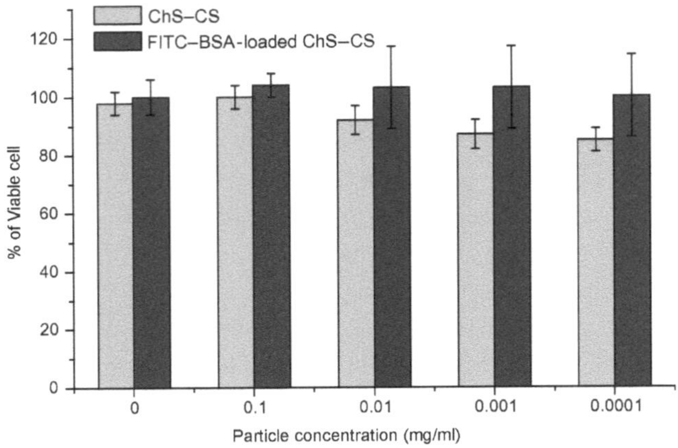

Figure 6 Cytotoxicity profile of ChS–CS and FITC–BSA-loaded ChS–CS NPs after 24 h incubation with Caco-2 cell. Percent viability of fibroblasts is expressed relative to control cells ($n = 6$). *Adapted with permission from Yeh, Cheng, Hu, Huang, and Young (2011).*

nanoparticles into the cells by a mechanism of endocytosis. It was further observed that positively charged nanoparticles had a more favorable transport than those exhibiting a negative charge (Hu, Chiang, Hong, & Yeh, 2012).

Similar nanoparticles were used for the delivery of platelet lysates in bone regenerative medicine. With a size around 200 nm, a first approach consisted in using BSA as model protein, which exhibited a controlled release in PBS pH 7.4 for up to 4 weeks (30–50% release, depending on formulation). The incubation with human adipose-derived stem cells (hASCs) evidenced an absence of toxicity and also demonstrated the ability of nanoparticles to be taken up by the cells. The latter effect is proportionally dependent on nanoparticle concentration and was maximized at 24 h, decreasing after that. Platelet lysates were successfully adsorbed onto the surface of nanoparticles. hASCs cultured in presence of these carriers showed enhanced cell viability, proliferation, and calcium deposition, suggesting increased osteogenic differentiation. The authors further proposed the possibility of using the carriers to replace conventional serum used in cell culture, as platelet lysate is as an autologous source of proteins, thereby permitting avoiding the risk of immunogenic responses (Santo, Gomes, Mano, & Reis, 2012).

Chondroitin sulfate was also used to form a nanocomplex (150–250 nm) with BMP-2. The nanocomplex was then incorporated in titanium-hydroxyapatite surface. Immobilized BMP-2 evidenced a controlled release up to 4 weeks. Titanium-hydroxyapatite surfaces provided faster proliferation of mouse osteoblast cells when the nanocomplex was present (Fig. 7). Additionally, the expression of osteocalcin and type I collagen, bone-specific markers, was significantly upregulated by the use of the nanocomplex. It was thus demonstrated that titanium surfaces can hold both osteoconductive apatite and osteoinductive growth factor and, therefore, the authors proposed the application of the system in dental and orthopedic implants (Bae et al., 2012).

3.12. Heparin

Heparin was associated to chitosan to produce nanoparticles by polyelectrolyte complexation. These were used as carriers of bFGF and were further incorporated inside a collagen membrane, in a strategy for tissue repair. Although bFGF was the target molecule for encapsulation and design of the system, human serum albumin (HSA) was also tested as model biomolecule, being used to perform the release assay. Nanoparticles displayed a size around 150 nm and a negative zeta potential (−39 mV). HSA exhibited a controlled release profile in PBS pH 7.4 for up to 20 days, in an assay performed directly from nanoparticles in absence of the membrane (Li et al., 2011). Although this is a work involving heparin nanoparticles, most of

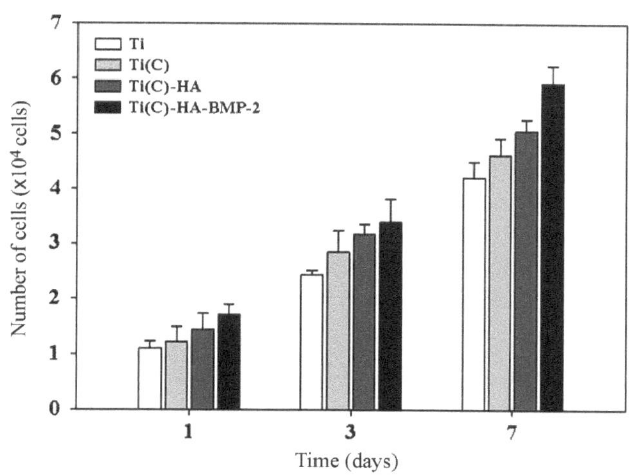

Figure 7 Cell proliferation on the different Ti substrates. As the number of cells was increasing with time, they were more proliferative on the Ti(C)-HA-BMP-2 than those on the other substrates. *Adapted with permission from Bae, Choi, Joung, Park, and Han (2012).*

the observed effects are ascribed to the nanoparticle associated collagen membranes.

Only one other work reports the use of heparin in the nanoparticle matrix, as a functionalization of magnetic nanoparticles. Owing to heparin negative charge, it acts as a linking moiety for cationic molecules, as is the case of protamine. Heparin-functionalized magnetic nanoparticles were pegylated and coated with starch, being proposed as a protein delivery platform. Protamine was used as model protein and a loading of 22.9 μg protamine/mg of Fe was achieved. These nanoparticles have shown a half-life of 9.37 h, 37.5-fold higher than commercially available pegylated magnetic nanoparticles, and they further demonstrated a superior tumor targeting ability (Zhang et al., 2013).

4. CONCLUSIONS AND FUTURE PERSPECTIVES

In the past decades, excipients have been assuming relevant roles on pharmaceutical formulation, endowing drug carriers with specialized functions and leaving aside the traditional concept that describes excipients as inert substances that simply aid on the formulation preparation. Polysaccharides are one of the material classes finding a greater number of applications in drug delivery and they are particularly envisaged in nanoparticle formulations. Nevertheless, a restrict number of these polymers (chitosan, hyaluronic acid, alginate, etc.) centralizes the attentions, while many other polymers remain quite unexplored regarding their abilities to compose the matrix of nanocarriers. This review shows that many of these less explored substances display great potential in protein delivery, when integrating nanoparticle matrix. Twelve polysaccharides were considered in this review, complying with the requisite of being a not very well-known material regarding the preparation of nanoparticles aimed at protein delivery. From this set of polymers, pullulan, chondroitin sulfate, and starch (in this order) were those having the higher numbers of reports on the mentioned objective. It was very interesting to observe that, by far, the specific application to which more works were dedicated is gene delivery, indicating that polysaccharides definitely might be an added value in this regard. The systemic delivery of proteins as a general approach was the second application in line, in this case BSA being the protein model tested in most cases, although insulin was used in several works as well. Some works approached strategies on tissue engineering, regarding the delivery of growth factors. Other applications gathering attention were immunization and the specific targeting of cells (cancer cells, macrophages, dendritic cells, etc.) that might be mediated by polysaccharides like pullulan or mannan.

As a whole, it is suggested that several of the presented polysaccharides worth a deeper exploration, particularly when specific niches of application can be found, which is the case for instance of cell targeting. Nevertheless, it is obviously considered that substantial amount of research remains to be conducted to unveil the real potentialities.

ACKNOWLEDGMENTS

The authors acknowledge the Fundação para a Ciência e Tecnologia through research projects PTDC/SAU-FCF/100291/2008 and PTDC/DTP-FTO/0094/2012, and strategic project PEst-OE/EQB/LA0023/2014.

REFERENCES

Akiyoshi, K., Kobayashi, S., Shichibe, S., Mix, D., Baudys, M., Wan Kim, S., et al. (1998). Self-assembled hydrogel nanoparticle of cholesterol-bearing pullulan as a carrier of protein drugs: Complexation and stabilization of insulin. *Journal of Controlled Release*, *54*(3), 313–320.

Ali, B. H., Ziada, A., & Blunden, G. (2009). Biological effects of gum arabic: A review of some recent research. *Food and Chemical Toxicology*, *47*(1), 1–8.

Alonso-Sande, M., Teijeiro-Osorio, D., Remuñán-López, C., & Alonso, M. J. (2009). Glucomannan, a promising polysaccharide for biopharmaceutical purposes. *European Journal of Pharmaceutics and Biopharmaceutics*, *72*(2), 453–462.

Alvarez-Lorenzo, C., Blanco-Fernandez, B., Puga, A. M., & Concheiro, A. (2013). Crosslinked ionic polysaccharides for stimuli-sensitive drug delivery. *Advanced Drug Delivery Reviews*, *65*(9), 1148–1171.

Anastyuk, S. D., Shevchenko, N. M., Nazarenko, E. L., Dmitrenok, P. S., & Zvyagintseva, T. N. (2009). Structural analysis of a fucoidan from the brown alga *Fucus evanescens* by MALDI-TOF and tandem ESI mass spectrometry. *Carbohydrate Research*, *344*(6), 779–787.

Antosova, Z., Mackova, M., Kral, V., & Macek, T. (2009). Therapeutic application of peptides and proteins: Parenteral forever? *Trends in Biotechnology*, *27*(11), 628–635.

Anwunobi, A., & Emeje, M. (2011). Recent applications of natural polymers in nanodrug delivery. *Journal of Nanomedicine & Nanotechnology*, *S4*, 2–7.

Aspinall, G. O., & Baillie, J. (1963). Gum tragacanth. Part I. Fractionation of the gum and the structure of tragacanthic acid. *Journal of the Chemical Society (Resumed)*, *318*, 1702–1714.

Avadi, M. R., Sadeghi, A. M., Dounighi, N., Dinarvand, R., Atyabi, F., & Rafiee-Tehrani, M. (2011). Ex vivo evaluation of insulin nanoparticles using chitosan and arabic gum. *International Scholarly Research Network*, *2011*, 1–6.

Avadi, M. R., Sadeghi, A. M., Mohammadpour, N., Abedin, S., Atyabi, F., Dinarvand, R., et al. (2010). Preparation and characterization of insulin nanoparticles using chitosan and Arabic gum with ionic gelation method. *Nanomedicine: Nanotechnology, Biology and Medicine*, *6*, 58–63.

Ayame, H., Morimoto, N., & Akiyoshi, K. (2008). Self-assembled cationic nanogels for intracellular protein delivery. *Bioconjugate Chemistry*, *19*(4), 882–890.

Bae, S. E., Choi, J., Joung, Y. K., Park, K., & Han, D. K. (2012). Controlled release of bone morphogenetic protein (BMP)-2 from nanocomplex incorporated on hydroxyapatite-formed titanium surface. *Journal of Controlled Release*, *160*(3), 676–684.

Balboa, E. M., Conde, E., Moure, A., Falqué, E., & Domínguez, H. (2013). In vitro antioxidant properties of crude extracts and compounds from brown algae. *Food Chemistry, 138*(2–3), 1764–1785.

Barbeyron, T., Michel, G., Potin, P., Henrissat, B., & Kloareg, B. (2000). ι-Carrageenases constitute a novel family of glycoside hydrolases, unrelated to that of κ-carrageenases. *Journal of Biological Chemistry, 275*, 35499–35505.

Beneke, C., Viljoen, A., & Hamman, J. (2009). Polymeric plant-derived excipients in drug delivery. *Molecules, 14*, 2602–2620.

Berteau, O., & Mulloy, B. (2003). Sulfated fucans, fresh perspectives: Structures, functions, and biological properties of sulfated fucans and an overview of enzymes active toward this class of polysaccharide. *Glycobiology, 13*(6), 29R–40R.

Bixler, H. (1994). The carrageenan connection IV. *British Food Journal, 96*, 12–17.

Buchala, A. J., Franz, G., & Meier, H. (1974). A glucomannan from the tubers of *Orchis morio*. *Phytochemistry, 13*(1), 163–166.

Bulmer, C., Margaritis, A., & Xenocostas, A. (2012). Encapsulation and controlled release of recombinant human erythropoietin from chitosan-carrageenan nanoparticles. *Current Drug Delivery, 9*, 527–537.

Carlsson, P., & Kjellén, L. (2012). Heparin biosynthesis. In R. Lever, B. Mulloy, & C. P. Page (Eds.), *Vol. 207. Heparin—A century of progress* (pp. 23–41). Berlin: Springer.

Casettari, L., & Illum, L. (2014). Chitosan in nasal delivery systems for therapeutic drugs. *Journal of Controlled Release, 190*, 189–200.

Chauhan, P., Puri, N., Sharma, P., & Gupta, N. (2012). Mannanases: Microbial sources, production, properties and potential biotechnological applications. *Applied Microbiology and Biotechnology, 93*, 1817–1830.

Chen, H., Leipprandt, J. R., Traviss, C. E., Sopher, B. L., Jones, M. Z., Cavanagh, K. T., et al. (1995). Molecular cloning and characterization of bovine β-mannosidase. *Journal of Biological Chemistry, 270*(8), 3841–3848.

Cheng, K.-C., Demirci, A., & Catchmark, J. (2011). Pullulan: Biosynthesis, production, and applications. *Applied Microbiology and Biotechnology, 92*(1), 29–44.

Chiellini, F., Piras, A. M., Errico, C., & Chiellini, E. (2008). Micro/nanostructured polymeric systems for biomedical and pharmaceutical applications. *Nanomedicine, 3*(3), 367–393.

Chourasia, M., & Jain, S. (2004). Polysaccharides for colon targeted drug delivery. *Drug Delivery, 11*, 129–148.

Copetti, G., Grassi, M., Lapasin, R., & Pricl, S. (1997). Synergistic gelation of xanthan gum with locust bean gum: A rheological investigation. *Glycoconjugate Journal, 14*, 951–961.

Csaba, N., Garcia-Fuentes, M., & Alonso, M. J. (2006). The performance of nanocarriers for transmucosal drug delivery. *Expert Opinion on Drug Delivery, 3*(4), 463–478.

Cui, Z., Hsu, C.-H., & Mumper, R. J. (2003). Physical characterization and macrophage cell uptake of mannan-coated nanoparticles. *Drug Development and Industrial Pharmacy, 29*(6), 689–700.

Cuña, M., Alonso-Sande, M., Remuñán-López, C., Pivel, J. P., Alonso-Lebrero, J. L., & Alonso, M. J. (2006). Development of phosphorylated glucomannan-coated chitosan nanoparticles as nanocarriers for protein delivery. *Journal of Nanoscience and Nanotechnology, 6*, 2887–2895.

Davé, V., & McCarthy, S. (1997). Review on konjac glucomannan. *Journal of Environmental Polymer Degradation, 5*(4), 237–241.

d'Ayala, G. G., Malinconico, M., & Laurienzo, P. (2008). Marine derived polysaccharides for biomedical applications: Chemical modification approaches. *Molecules, 13*, 2069–2106.

de la Fuente, M., Csaba, N., Garcia-Fuentes, M., & Alonso, M. J. (2008). Nanoparticles as protein and gene carriers to mucosal surfaces. *Nanomedicine, 3*, 845–857.

Desai, M. P., Labhasetwar, V., Amidon, G. L., & Levy, R. J. (1996). Gastrointestinal uptake of biodegradable microparticles: Effect of particle size. *Pharmaceutical Research, 13*(12), 1838–1845.

Ding, F., You, J., Weng, X., Zhou, J., Zhang, X., Zhou, X., et al. (2012). Exploring quaternized hydroxyethylcellulose as potential gene carriers. *Chinese Journal of Chemistry, 30*(9), 2212–2218.

Dionísio, M., Cordeiro, C., Remuñán-López, C., Seijo, B., Rosa da Costa, A., & Grenha, A. (2013). Pullulan-based nanoparticles as carriers for transmucosal protein delivery. *European Journal of Pharmaceutical Sciences, 50*, 102–113.

Dionísio, M., & Grenha, A. (2012). Locust bean gum: Exploring its potential for biopharmaceutical applications. *Journal of Pharmacy & Bioallied Sciences, 4*(3), 75–85.

Dombu, C. Y., & Betbeder, D. (2013). Airway delivery of peptides and proteins using nanoparticles. *Biomaterials, 34*(2), 516–525.

Du, J., Sun, R., Zhang, S., Zhang, L. F., Xiong, C. D., & Peng, Y. X. (2005). Novel polyelectrolyte carboxymethyl konjac glucomannan-chitosan nanoparticles for drug delivery. I. Physicochemical characterization of the carboxymethyl konjac glucomannan-chitosan nanoparticles. *Biopolymers, 78*, 1–8.

Du, J., Zhang, S., Sun, R., Zhang, L. F., Xiong, C. D., & Peng, Y. X. (2005). Novel polyelectrolyte carboxymethyl konjac glucomannan-chitosan nanoparticles for drug delivery. II. Release of albumin in vitro. *Journal of Biomedical Materials Research. Part B, Applied Biomaterials, 72*, 299–304.

Elsabahy, M., & Wooley, K. L. (2012). Design of polymeric nanoparticles for biomedical delivery applications. *Chemical Society Reviews, 41*(7), 2545–2561. http://dx.doi.org/10.1039/C2CS15327K.

Fattahi, A., Petrini, P., Munarin, F., Shokoohinia, Y., Golozar, M., Varshosaz, J., et al. (2013). Polysaccharides derived from tragacanth as biocompatible polymers and gels. *Journal of Applied Polymer Science, 129*(4), 2092–2102.

Fattahi, A., Sadrjavadi, K., Golozar, M. A., Varshosaz, J., Fathi, M.-H., & Mirmohammad-Sadeghi, H. (2013). Preparation and characterization of oligochitosan–tragacanth nanoparticles as a novel gene carrier. *Carbohydrate Polymers, 97*(2), 277–283.

Fayazpour, F., Lucas, B., Alvarez-Lorenzo, C., Sanders, N. N., Demeester, J., & De Smedt, S. C. (2006). Physicochemical and transfection properties of cationic hydroxyethylcellulose/DNA nanoparticles. *Biomacromolecules*, 7(10), 2856–2862.

FDA. (2014). *GRAS substances database*. Available from, http://www.fda.gov/Food/IngredientsPackagingLabeling/GRAS/SCOGS/default.htm.

Gao, S., & Nishinari, K. (2004). Effect of deacetylation rate on gelation kinetics of glucomannan. *Colloids and Surfaces. B, Biointerfaces, 38*, 241–249.

Ghorbani Gorji, S., Ghorbani Gorji, E., Mohammadifar, M. A., & Zargaraan, A. (2014). Complexation of sodium caseinate with gum tragacanth: Effect of various species and rheology of coacervates. *International Journal of Biological Macromolecules, 67*, 503–511.

Ghosh, D. (2007). Human sulfatases: A structural perspective to catalysis. *Cellular and Molecular Life Sciences, 64*, 2013–2022.

Gou, M., Dai, M., Li, X., Yang, L., Huang, M., Wang, Y., et al. (2008). Preparation of mannan modified anionic PCL–PEG–PCL nanoparticles at one-step for bFGF antigen delivery to improve humoral immunity. *Colloids and Surfaces. B, Biointerfaces, 64*(1), 135–139.

Grenha, A., Gomes, M. E., Rodrigues, M., Santo, V. E., Mano, J. F., Neves, N. M., et al. (2010). Development of new chitosan/carrageenan nanoparticles for drug delivery applications. *Journal of Biomedical Materials Research. Part A, 92A*(4), 1265–1272.

Gupta, A., Gupta, R., & Gupta, G. (2009). Targeting cells for drug and gene delivery: Emerging applications of mannans and mannan binding lectins. *Journal of Scientific and Industrial Research, 68*, 465–483.

Hägglund, P., Sabini, E., Boisset, C., Wilson, K., Chanzy, H., & Stålbrand, H. (2001). Degradation of aannan I and II crystals by fungal endo-β-1,4-mannanases and a β-1,4-mannosidase studied with transmission electron microscopy. *Biomacromolecules, 2*(3), 694–699.

Hu, C., Chiang, C., Hong, P., & Yeh, M. (2012). Influence of charge on FITC-BSA-loaded chondroitin sulfate-chitosan nanoparticles upon cell uptake in human Caco-2 cell monolayers. *International Journal of Nanomedicine*, 7, 4861–4872.

Huang, Y.-C., & Lam, U.-I. (2011). Chitosan/fucoidan pH sensitive nanoparticles for oral delivery system. *Journal of the Chinese Chemical Society, 58*, 779–785.

Huang, Y.-C., & Liu, T.-J. (2012). Mobilization of mesenchymal stem cells by stromal cell-derived factor-1 released from chitosan/tripolyphosphate/fucoidan nanoparticles. *Acta Biomaterialia, 8*(3), 1048–1056.

Huang, Y.-C., & Yang, Y.-T. (2013). Effect of basic fibroblast growth factor released from chitosan–fucoidan nanoparticles on neurite extension. *Journal of Tissue Engineering and Regenerative Medicine.* http://dx.doi.org/10.1002/term.1752 (in press).

Huang, Y., Liu, M., Gao, C., Yang, J., Zhang, X., Zhang, X., et al. (2013). Ultra-small and innocuous cationic starch nanospheres: Preparation, characterization and drug delivery study. *International Journal of Biological Macromolecules, 58*, 231–239.

Jain, A., Gupta, Y., & Jain, S. (2007). Perspectives of biodegradable natural polysaccharides for site-specific drug delivery to the colon. *Journal of Pharmacy and Pharmaceutical Sciences, 10*(1), 86–128.

Janes, K. A., Calvo, P., & Alonso, M. J. (2001). Polysaccharide colloidal particles as delivery systems for macromolecules. *Advanced Drug Delivery Reviews, 47*(1), 83–97.

Jing, F., Li, J., Liu, D., Wang, C., & Sui, Z. (2013). Dual ligands modified double targeted nano-system for liver targeted gene delivery. *Pharmaceutical Biology, 51*(5), 643–649.

Jobling, S. (2004). Improving starch for food and industrial applications. *Current Opinion in Plant Biology*, 7(2), 210–218.

Kaffashi, B., Zandieh, A., & Khadiv-Parsi, P. (2006). Drug release study of systems containing the tragacanth and collagen composite: Release characterization and viscoelastic measurements. *Macromolecular Symposia, 239*(1), 120–129.

Kaneiwa, T., Mizumoto, S., Sugahara, K., & Yamada, S. (2010). Identification of human hyaluronidase-4 as a novel chondroitin sulfate hydrolase that preferentially cleaves the galactosaminidic linkage in the trisulfated tetrasaccharide sequence. *Glycobiology, 20*(3), 300–309.

Karewicz, A., Bielska, D., & Nowakowska, M. (2012). Modified polysaccharides as versatile materials in controlled delivery of anti-degenerative agents. *Current Pharmaceutical Design, 18*, 2518–2535.

Kurosaki, T., Kitahara, T., Kawakami, S., Nishida, K., Nakamura, J., Teshima, M., et al. (2009). The development of a gene vector electrostatically assembled with a polysaccharide capsule. *Biomaterials, 30*(26), 4427–4434.

Lavoine, N., Desloges, I., Dufresne, A., & Bras, J. (2012). Microfibrillated cellulose—Its barrier properties and applications in cellulosic materials: A review. *Carbohydrate Polymers, 90*(2), 735–764.

Lee, I., & Akiyoshi, K. (2004). Single molecular mechanics of a cholesterol-bearing pullulan nanogel at the hydrophobic interfaces. *Biomaterials, 25*(15), 2911–2918.

Li, X., Anton, N., Arpagaus, C., Belleteix, F., & Vandamme, T. (2010). Nanoparticles by spray drying using innovative new technology: The Büchi Nano Spray Dryer B-90. *Journal of Controlled Release, 147*, 304–310.

Li, H., Bian, S., Huang, Y., Liang, J., Fan, Y., & Zhang, X. (2014). High drug loading pH-sensitive pullulan-DOX conjugate nanoparticles for hepatic targeting. *Journal of Biomedical Materials Research. Part A, 102*(1), 150–159.

Li, B., Lu, F., Wei, X., & Zhao, R. (2008). Fucoidan: Structure and bioactivity. *Molecules, 13*, 1671–1695.

Li, X., Wang, J., Su, G., Zhou, Z., Shi, J., Liu, L., et al. (2011). Spatiotemporal control over growth factor delivery from collagen-based membrane. *Journal of Biomedical Materials Research. Part A, 100A*(2), 396–405.

Li, H., Yu, Y., Faraji-Dana, S., Li, B., Lee, C., & Kang, L. (2013). Novel engineered systems for oral, mucosal and transdermal drug delivery. *Journal of Drug Targeting, 21*(7), 611–629.

Liang, Y., & Kiick, K. L. (2014). Heparin-functionalized polymeric biomaterials in tissue engineering and drug delivery applications. *Acta Biomaterialia, 10*(4), 1588–1600.

Linhardt, R. J. (2003). 2003 Claude S. Hudson Award address in carbohydrate chemistry. Heparin: Structure and activity. *Journal of Medicinal Chemistry, 46*(13), 2551–2564.

Liu, Z., Jiao, Y., Wang, Y., Zhou, C., & Zhang, Z. (2008). Polysaccharides-based nanoparticles as drug delivery systems. *Advanced Drug Delivery Reviews, 60*(15), 1650–1662.

Liu, Y., Yao, W., Wang, S., Geng, D., Zheng, Q., & Chen, A. (2014). Preparation and characterization of fucoidan-chitosan nanospheres by the sonification method. *Journal of Nanoscience and Nanotechnology, 14*, 3844–3849.

Malafaya, P., Silva, G., & Reis, R. (2007). Natural–origin polymers as carriers and scaffolds for biomolecules and cell delivery in tissue engineering application. *Advanced Drug Delivery Reviews, 59*, 207–233.

Mangalam, A. P., Simonsen, J., & Benight, A. S. (2009). Cellulose/DNA hybrid nanomaterials. *Biomacromolecules, 10*(3), 497–504.

Minimol, P. F., Paul, W., & Sharma, C. P. (2013). PEGylated starch acetate nanoparticles and its potential use for oral insulin delivery. *Carbohydrate Polymers, 95*(1), 1–8.

Mizrahy, S., & Peer, D. (2012). Polysaccharides as building blocks for nanotherapeutics. *Chemical Society Reviews, 41*, 2623–2640.

Moon, R. J., Martini, A., Nairn, J., Simonsen, J., & Youngblood, J. (2011). Cellulose nanomaterials review: Structure, properties and nanocomposites. *Chemical Society Reviews, 40*(7), 3941–3994.

Moura, L. I. F., Dias, A. M. A., Carvalho, E., & de Sousa, H. C. (2013). Recent advances on the development of wound dressings for diabetic foot ulcer treatment—A review. *Acta Biomaterialia, 9*(7), 7093–7114.

Müller, A., Ni, Z., Hessler, N., Wesarg, F., Müller, F. A., Kralisch, D., et al. (2013). The biopolymer bacterial nanocellulose as drug delivery system: Investigation of drug loading and release using the model protein albumin. *Journal of Pharmaceutical Sciences, 102*(2), 579–592.

Namazi, H., Fathi, F., & Heydari, A. (2011). Nanoparticles based on modified polysaccharides. In A. Hashim (Ed.), *The delivery of nanoparticles* (pp. 149–184). Croatia: Intech.

Necas, J., & Bartosikova, L. (2013). Carrageenan: A review. *Veterinární Medicína, 58*(4), 187–205.

Nochi, T., Yuki, Y., Takahashi, H., Sawada, S.-I., Mejima, M., Kohda, T., et al. (2010). Nanogel antigenic protein-delivery system for adjuvant-free intranasal vaccines. *Nature Materials, 9*(7), 572–578.

Noga, M., Edinger, D., Rödl, W., Wagner, E., Winter, G., & Besheer, A. (2012). Controlled shielding and deshielding of gene delivery polyplexes using hydroxyethyl starch (HES) and alpha-amylase. *Journal of Controlled Release, 159*(1), 92–103.

Opanasopit, P., Apirakaramwong, A., Ngawhirunpat, T., Rojanarata, T., & Ruktanonchai, U. (2008). Development and characterization of pectinate micro/nanoparticles for gene delivery. *AAPS PharmSciTech, 9*(1), 67–74.

Oyarzun-Ampuero, F., Brea, J., Loza, M., Torres, D., & Alonso, M. J. (2009). Chitosan-hyaluronic acid nanoparticles loaded with heparin for the treatment of asthma. *International Journal of Pharmaceutics, 381*, 122–129.

Padalkar, S., Capadona, J. R., Rowan, S. J., Weder, C., Won, Y.-H., Stanciu, L. A., et al. (2010). Natural biopolymers: Novel templates for the synthesis of nanostructures. *Langmuir, 26*(11), 8497–8502.

Pauly, M., Gille, S., Liu, L., Mansoori, N., de Souza, A., Schultink, A., et al. (2013). Hemicellulose biosynthesis. *Planta, 238*, 627–642.

Pinto Reis, C., Neufeld, R. J., Ribeiro, A. J., & Veiga, F. (2006). Nanoencapsulation I. Methods for preparation of drug-loaded polymeric nanoparticles. *Nanomedicine: Nanotechnology, Biology and Medicine, 2*(1), 8–21.

Pollard, M., Kelly, R., Fischer, P., Windhab, E., Eder, B., & Amadò, R. (2008). Investigation of molecular weight distribution of LBG galactomannan for flours prepared from individual seeds, mixtures, and commercial samples. *Food Hydrocolloids, 22*, 1596–1606.

Pouliot, J. M., Walton, I., Nolen-Parkhouse, M., Abu-Lail, L. I., & Camesano, T. A. (2005). Adhesion of *Aureobasidium pullulans* is controlled by uronic acid based polymers and pullulan. *Biomacromolecules, 6*(2), 1122–1131.

Rekha, M., & Chandra, P. (2007). Pullulan as a promising biomaterial for biomedical applications: A perspective. *Trends in Biomaterials and Artificial Organs, 20*(2), 116–121.

Rekha, M. R., & Sharma, C. P. (2011). Hemocompatible pullulan–polyethyleneimine conjugates for liver cell gene delivery: In vitro evaluation of cellular uptake, intracellular trafficking and transfection efficiency. *Acta Biomaterialia, 7*(1), 370–379.

Ridley, B. L., O'Neill, M. A., & Mohnen, D. (2001). Pectins: Structure, biosynthesis, and oligogalacturonide-related signaling. *Phytochemistry, 57*(6), 929–967.

Rodrigues, S., Cordeiro, C., Seijo, B., Remuñán-López, C., & Grenha, A. (2014). Hybrid nanosystems based on natural polymers as protein carriers for respiratory delivery: Stability and toxicological evaluation. *Carbohydrate Polymers,* (submitted).

Rodrigues, S., Rosa da Costa, A., & Grenha, A. (2012). Chitosan/carrageenan nanoparticles: Effect of cross-linking with tripolyphosphate and charge ratios. *Carbohydrate Polymers, 89*, 282–289.

Rönnberg, E., Melo, F. R., & Pejler, G. (2012). Mast cell proteoglycans. *The Journal of Histochemistry and Cytochemistry: Official Journal of the Histochemistry Society, 60*(12), 950–962.

Rowe, R., Sheskey, P., & Owen, S. (2009). *Handbook of pharmaceutical excipients* (6th ed.). London: Pharmaceutical Press.

Sankalia, M. G., Mashru, R. C., Sankalia, J. M., & Sutariya, V. B. (2006). Stability improvement of alpha-amylase entrapped in kappa-carrageenan beads: Physicochemical characterization and optimization using composite index. *International Journal of Pharmaceutics, 312*(1–2), 1–14.

Santo, V. E., Gomes, M. E., Mano, J. F., & Reis, R. L. (2012). Chitosan–chondroitin sulphate nanoparticles for controlled delivery of platelet lysates in bone regenerative medicine. *Journal of Tissue Engineering and Regenerative Medicine, 6*(S3), s47–s59.

Shiku, H., Wang, L., Ikuta, Y., Okugawa, T., Schmitt, M., Gu, X., et al. (2000). Development of a cancer vaccine: Peptides, proteins, and DNA. *Cancer Chemotherapy and Pharmacology, 46*(Suppl.), S77–S82.

Siahi, M., Barzegar-Jalali, M., Monajjemzadeh, F., Ghaffari, F., & Azarm, S. (2005). Design and evaluation of 1- and 3-layer matrices of verapamil hydrochloride for sustaining its release. *AAPS PharmSciTech, 6*(4), E626–E632.

Sihorkar, V., & Vyas, S. P. (2001). Potential of polysaccharide anchored liposomes in drug delivery, targeting and immunization. *Journal of Pharmacy and Pharmaceutical Sciences, 4*(2), 138–158.

Song, Y., Sun, Y., Zhang, X., Zhou, J., & Zhang, L. (2008). Homogeneous quaternization of cellulose in NaOH/urea aqueous solutions as gene carriers. *Biomacromolecules, 9*(8), 2259–2264.

Song, Y., Wang, H., Zeng, X., Sun, Y., Zhang, X., Zhou, J., et al. (2010). Effect of molecular weight and degree of substitution of quaternized cellulose on the efficiency of gene transfection. *Bioconjugate Chemistry*, *21*(7), 1271–1279.

Song, Y., Zhou, Y., & Chen, L. (2012). Wood cellulose-based polyelectrolyte complex nanoparticles as protein carriers. *Journal of Materials Chemistry*, *22*(6), 2512–2519.

Sungthongjeen, S., Sriamornsak, P., Pitaksuteepong, T., Somsiri, A., & Puttipipatkhachorn, S. (2004). Effect of degree of esterification of pectin and calcium amount on drug release from pectin-based matrix tablets. *AAPS PharmSciTech*, *5*(1), 1–8.

Teramoto, N., & Shibata, M. (2006). Synthesis and properties of pullulan acetate. Thermal properties, biodegradability, and a semi-clear gel formation in organic solvents. *Carbohydrate Polymers*, *63*(4), 476–481.

Toita, S., Hasegawa, U., Koga, H., Sekiya, I., Muneta, T., & Akiyoshi, K. (2008). Protein-conjugated quantum dots effectively delivered into living cells by a cationic nanogel. *Journal of Nanoscience and Nanotechnology*, *8*(5), 2279–2285.

Valo, H., Arola, S., Laaksonen, P., Torkkeli, M., Peltonen, L., Linder, M. B., et al. (2013). Drug release from nanoparticles embedded in four different nanofibrillar cellulose aerogels. *European Journal of Pharmaceutical Sciences*, *50*(1), 69–77.

Verbeken, D., Dierckx, S., & Dewettinck, K. (2003). Exudate gums: Occurrence, production, and applications. *Applied Microbiology and Biotechnology*, *63*(1), 10–21.

Wang, Y., Chen, H., Liu, Y., Wu, J., Zhou, P., Wang, Y., et al. (2013). pH-sensitive pullulan-based nanoparticle carrier of methotrexate and combretastatin A4 for the combination therapy against hepatocellular carcinoma. *Biomaterials*, *34*(29), 7181–7190.

Wolfrom, M. L., Laver, M. L., & Patin, D. L. (1961). Carbohydrates of the coffee bean. II. Isolation and characterization of a mannan. *The Journal of Organic Chemistry*, *26*(11), 4533–4535.

Xiao, C., Gao, S., & Zhang, L. (2000). Blend films from konjac glucomannan and sodium alginate solutions and their preservative effect. *Journal of Applied Polymer Science*, *3*, 617–626.

Yeh, M.-K., Cheng, K.-M., Hu, C.-S., Huang, Y.-C., & Young, J.-J. (2011). Novel protein-loaded chondroitin sulfate–chitosan nanoparticles: Preparation and characterization. *Acta Biomaterialia*, 7(10), 3804–3812.

Yu, W., Liu, C., Liu, Y., Zhang, N., & Xu, W. (2010). Mannan-modified solid lipid nanoparticles for targeted gene delivery to alveolar macrophages. *Pharmaceutical Research*, *27*(8), 1584–1596.

Zhang, H., Ma, Y., & Sun, X. (2010). Recent developments in carbohydrate-decorated targeted drug/gene delivery. *Medicinal Research Reviews*, *30*(2), 270–289.

Zhang, J., Shin, M. C., David, A. E., Zhou, J., Lee, K., He, H., et al. (2013). Long-circulating heparin-functionalized magnetic nanoparticles for potential application as a protein drug delivery platform. *Molecular Pharmaceutics*, *10*(10), 3892–3902.

CHAPTER EIGHT

Bioconjugation of Silk Fibroin Nanoparticles with Enzyme and Peptide and Their Characterization

Fei Wang, Yu-Qing Zhang[1]
Silk Biotechnology Laboratory, School of Biological and Basic Medical Sciences, Soochow University, Suzhou, PR China
[1]Corresponding author: e-mail address: sericult@suda.edu.cn

Contents

Abstract

Bombyx mori silk fibroin is a type of protein-based polymer with unique characteristics that is widely used in the research and development of medical biomaterials. The

ISSN 1876-1623
http://dx.doi.org/10.1016/bs.apcsb.2014.11.005

degummed filament of silk fibroin can be dissolved in a highly concentrated salt solution. After desalination, the regenerated liquid silk fibroin (LSF) solution could be made into various forms of silk biomaterials, such as powder, fiber, film, porous matrix, 3D scaffold, and hydrogel, depending on its application. In this study, we mixed the liquid silk solution with enzymes, including oxidase and hydrolase, and rapidly injected the mixture into an excess of acetone. The enzyme retained most of its enzymatic activity and was also captured in silk fibroin nanoparticles (SFNs), which instantly formed via a configuration transition of the regenerated silk protein from a random coil and α-helix to a β-sheet. The resulting enzyme-captured SFNs displayed a fine crystal structure with a high activity recovery and good thermal stability. Moreover, the affinities of these modified enzymes to their substrate did not evidently suffer from the capture. When only the liquid silk solution was rapidly injected into acetone, the resulting globular SFNs with the same crystallinity were also a good carrier that was covalently conjugated to enzymes and insulin. Thus, silk protein nanoparticles are of potential value as an enzyme or peptide delivery system for the research and development of medical biomaterials. In this report, the bioconjugation of SFNs with glucose oxidase, superoxidase, β-glucosidase, L-asparaginase, neutral protease, and insulin and their characterization are described in detail.

1. INTRODUCTION

The cocoon silks spun by the domestic silkworm *Bombyx mori* have been used in textiles for thousands of years (at least 5000 years). Silk has long been an important textile stock due to its unique feel, luster, dyeability, tensile strength and elasticity, good thermal stability, hygroscopic nature, and microbial resistance. For almost two decades, the regenerated liquid silk fibroin (LSF) solution has been extensively applied to the research and development of various forms of biomaterials, such as drug/enzyme delivery systems, regenerated fibers (Trabbic & Yager, 1998), artificial skin (Li, Masayo, & Minoura, 2003; Li, Wu, & Zhang, 2001), porous matrix or 3D scaffolds (Tamada, 2005), biomimetic nanofibrous scaffolds (Park, Jungb, Leec, Min, & Park, 2006), and a platform for transistors (Kim et al., 2009), and various classes of photonic devices (Amsden et al., 2009; Kim et al., 2011; Parker et al., 2009), due to its biocompatibility. Recently, it is often used as a model animal to be studied and utilized in biology, especially in molecular biology.

Two parallel monofilaments spun by matured silkworm larvae are often referred to as silk fibers. They are composed of 65–75% fibroin, 20–30% sericin, and 5% wax, pigments, sugars, and other impurities. Silk fibroin (SF) is a crystalline protein-based fiber surrounded by layers of the gum sericin protein, which reinforces the cocoon structure, where it is very suitable for the subsequent development and metamorphosis of the matured larvae to pupae. The processing procedures of modern silk fibers, including textile fibers, are exactly the opposite of those of biomedical materials: the silk industry removes the sericin that surrounds the natural fibroin fibers.

The glue protein sericin shows a layered structure and is synthesized and excreted in the middle silk gland cells. The inner layer of sericin that closely covers the fibroin fiber is excreted in the posterior of the middle silk gland. The outer sericin is excreted in the anterior of the middle silk gland. This sericin has a wide range of molecular weights ranging from 40 to over 400 kDa. It consists of 17 amino acids and is primarily composed of 25.50% serine (mol%), 17.85% glycine, 18.38% aspartic acid, and 7.47% threonine (Appendix B). A few years ago, some authors reported that the sericin that coats SF fiber presents a layered structure (Wang & Zhang, 2011). The outermost layer of sericin could be removed by immersing the fiber in boiling water for 30 min, while the middle layer of sericin could be stripped via immersion in 120 °C water for 1 h, and the innermost layer of sericin could be degummed by boiling in 0.2–0.5% Na_2CO_3 solution for 20 min. However, sericin is well-known to completely dissolve in boiling Na_2CO_3 solution after 30 min. According to the solubility in water or solutions at different temperatures and pressures, sericin is divided into three layers, i.e., the outermost, middle, and innermost layers of sericin, which account for 15%, 10.5%, and 4.5% to the total silk protein (whole cocoon shell) by mass, respectively (Fig. 1). Furthermore, the differences in the amino acid composition of the three layers are very significant (Appendix B). The content of nonpolar amino acids in the outer, middle, and inner sericins gradually increase, that of hydrophobic amino acids gradually

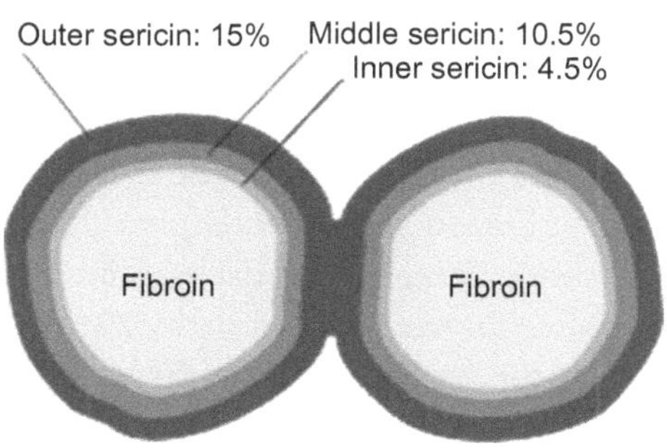

Figure 1 The three layers of sericin surrounded the fibroin fiber and their rates.

increase, and that of hydrophilic amino acids decrease. These results indicate that the amino acid compositions of the outer and middle layers are similar, while those of the inner sericin differ. The closer the sericin layer is to the SF fiber, the closer its amino acid composition is to that of the fibroin fiber, and the poorer its dissolution is in water. These results constitute important reference values for the degumming and processing of silk fiber.

The SF secreted in the lumen of the posterior silk gland of *B. mori* consists of three protein components, namely heavy chain (H-chain), light chain (L-chain), and glycoprotein, P25 (Chevillard, Couble, & Prudhomme, 1986; Chevillard, Deleage, & Couble, 1986; Shimura, Kikuchi, Katagata, & Ohtomok, 1982; Yamaguchi et al., 1989). The H- and L-chains are linked by a disulfide bond, and P25 associates with the disulfide-linked H- and L-chains via noncovalent interactions. A large protein complex of approximately 2.3 MDa, which acts as an elementary unit of fibroin and has 6:6:1 molar ratios of the H-chain, L-chain, and P25, respectively, is synthesized in posterior silk gland cells and then secreted into its cocoon and fibroin fiber (Inoue, Tanaka, Arisaka, Kimura, & Ohtomo, 2000). The amino acid sequence of the H-chain was deduced to be 5263 amino acid residues long, with a molecular weight of 391,367 Da. Its primary structure is rich in glycine, alanine, and serine (approximately 85 mol%) at a molar ratio of 3:2:1, respectively, and these residues form typical repeating motifs: 432 copies of -(GAGAGS)$_n$- and 120 copies of -(GAGAGY)$_n$-. The two hexapeptides together account for 64.8% of the H-chain (Zhou et al., 2000). The amino acid sequence of the L-chain was deduced to be 244 amino acid residues long, with a molecular weight of 25,800 Da (Yamaguchi et al., 1989). Two crystalline forms have been identified, Silk II and Silk I, which are reported as the dimorphs of SF. The former mainly consists of the two repeating motifs present in the H-chain that form a prominent β-sheet structure and is responsible for the mechanical properties of the fiber. The structure of the fiber after spinning consists mainly of Silk II, i.e., an antiparallel β-sheet (Fraser & MacRae, 1973; Lotz & Cesari, 1979; Marsh, Corey, & Pauling, 1995; Mita, Ichimura, & James, 1994; Takahashi, Gehoh, & Yuzuriha, 1999). The latter is composed mainly of amino residues with larger side groups, such as Phe, Tyr, and Try, which mainly comprise the L-chain and H-chain hydrophilic regions and are responsible for the hygroscopicity and resistance of the fiber to UV radiation. Silk I is the conformation of fibroin in the solid state prior to spinning; conformations similar to Silk I are obtained as a film from the liquid silk of the silk gland or from an aqueous solution of regenerated SF

(Kim et al., 2009). The physical and mechanical properties of regenerated SF depend on the conformation of the molecular chain and the crystal structure (Asakura, Suita, Kameda, Afonin, & Ulrich, 2004).

2. THE REGENERATION OF LSF

Silkworm cocoons or silk filaments were processed into a liquid aqueous SF via a series of processing, degumming, dissolution, purification, and concentration steps that are often referred to as silk regeneration. The resulting LSF is often referred to as regenerated liquid silk. The regenerated LSF is very unstable in aqueous solution, and its molecular structure changes from a random coil and α-helix form (Silk I) into an antiparallel β-sheet form (Silk II) due to environmental factors, such as physics and chemistry. Due to the properties of structure transition, the regenerated LSF is used in various forms, such as hydrogel, fiber, powder, membrane, and porous matrix, depending on the application. The physical and mechanical properties of silk biomaterials derived from the regenerated LSF depend on the conformation of the molecular chain and the crystal structure (Wen, Wang, Zhu, & Sun, 2005). Therefore, the regeneration of LSF and subsequent processing and solidification evidently affected the structure and properties of the final forms of silk biomaterials.

2.1. Silk degumming

Of the sericin layers that surround the fibroin fiber, the outermost sericin is easily soluble in boiling water, the middle sericin dissolves at exactly 120 °C in water, and the innermost sericin is only soluble in alkaline boiling water. The partly degummed silk fiber does not dissolve well in highly concentrated salt solutions, such as LiBr and $CaCl_2$, due to some sericin residues that surround the fibroin fiber. First, an appropriate method must be selected for silk degumming according to the degumming efficiency and its effect on silk fiber. Several degumming/refining methods have been developed, such as the use of 25 mM NaOH (Miyaguchi & Hu, 2005), 0.2–0.5% Na_2CO_3 (Zhang, Tao, Shen, Zhou, & Ding, 2004), neutral soap–alkali solution (Tsunokaye, 1932), highly concentrated urea buffer (Yamada, Nakao, & Takasu, 2001), protease (Freddi, Mossotti, & Innocenti, 2003), strongly alkaline electrolyzed water (Cao, Wang, & Zhang, 2013), and HTHP water (Zhang, 1997). In addition, degumming with acid (Gulrajani, Sethi, Gupta, & Brezizinnski, 1992), alkali (Walters & Hougen, 1934), and microwave-assisted treatments (Haggag, El-Sayed, & Allam, 2007) might impact the surface characteristics and mechanical properties of SF fiber.

The degumming methods also affect the molecular structure of the degummed fiber of SF. Recently, several authors have investigated the effect of the regeneration of LSF on its structure and characterization by using the following three degumming methods (Wang & Zhang, 2013). The thermal behavior, mechanical properties, and XRD analysis showed that degumming in 8.0 M urea had the smallest effect on SF, followed by degumming with electrolyzed water, while Na_2CO_3 degumming had the greatest impact on the thermal properties of SF; the thermal decomposition temperature decreased by 6.34 °C. SDS-PAGE showed boiling silk filaments twice in Na_2CO_3 for 0.5 h resulted in serious breakage of the peptide chains of SF, leading to a sharp decline in the mechanical properties of the fiber.

The present experiment used the Na_2CO_3 method for the degumming treatment of *B. mori* silk. The advantage of this method lies in its ability to remove all of the sericin that covers the fiber. Moreover, the liquid silk that results from the easily dissolved broken peptide chains of the fibroin fiber did not affect the formation of silk fibroin nanoparticles (SFNs) or the production of enzyme bioconjugates. The detailed degumming steps are as follows.

The experimental materials are the cocoons of silkworm *B. mori* (available strain: QingSong × HaoYue) provided by the Sericulture Department, Soochow University. Five hundred grams of clean cocoon shell was added to a 0.2% (w/v) Na_2CO_3 solution, and the mixed solution was then heated to 98 °C while stirring, i.e., the solution was boiled for 30 min. After removing the silk fiber, the procedure again was repeated in fresh Na_2CO_3 solution. The resulting SF fiber was washed repeatedly in deionized water and then air-dried.

2.2. Dissolution of degummed SF fiber

The dissolution of SF fiber affects the molecular chains and structure of regenerated LSF. Like the degumming process, the dissolution results in different lengths of broken peptide chains in the regenerated LSF, and these chains then directly affect the structure and properties of various silk biomaterials.

Like cellulose, crystalline silk fiber is very difficult to dissolve in general solvents or in common organic solvents due to its intermolecular hydrogen bonding. Nevertheless, it can be dissolved only three groups, including Cuoxam complex systems (Kiyoshi, Hisao, & Yukio, 1979), saturated aqueous solutions of inorganic salts (von Weimarn, 1927), and organic solvents, such as HFIP (Lock & Newark, 1993), DMSO (Shimizu, 1978), MMNO

(Freddi, Pessina, & Tsukada, 1999), and AmimCl (Wang, Yang, Chen, & Shao, 2012). At present, the most common dissolving systems are highly concentrated or mixed salt solutions, such as LiSCN, LiBr, and the $CaCl_2$–EtOH–H_2O ternary system (shortly $CaCl_2$-TS), in which the latter two components are the most useful. Ordinarily, the degummed silk fiber can easily be dissolved by stirring in 9.3 M LiBr at $\geq$50 °C or $CaCl_2$-TS at $\geq$70 °C for more than 2 h. The former solvent is much more expensive than the latter. Regarding the effects of the two solvents on the molecular chains, our results are similar to the conclusions of Shao's research group: the effects of silk-degumming approaches on the breakage of the molecular chain and mechanical properties of SF are much greater than those of the dissolving process (Wang, Chen, Yang, & Shao, 2013; Wang & Zhang, 2013). The alkalinity of the cocoon-degumming reagent positively correlates with the magnitude of the effect on the SF molecular chain. The magnitude of the effect of the two inorganic salts on the degradation degree of SF molecular chains follows the order LiBr < $CaCl_2$–EtOH–H_2O. The detailed procedures of the dissolution experiment are as follows:

The degummed silk fibers were added to 20 volumes (v/w) of $CaCl_2$-TS ($CaCl_2$/ethanol/water, 1:2:8 mol) and stirred at 70 °C to form a clear solution. After the SF/salt solution was filtered through deionized water or centrifuged at 8000 rpm for 10 min, the filtered solution or supernatant was dialyzed continuously for 48 h with running water to remove $CaCl_2$, smaller molecules, and some impurities using a cellulose semi-permeable membrane (cutoff 5 kDa). The aqueous solutions of regenerated SF were used to prepare pure SFN and its bioconjugates with enzymes or insulin.

2.3. Purification

The purification of the fibroin–salt solution included two steps, dialysis and concentrating. If the protein was not immediately used, the $CaCl_2$-TS of the degummed SF could be stored at 4 °C or frozen for long periods and subsequent experimental use.

Prior to dialysis, the protein–salt solution was centrifuged at 8000 rpm for 10 min to remove the insoluble particles, and the supernatant was dialyzed continuously for 48 h with running pure water to remove $CaCl_2$, smaller molecules, and some impurities using a cellulose semi-permeable membrane (cutoff 5 kDa) or Slide-A-Lyzer® 5 K Dialysis Cassettes G2 (Thermo Scientific, USA). The dialyzed SF solution could again be centrifuged at 8000 rpm for 10 min to remove the insoluble particles. The supernatant volumes were diluted to a final concentration with purified water or concentrated by immersing the dialyzed tube containing SF solution in polyethylene glycol (PEG) powder. A simpler method consists of suspending the dialyzed tube containing silk in a 4 °C refrigerator and spinning the tube continuously until a certain volume is reached. The resulting regenerated LSF solution was used to prepare the silk nanoparticles and their bioconjugates.

3. PROCESSING OF SFN AND ITS CHARACTERISTICS

3.1. Background

Regenerated SF films are the first and most extensively studied carriers for enzyme/drug immobilization via physical treatment other than chemical modification (Demura, Asakura, & Kuroo, 1989; Demura, Asakura, Nakamura, & Tamura, 1989; Zhang, 1998; Zhang, Shen, Gu, Zhu, & Xue, 1998). This application is primarily attributable to its unique characteristics, i.e., the configuration transition from a random coil and α-helix to a β-sheet. Powdered silk has been most extensively utilized in cosmetics, surface-improving materials, health foods, and industrial materials (Tajima & Tanaka, 1994; Toshio, Kazuo, & Hiroyuki, 2000). Silk powder can be produced via physical/chemical processing. Chemical processing refers to the addition of salts or alcohol to regenerated LSF. The powders are characterized by poor crystallinity, irregular morphology, large sizes, and high cost. Physical processing refers to the use of mechanical processing to generate powdered silk with fine crystallinity. The powders have numerous applications in many fields and have been most extensively utilized in cosmetics and surface-improving materials (Sano et al., 1998; Tsubouchi, 1998). However, these crystalline particles also suffer from three flaws: a larger particle size, a broader size range, and an irregular morphology. To date, few studies have investigated mechanically powdered silk as a vehicle for enzyme/drug delivery. The further development of silk powder is restricted by the current limitations in the preparation of finer particles of SF that are less than 1 μm in diameter. However, pulverized silk may not be desirable for biomedical purposes, especially for enzyme/drug delivery systems.

The third processing technique consists of chemical–physical treatments (Minoura & Tsukada, 1990; Otoi & Horikawa, 1980), such as adding coagulating reagents, aeration, coagulation at the isoelectric point, exposure to ultrasonic waves, and agitation at high shear rates, which often results in the reconstituted LSF aggregating or precipitating due to the denaturing of silk protein, i.e., the above-stated configuration transition. The addition of alcohol to LSF is also a conventional method to induce it to gel or precipitate. Methanol and ethanol are the first and most frequently used agents to induce a configuration change of regenerated silk film from a random coil and α-helix to a β-sheet (Kuzuhara, Asakura, Tomoda, & Matsunaga, 1987). The volumes of these alcohols added to the silk solution were small, mostly less than 150% (v/v) of the final mixture. The addition of these alcohols could only transform the liquid silk into a viscous silk precipitate in gel form for use as a tissue-engineering scaffold (Nazarov, Jin, & Kaplan, 2004) or slowly induce it to gel to form a micrometric spherical particle of SF (Nam & Park, 2001).

The experiments in this study indicated that the polar protonic and aprotonic organic solvents methanol, ethanol, propanol, isopropanol, tetrahydrofuran, and acetone could induce reconstituted LSF to form finer globular microparticles with fine crystallinity. Unlike the methods mentioned above, when a small amount of LSF solution was rapidly injected into an excess of polar protonic solvent, a milky white suspension of silk substances, i.e., SFNs, immediately formed (Zhang et al., 2007). Here, we rapidly prepared globular SFNs with a fine crystallinity from regenerated LSF using acetone. Because of the defined primary structure, the SFNs may offer various possibilities for surface modification and covalent drug attachment.

3.2. Preparation

The regenerated LSF containing 2.5% by weight was rapidly injected into at least 70% (v/v) of the final mixture volume of acetone using a sample pipette at room temperature. The silk particles suspended in the mixture consisted of water and organic solvent were water-insoluble and slowly precipitated due to nanoparticles agglomeration. The SFN precipitates were collected and purified from the mixture via repeated centrifugation at 12,000 rpm for 20 min or filtering to separate these particles from the solvent. When the precipitate transferred in water was subjected to a supersonic treatment for 1–5 min, the SFNs dispersed evenly in water or 50 mmol/l phosphate buffer. The resulting lyophilized or liquid SFNs were used for all experiments in the section.

3.3. Characterization

The globular SFNs were rapidly prepared from an aqueous solution of regenerated LSF using water-miscible acetone (Fig. 2A). The globules were approximately 35–125 nm in diameter, insoluble but well dispersed and stable in aqueous solution, and characterized by a β-sheet conformation and crystallinity close to that of native fibers (Fig. 2B). The SFNs formed in various solvents, including methanol, ethanol, propanol, isopropanol, tetrahydrofuran, and acetone, of which acetone and isopropanol were the most preferable.

The globular nanoparticles were prepared with the regenerated LSF from the degummed silk filaments solubilized in $CaCl_2$-TS at 90 °C for

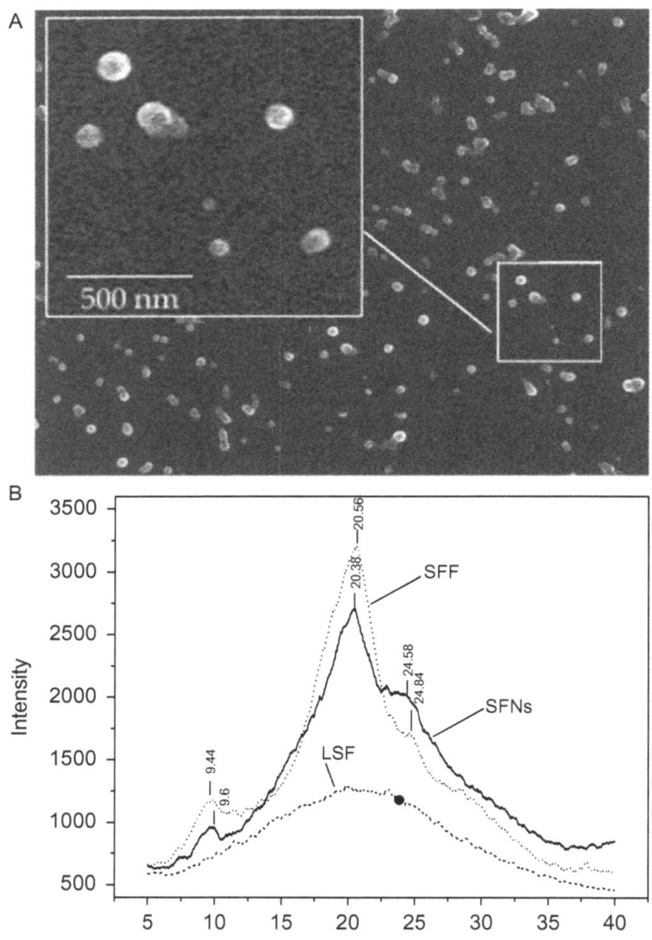

Figure 2 SEM image (×10 K, Hitachi S-570 SEM) and X-ray diffraction curves (D/MAX-IIIC, Rigaku Co., JP) of SFNs. SFNs were made with aqueous silk (LSF) derived from the dissolution of $CaCl_2$-TS at 90 °C for 12 h. SFF: powdered natural silk fibroin filaments.

12 h. The regenerated LSF was a mixture of gradually degraded SF polypeptides that broadly ranged in relative molecular mass from approximately 8 to 100 kDa. LSF with a broader range of MW (8–200 kDa) can also be made into SFNs. The ability of SFNs formed in acetone to resist UV radiation is approximately 20 times higher than that of the liquid silk. Both *Escherichia coli* and *Bacillus subtilis* could not propagate normally in the aqueous solution containing SFNs due to a lack of essential amino acids. Over half of the ε-amino groups exist on the surface of the SFNs, where the tyrosine residues are more exposed than on the fibers.

The globular SFNs formed from the LSF rapidly denatured in excessive acetone, and the configuration of LSF molecules was simultaneously instantly converted from Silk I into Silk II. The SFNs also displayed a β-sheet form based on an electron diffraction photograph. The XRD curves indicate that the crystallinity of the SFNs is approximately four-fifths of that of the native SF fiber. DSC showed that the endothermic peaks of the maximum decomposition of both aqueous SF and nanoparticles shift to a lower temperature, and the exothermic peaks attributed to the recrystallization of amorphous SF almost disappear from the DSC curves of the nanoparticles. The DSC curve of native silk fiber showed a broad and large endothermic peak at approximately 316.6 °C, which was attributed to the thermal degradation of the β-sheet material. The SFN endothermic peak is almost the same as that of the lyophilized LSF (288.6 °C) and also shifted to a lower temperature of 288.7 °C. This shift indicates that the decrease in the thermal stability of SF was related to the breakage of amide bonds and nonoriented β-sheet arrangement in SF molecules.

These results show that the degraded polypeptide fragments of these regenerated LSFs homogeneously or heterogeneously aggregated to form a looser globular structure. The formation of crystalline SFNs results from the rapid dispersion, instant dehydration, and internal and external denaturing in the solvent when looser silk globules are introduced into an excess of solvent. In addition, this simple processing for SFN preparation provides a novel protocol that is useful to illustrate the advanced structure and processing mechanism of silk protein *in vivo* in insects.

The silk protein nanoparticles with a good biocompatibility and crystallinity, anti-UV radiation, and lots of active amino groups and their surrounding tyrosine residues favor attachment and bioconjugation with other molecules. This attribute is potentially useful in medical biomaterials, especially in enzyme/drug delivery systems.

4. ENZYME-CAPTURED SFNs AS AN ENZYME DELIVERY SYSTEM

4.1. Background

B. mori silk fiber has long been used as a surgical suture due to its biodegradability and biocompatibility (Sadayuk et al., 1999; Sakabe, Ito, Miyamoto, Noishiki, & Ha, 1989; Santin, Motta, Freddi, & Cannas, 1999). Silk protein has also long been studied as a vector for enzyme modification. As early as 1976, Inada and Hirose applied for a Japanese patent on regenerated silk fiber from liquid silk used as a carrier for enzyme absorption *in vivo* (Inada & Hirose, 1976). Subsequently, several reports were published on alkaline phosphatase and ribonuclease immobilized on woven silk and diazotized silk fiber (Cordier, Counturier, Grasset, & Ville, 1982; Grasset, Cordier, & Ville, 1977, 1979). These enzymes were immobilized on the protein using chemically covalent conjugation. In 1978, Miyairi et al. attempted to immobilize β-glucosidase (βG) into a fibroin membrane by immersing it in alcohol solution for 1 min (Miyairi, Sugiura, & Fukui, 1978)—this study is the earliest report on enzyme immobilization in a silk membrane based on the configuration transition from Silk I to Silk II. Since then, many enzymes have been immobilized in silk membranes using the conformation transition (Asakura, Yoshimizu, Kuzuhara, & Matsunaga, 1988; Koichiro, 1980; Kuzuhara et al., 1987; Miyairi, 1977; Zhang, 1998; Zhang & Zhou, 1995). The above configuration transition and simultaneous enzyme immobilization can also be induced by physical treatments, i.e., stretching, compressing, and incubation at high humidity (Demura & Asakura, 1989; Demura, Asakura, & Kuroo, 1989). The immobilization of enzymes entrapped in silk membranes is unique, and such immobilized enzymes have potential applications in biosensors (Demura, Asakura, & Kuroo, 1989; Demura, Asakura, Nakamura, et al., 1989; Demura, Komura, & Asakura, 1991).

Early in 1981, both alkali and neutral proteases (NPs) were immobilized in powdered SF by adjusting the pH to 5.0–5.5 (Inoue, Matsunaga, & Iwane, 1986; Nakayama, 1981). These proteases were again captured in silk powder by stirring at a high shear rate and/or combined addition of at least one of the following water-miscible organic solvents: methanol, ethanol, propanol, ethylene glycol, propylene glycol, butane diol, glycerol, and acetone (Yoshimizu & Asakura, 1990; Yukio & Hiroshi, 1981). However, these enzymes entrapped in these silk powders were characterized by poor crystallinity, irregular morphology, and inconsistent sizes, which limited their applications.

The studies of protein-based nanoparticles as a vehicle for drug delivery systems focus mainly on the albumin that is primarily derived from human or bovine serum. In essence, three different methods for particle preparation have been described, i.e., emulsion formation, coacervation, or desolvation (Lin, Coombes, Davies, Davis, & Illum, 1993; Müller, Leuenberger, & Kissel, 1996; Weber, Coester, Kreuter, & Langer, 2000). The shortcomings of these methods are the need to apply surfactants to stabilize emulsions or glutaraldehyde for to cross-link particles. In fact, the protein-based nanoparticles were colloidal and cross-linked albumin nanoparticles. The advantage of albumin nanoparticles is that they may be biocompatible, but they are also very expensive.

In this section, we reported another novel and highly efficient method to process fine-crystalline enzyme-entrapped SFNs in acetone based on the configuration transition mentioned above. The biosynthesis, kinetics, *in vitro* activities and biostability of the modified oxidases, such as glucose oxidase (GOD) and superoxidase (SOD), and hydrolases, such as βG, L-asparaginase (ASNase), and NP, are described in detail.

4.2. Preparation of oxidase-captured SFNs

4.2.1 Compounds

GOD from *Aspergillus niger* (EC 1.1.3.4.), which had a specific activity of 23,900 U/g, βG (37 U/mg), and trypsin (1000 U/mg) were obtained from Sigma Chemicals. Peroxidase (POD) from horseradish (EC 1.11.1.7) was purchased from Shanghai Orient Chemicals Factory (China) and had a specific activity of 250 U/mg. *E. coli* ASNase (MW 140 kDa) with a specific activity of 271.8 IU/mg protein was obtained from Changzhou Qianhong Bio-Pharma Co. Ltd (China). *p*-Nitrophenyl-β-D-glucopyranoside (*p*-NPG), asparagine, glutaraldehyde, glycine, and the electrophoresis reagents were purchased from Sigma. The protein ladder was purchased from Fermentas Co. Ltd (Lithuania). Other chemicals used in the experiment, such as 4-aminoantipyrine, phenol, disodium hydrogen phosphate, methanol, citric acid, calcium chloride, LiBr, acetone, KI, HgI_2, and trichloroacetic acid, were all of analytical reagent grade. The water was distilled twice.

4.2.2 Preparation of oxidase-captured SFNs

A given volume of LSF (approximately 2.5% by weight) was mixed with βG or ASNase solution in a plastic flask at 4 °C. After mild homogenization, the mixed solution containing the oxidase was rapidly injected into a container of acetone whose volume was at least 70% (v/v) of the final mixture volume at room temperature. After injection, the acetone in the container was stirred with a magnetic stirrer. Milk-like silk particles containing oxidase immediately formed and were suspended in the mixture of water and acetone. These oxidase-entrapped particles were water-insoluble and precipitated very slowly due to the agglomeration of nanoparticles. These precipitates were collected with water centrifuging three times at 30,000 rpm (Beckman Avanti J30I, USA) to separate the particles from acetone. When the precipitate in water was subjected to five ultrasonic treatments (Output: 16 W, Ultrasonic Processor, Sonics & Materials Inc., Newtown, CT) for 2-min each, the oxidase-captured SFNs dispersed evenly in water and were stored at 4 °C for the following experiments.

4.3. The properties of oxidase-captured SFNs

4.3.1 Glucose oxidase

GOD has been often used as a test enzyme for delivery systems. Here, GOD was captured into silk protein nanoparticles and transferred immediately from the regenerated LSF via the configuration transition. The activities of free and captured GOD in SFNs were determined using 4-aminoantipyrine as an oxidation–reduction indicator, as reported by Kawahara (1993) with a slight modification. The enzyme reaction was carried out at 37 °C in a reaction mixture of 3.1 ml that contained 2.4 ml dye-buffer solution at pH 5.5, 0.5 ml of 10% glucose solution, 0.1 ml POD solution (12 U/ml), and 0.1 ml free GOD or GOD-captured SFNs. The GOD activity was calculated based on the change in the optical density over 5 min at 500 nm using a spectrophotometer (Hitachi 200-20, Tokyo, Japan) with a 1-cm quartz cell. Each sample was measured five times to obtain a mean value.

Table 1 shows that the modified GOD has a broad range of optimal pH values (5.5–6.5) when captured in SFNs compared to native GOD (pH 6.5). The optimal reaction temperature range (30–40 °C) of the captured GOD is also wider than that (30 °C) of the native enzyme. The activity recovery is maximized when GOD is loaded in SFNs at a concentration of 0.1 U/mg. In addition, the captured GOD could resist to the hydrolysis of protease. Native GOD completely lost its bioactivity in 1% pepsin after 2 min, while the GOD captured in SFNs retained 60% of its original activity in the same pepsin solution for 60 min. The above results demonstrate that GOD retained most of its activity and better resisted external environmental conditions when it was entrapped in silk protein nanoparticles transferred instantly in acetone from the regenerated liquid silk state. These results show that silk protein nanoparticles are an excellent biomaterial in the film or membrane state for enzyme vectors or delivery systems.

Table 1 The kinetic characteristics of the two oxidases before or after being captured in SFNs

Enzyme species	**Glucose oxidase (GOD)**		**Superoxide dismutase (SOD)**	
Free/immobilization	**Entrapped in SFNs**	**Native state**	**Entrapped in SFNs**	**Native state**
Optimal pH	5.5–6.5	6.5	7.0–9.0	7.0–9.0
Optimal reaction temperature	30–40 °C	30 °C	40 °C	40 °C
Optimal entrapped amount	0.1 U/mg SFNs	–	0.1 U/mg SFNs	–
Recovery of entrapped enzyme	69.2%	–	79.4%	–
Wet thermostability ($T_{1/2}$)	–	–	84 °C	73 °C
Stability in 1.0% pepsin	60% (60 min)	0% (2 min)	97.8% (48 h)	57.4% (24 h)
Half-life in rat ($T_{1/2}$)	–	–	36 h	18 h
Storage stability (4 °C/ RT)	No loss (1 M)	–	No less (6 M)	–

Storage stability test: The activity of immobilized enzymes stored at 4 °C and room temperature for 6 months was measured at 1 month of intervals; $T_{1/2}$: the half-life of entrapped enzyme or native enzyme in rat.

4.3.2 Superoxidase

SOD is the only antioxidant enzyme that scavenges the superoxide anion by converting this free radical to oxygen and hydrogen peroxide, thus preventing peroxynitrite production and further damage. Free radicals are strongly associated with many pathological processes in the body. Due to this scavenging ability, SOD has garnered significant attention for therapeutic use. SOD is extensively researched and used in anti-inflammatory, antitumor, radiation protection, and antisenility applications (Luisa et al., 2002; Morikawa & Morikawa, 1996). However, the clinical application of SOD as a therapeutic agent has been limited due to its extremely rapid plasma clearance time, instability, and immunogenicity *in vivo*. Several strategies have been proposed to overcome these problems, including the use of SOD conjugates with alginate–chitosan (Mi, Shyu, Chen, & Schoung, 1999) or hydrophilic polymers, such as PEG (Yang et al., 2007), as appropriate delivery systems.

Here, we prepared novel SOD-entrapped silk nanoparticles in acetone from the regenerated LSF. The activity of the SOD-entrapped SFNs was determined with a SOD typed assay kit (hydroxylamine method) (Oyanagui, 1984) provided by Nanjing Jiancheng Bioengineering Institute (Nanjing, China). The results showed that a SOD-loading amount of 1.0 U/10 mg SFNs is the optimum concentration at which the activity recovery of SOD-captured SFNs reaches 79.4% of that of the free SOD (Table 1). The optimum pH of the SOD-captured SFNs varied widely from 7.0 to 9.5, and these values are similar to those obtained for native SOD. The optimal reactive temperature (40 °C) of the entrapped SOD is also the same as that of native enzyme. The wet thermostability of the modified SOD is higher than that of free SOD. SOD did not leak from the entrapped enzyme after being stored in solution for 6 months. The entrapped SOD also remained stable *in vitro* and *in vivo*. Modified SOD incubated in 1.0% pepsin for 48 h retained 97.8% of its original activity, while native SOD incubated in 1.0% pepsin for 24 h only retained 57.4% of its activity. When the captured SOD (200 U/kg BW) was subcutaneously injected into rats, the IC_{50} of SOD detected in the serum of the rat was 36 h, while the value of free SOD in the rat was only 18 h. These results indicate that the biological stability of the enzyme increased after being entrapped in the SFNs both *in vitro* and *in vivo*. These results for SOD modified in SFNs are the almost same as those obtained for GOD in SFNs.

4.4. The properties of hydrolase-captured SFNs

4.4.1 β-Glucosidase (Cao, Zhou, & Zhang, 2014)

Based on five different loading amounts of the enzyme (5–50 U) into 10 mg of SFNs, 10 U/10 mg SFNs was determined to be the optimum loading concentration; this concentration yielded an activity recovery for βG-captured SFNs of 59.2% of the free enzyme (Table 2). The optimum pH of the free enzyme and the βG-captured SFNs was ~5.0. The optimum temperature and the thermal stability of the SFNs were the same as those of the native enzyme. When the βGs were entrapped and fixed in SFNs, its affinity to substrate was not significantly reduced, and the enzyme did not leak. These results indicate that the microenvironmental factors did not affect these kinetic characteristics of the enzyme before or after being entrapped. These results confirmed that silk protein nanoparticles were good carriers and acted as bioconjugates for the modification of enzymes with a potential value for research and development. The method used in this study has potential applications in

Table 2 The kinetic characteristics of the two enzymes before or after being entrapped in SFNs

Enzyme species	**β-Glucosidase**		**L-Asparaginase (ASNase)**	
Enzyme states	**Entrapped in SFNs**	**Native state**	**Entrapped in SFNs**	**Native state**
Optimal concentration	1.0 U/mg SFNs	–	10.0 U/mg SFNs	–
Optimal pH	~5.0	~5.0	7.0–8.0	7.0
Optimal reaction temperature	60 °C	60 °C	50 °C	50 °C
Wet thermostability	60% RA at 60 °C	60% RA at 60 °C	50% at 55 °C	50% at 55 °C
Dry thermostability	–	–	>80% at 110 °C	–
Recovery of entrapped enzyme	59.2%	–	90%	–
Operation stability/10 times	>80% RA	–	>93% RA	–
K_m (app) (mol/l)	1.01×10^{-2}	7.26×10^{-3}	2.42×10^{-3}	1.0×10^{-3}
Half-life in 1.0% trypsin ($T_{1/2}$)	–	–	50 min	5 min
Half-life in human serum ($T_{1/2}$)	–	–	75 h	32 h
Storage stability (RA)	100% at 37 °C for 2 days	–	100% for 10 days	60% for 10 days

RA: residual activity; $T_{1/2}$: the half-life of entrapped enzyme or native enzyme in human serum.

food processing and the production of flavor agents. This immobilization of biocatalysts has many advantages, namely the reuse of the biocatalyst, ease of separation of the biocatalyst from reaction media, and a continuous mode of operation.

4.4.2 L-Asparaginase (Zhang, Wang, Wang, Zhu, & Zhou, 2011)

Based on five different amounts of the enzyme (2, 10, 15, 20, 40, and 100 U) loaded to four concentration of SFNs (0.5–5.0 mg), a loading amount of 10 U/1.0 mg SFNs was determined to be the optimum concentration; this concentration yielded an activity recovery for βG-captured SFNs of 90% of that of the free enzyme. The recovery was higher than that of βG-entrapped SFNs (Table 2). The optimum pH range of the ASNase-captured SFNs was pH 7.0–8.0, which is a broader range than that of the free enzyme (pH 7.0). The optimum temperature and the wet thermostability of the restructured SFNs were the same as those of the native enzyme. However, the dry thermostability of the ASNase-captured SFNs was much higher than that of the native enzyme. When incubated in an oven at 110 °C for 30 min, the modified enzyme retained 80% of its original activity, which shows that the powdered ASNase entrapped in SFNs could tolerate 110 °C in dry heat conditions due to the protection provided by the fine-crystalline SFNs to the active center of ASNase. When the enzyme was entrapped in SFNs, its affinity to the substrate slightly decreased to a value approximately 2.4 times lower than that of the free enzyme. This decrease suggested that immobilization in SFNs influenced the affinity between ASNase and its substrate. The other parameters, such as the stabilities (IC_{50}) of the immobilized ASNase in 1.0% trypsin and human serum and storage state, were much higher than those of free ASNase.

FT-IR and XRD demonstrated that the configuration transition of ASNase-captured SFNs from a random coil and α-helix of LSF containing ASNase to a β-sheet occurred instantaneously in acetone, and ASNase molecules were simultaneously tightly entrapped in the globular granules of the silk protein with a final crystallinity. The ASNase that was uniquely and efficiently captured and immobilized in the SFNs is mainly attracted to the above configuration transition of LSF. Thus, the highly efficient processing technology and use of silk nanoparticles as a novel drug delivery system has potential applications in research and development.

5. COVALENT BIOCONJUGATION OF SFNs WITH ENZYME/POLYPEPTIDE

During the above investigation, the authors found that two types of oxidase and hydrolase as well as isomerases, such as glucose isomerase (data not shown), could be entrapped and immobilized at the SFNs transferred instantly in acetone from a liquid silk solution. The resulting immobilized enzymes could retain most of their original activities. However, when insulin was used for this entrapping test, the results drastically differed from those described above. The insulin-entrapped SFNs generated using the same method described above completely lost its activity. Thus, we speculated that the enzyme captured in the crystalline silk nanoparticles did not affect its activity center, which continued to enzymatically react when being bound by the substrate. While the insulin was captured in crystalline SFNs, the capturing directly affected the combination of the insulin surface with its receptor. This finding suggested that the covalent binding of this nanoparticle with insulin and enzyme molecules made of SFN bioconjugates could serve as a new drug/enzyme delivery system.

5.1. Preparation and characterization of βG–SFN bioconjugates

According to the processing procedure in Section 3.2, the SFNs were first prepared via the instant configuration transition of silk protein by injecting the regenerated LSF solution into acetone. The SFNs were then covalently conjugated with a hydrolase or insulin using glutaraldehyde as the cross-linking reagent according to the method reported previously by other authors (Zhu, Hu, Wang, Wang, & Zhang, 2011). A 1.0-ml sample of SFN (10 mg/ml) was introduced into a plastic flask and mixed with a solution of hydrolase or insulin at 4 °C. After mild homogenization, the required volume of 25% glutaraldehyde was slowly added to the mixture. Finally, phosphate-buffered saline (PBS, pH 7.0) was added to a final volume of 5.0 ml. The flask was tightly stoppered and then placed onto an orbital shaker for 30 min at 4 °C to allow the cross-linking reaction to proceed. The reaction was stopped by the addition of 200 mg of glycine. The mixture was then centrifuged twice at 30,000 rpm (Beckman Avanti J-30I) and 4 °C for 30 min to remove impurities. The supernatant was discarded, and the precipitate was concentrated to 10 IU/ml in PBS.

The activities of βG–SFN bioconjugates and free βG were determined using *p*-NPG as a substrate according to the method reported previously by Liu, Yang, Zhang, and Xiong (2005) with a slight modification (Zhou & Zhang, 2011). The optimal conditions for the biosynthesis of βG–SFNs bioconjugates were investigated (Table 3). The analytical results showed that the βG–SFN obtained after 5 h of covalent cross-linking at 0.25% cross-linking reagent, 37 °C and a proportion of βG and SFNs of 75:100 U/mg yielded a 46% activity recovery. The results showed that the kinetic parameters of βG–SFNs were the same as those of the free βG. Its optimal pH and optimal temperature were pH 5.0 and 60 °C, respectively. When βG was covalently coupled with silk nanoparticles, the thermal stability of βG–SFNs was slightly enhanced compared with that of free βG. The apparent K_m of βG–SFNs (1.41×10^{-3} mol/l) was almost fivefold less

Table 3 The kinetic characteristics of the two oxidases/insulin before or after being conjugated with SFNs

Enzyme species	**β-Glucosidase (βG)**		**Neutral protease (NP)**		**Insulin (Ins)**	
Enzyme states	**βG-SFNs**	**Native state**	**NP–SFNs**	**Native state**	**Ins–SFNs**	**Native state**
Optimal enzyme:SFNs (U:mg)	75:100	–	1:5–8	–	10:5	–
Optimal pH	5.0	5.0	7.5	7.0	–	–
Optimal reaction temperature	60 °C	60 °C	40–50 °C	40 °C	–	–
Recovery of conjugated enzyme	46.0%	–	31.5%	–	90–115%	–
K_m (app) (mol/l)	1.41×10^{-3}	7.26×10^{-3}	6.97×10^{-4}	3.55×10^{-4}	–	–
Wet thermostability (RA)	60%/60 °C	60%/60 °C	55%/50 °C	35%/50 °C	–	–
Operation stability/10 times	>60% RA	–	–	–	–	–
Resistance to trypsin (30 min)	–	–	–	–	85% RA	50% RA
Half-life in human serum ($T_{1/2}$)	–	–	–	–	42 h	17 h

RA: residual activity; $T_{1/2}$: the half-life of Ins–SFNs or native insulin in human serum.

than that of the free enzyme (7.26×10^{-3} mol/l), which clearly demonstrates that the enzyme affinity for the substrate increased after the free enzyme was conjugated with SFNs. These results showed that the SFNs were also a good carrier for the bioconjugation or modification of enzymes. Moreover, the βG–SFN bioconjugates have potential value in the research and development of food processing and flavor processing agents.

5.2. Preparation and characterization of NP–SFN bioconjugates (Zhu et al., 2011)

The objective of this study was to determine and investigate the dynamic characteristics of the NP–SFN bioconjugates and explore their possible application as a hydrolyzing sericin protein. The enzymatic activities of native NP and NP–SFN were determined with a UV spectrophotometer (U-3000; Hitachi Co. Ltd., Japan). First, the SFNs were covalently conjugated with a NP using glutaraldehyde as the cross-linking reagent according to the above-stated procedure. The NP–SFN bioconjugates were obtained via the covalent cross-linking of NP and SFN at an NP/SFN ratio of 1 IU:5–8 mg with 0.75% glutaraldehyde for 6 h at 25 °C. Adding 50 IU of the enzyme increased the residual activity of biological conjugates to 31.5%. A kinetic analysis showed that the NP–SFN bioconjugates were more stable than the free enzyme, the optimum reaction temperature increased by 10 °C, and the optimum pH value increased to pH 7.5. Furthermore, the thermal stability improved to some extent (Table 3). A controlled hydrolysis test using the poorly water-soluble protein sericin (30–200 kDa) as a substrate and NP–SFN as the enzyme showed that the reaction time (within 1 h) negatively correlated with the molecular mass (<30 kDa) of the produced sericin peptide. The NP–SFN bioconjugates are easily recovered by centrifugation and can be used repeatedly. Therefore, the highly efficient hydrolysis technology and the use of SFNs as a novel vector for protease have great potential in the research and the development of food processing.

5.3. Preparation and characterization of Ins–SFN bioconjugates

Insulin is an important polypeptide drug used to lower the levels of blood glucose in diabetics. The drug is subject to serious hydrolysis of protease and the loss of its activity in blood within a very short period of time (2–3 h generally). Thus, many studies have examined the delivery methods of insulin for diabetic treatment. Here, we attempted to modify the polypeptide drug with SFNs using glutaraldehyde as a cross-linking reagent. The bioconjugation and preparation of insulin and SFNs were carried out according to the same method used to conjugate βG–SFNs and NP–SFNs. The bioactivities of the modified insulin and native insulin were determined with an enzyme-linked immunosorbent assay (ELISA) according to the method stated previously by the authors (Yan, Zhang, Ma, & Zhou, 2009).

The optimal conditions for the biosynthesis of Ins–SFN bioconjugates were investigated (Table 3). The Ins–SFN constructs obtained after 8 h of covalent cross-linking with 0.7% cross-linking reagent at an insulin:SFN ratio of 10 IU:5 mg showed much higher recoveries (90–115%). When insulin was coupled covalently with silk nanoparticles, the resistance of the modified insulin to trypsin digestion and *in vitro* stability in human serum were greatly enhanced compared with native insulin. After being hydrolyzed for 30 min by trypsin, native insulin only retained approximately half of its original activity, while Ins–SFN bioconjugates retained approximately 85% of the original activity. The results in human serum indicated that the *in vitro* half-life of Ins–SFN derivatives was 42 h, which is approximately 2.5 times higher than that of native insulin. When incubated in the same manner as the physiological solution at 37 °C for 60 h, native insulin lost almost all of its activity, while the insulin modified with SFN retained approximately 40% of its initial activity levels. Therefore, the silk protein nanoparticles have potential value for being studied and developed as a new bioconjugate for enzyme/polypeptide drug delivery systems.

ACKNOWLEDGMENTS

The authors gratefully acknowledge the earmarked fund (CARS-22-ZJ0504) for China Agriculture Research System (CARS) and a project funded by the Priority Academic Program Development of Jiangsu Higher Education Institutions, P. R. China.

APPENDIX A. ABBREVIATION TABLE

No.	Abbreviation	Full name	No.	Abbreviation	Full name
1.	3D	Three-dimensional	21.	LSF	Liquid silk fibroin
2.	AmimCl	1-Allyl-3-methylimidazolium chloride	22.	L-chain	Light chain
3.	ASNase	L-Asparaginase	23.	LiBr	Lithium bromide

No.	Abbreviation	Full name	No.	Abbreviation	Full name
4.	βG	β-Glucosidase	24.	MMNO	*N*-methyl morpholine N-oxide
5.	$CaCl_2$	Calcium chloride	25.	MW	Molecular weight
6.	$CaCl_2$-TS	Calcium chloride ternary system	26.	Na_2CO_3	Sodium carbonate
7.	Cuoxam	Copper hydroxide–ammonia solution	27.	NP	Neutral protease
8.	DMSO	Dimethylsulfoxide	28.	PAGE	Polyacrylamide gel electrophoresis
9.	DMA	*N*,*N*-dimethylacetamide	29.	PEG	Polyethylene glycol
10.	DSC	Differential scanning calorimetry	30.	*p*-NPG	*p*-Nitrophenyl-β-D-glucopyranoside
11.	E	Enzyme	31.	RA	Residual activity
12.	ELISA	Enzyme-linked immunosorbent assay	32.	SDS	Sodium dodecyl sulfate
13.	GOD	Glucose oxidase	33.	SEM	Scanning electronic microscope
14.	H-chain	Heavy chain	34.	SF	Silk fibroin
15.	HFIP	Hexafluoro-2-propanol	35.	SFF	Silk fibroin filaments
16.	HTHP	High temperature and high pressure	36.	SFN	Silk fibroin nanoparticles
17.	Ins	Insulin	37.	SOD	Superoxidase
18.	IC_{50}	Half-inhibition concentration	38.	$T_{1/2}$	Half-life
19.	IP	Isoelectric point	39.	UV	Ultraviolet
20.	K_m (app)	Apparent Michaelis constant	40.	XRD	X-ray diffraction

APPENDIX B. AMINO ACID COMPOSITIONS OF WHOLE (S), OUTER (S*O*), MIDDLE (S*M*), AND INNER (S*I*) SERICIN AND FIBROIN (MOL%)

Amino acids	S	S*o*	S*m*	S*i*	Fibroin
Gly	17.85	16.29	16.35	17.87	42.62
Ala	6.70	5.20	6.13	11.58	33.38
Val*	4.05	3.77	4.27	5.43	2.58
Leu*	1.49	1.21	1.77	4.06	0.54
Ile*	1.02	0.79	1.17	3.76	0.72
Phe*	0.67	0.64	0.66	2.49	0.81
Met*	0.31	0	0	0.83	0.15
Pro	0.81	0	0.64	0	0.47
Tyr	3.10	2.87	3.98	4.09	5.84
Cys	0.38	0.69	0.95	0.75	0.26
Ser	25.50	28.00	25.57	13.32	7.65
Thr*	7.47	7.78	8.13	5.66	0.85
Asp	18.38	17.97	17.08	15.83	1.79
Glu	5.74	6.25	4.65	7.34	1.36
His	1.32	1.32	1.69	1.38	0.21
Lys*	2.08	3.72	3.16	2.18	0.33
Arg	3.12	3.52	3.83	3.41	0.44
Aa	S	S*o*	S*m*	S*i*	Fibroin

Note: "*" means essential amino acids for human body.

REFERENCES

Amsden, J. J., Perry, H., Boriskina, S. V., Gopinath, A., Kaplan, D. L., Dal Negro, L., et al. (2009). Spectral analysis of induced color change on periodically nanopatterned silk films. *Optics Express*, *17*, 21271–21279.

Asakura, T., Suita, K., Kameda, T., Afonin, S., & Ulrich, A. S. (2004). Structural role of tyrosine in *Bombyx mori* silk fibroin, studied by solid-state NMR and molecular mechanics on a model peptide prepared as silk I and II. *Magnetic Resonance in Chemistry*, *42*, 258–266.

Asakura, T., Yoshimizu, H., Kuzuhara, A., & Matsunaga, T. (1988). Mechanism of glucose oxidase immobilization with silk fibroin. *Journal of Sericulture Science of Japan*, *57*, 203–209.

Cao, T. T., Wang, Y. J., & Zhang, Y. Q. (2013). Effect of strongly alkaline electrolyzed water on silk degumming and the physical properties of the fibroin fiber. *PLoS One*, *8*(6), e65654. http://dx.doi.org/10.1371/journal.pone.0065654.

Cao, T. T., Zhou, Z. Z., & Zhang, Y. Q. (2014). Processing of β-glucosidase-silk fibroin nanoparticle bioconjugates and their characteristics. *Applied Biochemistry and Biotechnology*, *173*, 544–551. http://dx.doi.org/10.1007/s12010-014-0861-y.

Chevillard, M., Couble, P., & Prudhomme, J. C. (1986). Complete nucleotide sequence of the gene encoding the *Bombyx mori* silk protein P25 and predicted amino acid sequence of the protein. *Nucleic Acids Research*, *14*, 6341–6342.

Chevillard, M., Deleage, G., & Couble, P. (1986). Amino acid sequence and putative conformational characteristic of the P25 silk protein of *Bombyx mori*. *Sericologia*, *26*, 435–449.

Cordier, D., Counturier, R., Grasset, L., & Ville, A. (1982). Ribonuclease insolubilization using diazotized silk. *Enzyme and Microbial Technology*, *4*, 249–255.

Demura, M., & Asakura, T. (1989). Immobilization of glucose oxidase with Bombyx mori silk fibroin membrane by only stretching treatment and its application to glucose sensor. *Biotechnology and Bioengineering*, *33*, 598–603.

Demura, M., Asakura, T., & Kuroo, T. (1989). Immobilization of biocatalysts with *Bombyx mori* silk fibroin by several kinds of physical treatment and its application to glucose sensors. *Biosensors*, *4*, 361–372.

Demura, M., Asakura, T., Nakamura, E., & Tamura, H. (1989). Immobilization of peroxidase with a *Bombyx mori* silk fibroin membrane and its application to biophotosensors. *Journal of Biotechnology*, *10*, 113–120.

Demura, M., Komura, T., & Asakura, T. (1991). Membrane potential of *Bombyx mori* silk fibroin membrane induced by an immobilized enzyme reaction. *Bioelectrochemistry and Bioenergetics*, *26*, 167–175.

Fraser, R. D. B., & MacRae, T. P. (1973). *Conformations of fibrous proteins and related synthetic polypeptides*. New York: Academic Press.

Freddi, G., Mossotti, R., & Innocenti, R. (2003). Degumming of silk fabric with several proteases. *Journal of Biotechnology*, *106*, 101–112.

Freddi, G., Pessina, G., & Tsukada, M. (1999). Swelling and dissolution of silk fibroin (*Bombyx mori*) in N-methyl morpholine N-oxide. *International Journal of Biological Macromolecules*, *24*, 251–263.

Grasset, L., Cordier, D., & Ville, A. (1977). Woven silk as a carrier for the immobilization of enzyme. *Biotechnology and Bioengineering*, *19*, 611–618.

Grasset, L., Cordier, D., & Ville, A. (1979). Silk: A natural protein for enzyme immobilization. *Process Biochemistry*, *14*, 2–5.

Gulrajani, M. L., Sethi, S., Gupta, & Brezizinnski, S. (1992). Some studies in degumming of silk with organic acids. *Journal of Society Dyers Coloration*, *108*(2), 79–86.

Haggag, K., El-Sayed, H., & Allam, O. G. (2007). Degumming of silk using microwave-assisted treatments. *Journal of Natural Fibers*, *4*(3), 1–23. http://dx.doi.org/10.1300/J395v04n03.

Inada, J., & Hirose, S. (1976). Fuyokakosono Seizohoho. *Japan Patent JP51067785*.

Inoue, S., Matsunaga, Y., & Iwane, H. (1986). Entrapment of phenylalanine ammonia-lyase in silk fibroin for protection from proteolytic attack. *Biochemical and Biophysical Research Communications*, *141*, 165–170.

Inoue, S., Tanaka, K., Arisaka, F., Kimura, S., & Ohtomo, K. (2000). Silk Fibroin of *Bombyx mori* is secreted, assembling a high molecular mass elementary unit consisting of H-chain, L-chain, and P25, with a 6:6:1 molar ratio. *Journal of Biological Chemistry*, *275*(51), 40517–40528.

Kawahara, Y. (1993). Immobilization of glucose oxidase in wild silk fiber (*Antheraea pernyi*) treated with an alkaline solution. *Journal of Sericulture Science of Japan, 62*(4), 272–275.
Kim, D. H., Kim, Y. S., Amsden, J., Panilaitis, B., Kaplan, D. L., Omenetto, F. G., et al. (2009). Silicon electronics on silk as a path to bioresorbable, implantable devices. *Applied Physics Letters, 95*, 133701–133703.
Kim, D., Viventi, J., Amsden, J. J., Xiao, J., Kim, Y., Blanco, J. A., et al. (2011). Dissolvable films of silk fibroin for ultrathin, conformal bio-integrated electronics. *Nature Materials, 9*(6), 511–517. http://dx.doi.org/10.1038/nmat2745.
Kiyoshi, O., Hisao, N., & Yukio, H. (1979). Preparation of fibroin dope. *Japan Patent JP54096126A*.
Koichiro, O. (1980). Preparation of making film containing enzyme. *Japan Patent JP55082128*.
Kuzuhara, K., Asakura, T., Tomoda, R., & Matsunaga, T. (1987). Use of silk fibroin for enzyme membrane. *Journal of Biotechnology, 5*, 199–207.
Li, M., Masayo, O., & Minoura, N. (2003). Enzymatic degradation behavior of porous silk fibroin sheets. *Biomaterials, 24*, 357–365.
Li, M., Wu, Z., & Zhang, C. J. (2001). Study on porous silk fibroin materials. II. Preparation and characteristics of spongy porous silk fibroin materials. *Applied Polymer Science, 79*, 2192–2199.
Lin, W., Coombes, A. G. A., Davies, M. C., Davis, S. S., & Illum, L. J. (1993). Preparation of sub-100 nm human serum albumin nanospheres using a pH-coacervation method. *Journal of Drug Targeting, 1*, 237–243.
Liu, D. C., Yang, G. Y., Zhang, L., & Xiong, X. T. (2005). Studies on immobilization of Neutrase. *Food Sciences, 26*, 61–65 (in Chinese).
Lock, R. L., & Newark, D. (1993). Process for making silk fibroin fibers. *US Patent US5252285*.
Lotz, B., & Cesari, F. C. (1979). The chemical structure and the crystalline structures of *Bombyx mori*. *Biochimie, 61*, 205–214.
Luisa, C. M., Jorge, J. C., van't Hof, R., Cruz, M. E., Crommelin, D. J., & Storm, G. (2002). Superoxide dismutase entrapped in long-circulating liposomes: Formulation design and therapeutic activity in rat adjuvant arthritis. *Biochimica et Biophysica Acta, 1564*, 227–236.
Marsh, R. E., Corey, R. B., & Pauling, L. (1995). An investigation of the structure of silk fibroin. *Biochimica et Biophysica Acta, 16*, 1–34.
Mi, F. L., Shyu, S. S., Chen, C. T., & Schoung, J. Y. (1999). Porous chitosan microsphere for controlling the antigen release of Newcastle disease vaccine: Preparation of antigen-adsorbed microsphere and in vitro release. *Biomaterials, 20*, 1603–1612.
Minoura, N., & Tsukada, M. (1990). Coagulation of silk fibroin. *Japan Patent JP02084503*.
Mita, K., Ichimura, S., & James, T. C. (1994). Highly repetitive structure and its organization of the silk fibroin gene. *Journal of Molecular Evolution, 38*, 583–592.
Miyaguchi, Y., & Hu, J. (2005). Physicochemical properties of silk fibroin after solubilization using calcium chloride with or without ethanol. *Food Science and Technology Research, 11*(1), 37–42.
Miyairi, Y. (1977). Immobilized enzyme using fibroin as carrier and method of preparing the same. *Japan Patent JP52057392*.
Miyairi, S., Sugiura, M., & Fukui, S. (1978). Immobilized of β-glucosidase in fibroin membrane. *Agricultural and Biological Chemistry, 42*, 1661–1667.
Morikawa, K., & Morikawa, S. (1996). Immunomodulatory effect of recombinant human superoxide dismutase (SOD) on human B lymphocyte function in vitro. *Cellular Immunology, 172*, 70–76.
Müller, G. M., Leuenberger, H., & Kissel, T. (1996). Albumin nanospheres as carriers for passive drug targeting: An optimized manufacturing technique. *Pharmaceutical Research, 13*, 32–37.

Nakayama, H. (1981). Immobilized protease and its preparation. *Japan Patent JP56015687.*
Nam, J., & Park, Y. H. J. (2001). Morphology of regenerated silk fibroin: Effects of freezing temperature, alcohol addition, and molecular weight. *Journal of Applied Polymer Science, 81*(12), 3008–3021.
Nazarov, R., Jin, H. J., & Kaplan, D. L. (2004). Porous 3-D scaffolds from regenerated silk fibroin. *Biomacromolecules, 5*, 718–726.
Otoi, K., & Horikawa, Y. (1980). Process for producing a fine powder of silk fibroin. *US Patent US4233212.*
Oyanagui, Y. (1984). Reevaluation of assay methods and establishment of kit for superoxide dismutase activity. *Analytical Biochemistry, 142*(2), 290–296.
Park, K. E., Jungb, S. Y., Leec, S. J., Min, B. M., & Park, W. H. (2006). Biomimetic nanofibrous scaffolds preparation and characterization of chitin silk fibroin blend nanofibers. *International Journal of Biological Macromolecules, 38*, 165–173.
Parker, S. T., Domachuk, P., Amsden, J., Bressner, J., Lewis, J. A., Kaplan, D. L., et al. (2009). Biocompatible silk printed optical waveguides. *Advanced Materials, 21*, 2411–2415.
Sadayuk, K., Hirokuni, O., Mayumi, K., Mamor, K., Ruby, P., & Koshi, M. (1999). Fibroin allergy: IgE mediated hypersensitivity to silk suture materials. *Journal of Nippon Medical School, 66*, 41–44.
Sakabe, H., Ito, H., Miyamoto, T., Noishiki, Y., & Ha, W. S. (1989). In vivo blood compatibility of regenerated silk fibroin. *Sen-i Gakkaishi, 45*, 487–490.
Sano, M., Mikami, S., Sasaki, N., Kusamoto, N., Fukatsu, F., Ubara, A., et al. (1998). Process for producing fine silk fibroin powder. *European Patent EP0875523A1.*
Santin, M., Motta, A., Freddi, G., & Cannas, M. (1999). In vitro evaluation of the inflammatory potential of the silk fibroin. *Journal of Biomedical Materials Research, 46*, 382–389.
Shimizu, Y. (1978). The swelling and dissolution of silk with organic solvents. *Journal of Sericulture Science of Japan, 47*, 417–420.
Shimura, K., Kikuchi, A., Katagata, Y., & Ohtomok, K. (1982). The occurrence of smallest component protein in the cocoon of *Bombyx mori*. *Journal of Sericulture Science of Japan, 51*, 20–26.
Tajima, M., & Tanaka, T. (1994). Silk powder and its production, coating agent using silk powder and its production, and cleaning finishing sizing agent using silk powder and its use. *Japan Patent JP06306772.*
Takahashi, Y., Gehoh, M., & Yuzuriha, K. (1999). Structure refinement and diffuse streak scattering of silk (*Bombyx mori*). *International Journal of Biological Macromolecules, 24*, 127–138.
Tamada, Y. (2005). New process to form a silk fibroin porous 3-D structure. *Biomacromolecules, 6*, 3100–3106.
Toshio, U., Kazuo, K., & Hiroyuki, A. (2000). Properties of silk pigment and it's application for cosmetics. *Fragrance Journal, 28*(4), 15–21.
Trabbic, K. A., & Yager, P. (1998). Comparative structural characterization of naturally- and synthetically-spun fibers of *Bombyx mori* fibroin. *Macromolecules, 31*, 462–471.
Tsubouchi, K. (1998). Process for preparing fine powder of silk fibroin. *US Patent US5853764.*
Tsunokaye, R. (1932). Degumming silk. *Journal of Society Dyers Coloration, 48*, 20.
von Weimarn, P. P. (1927). *Report of Osaka Institute*, 17, 3.
Walters, R. H., & Hougen, O. A. (1934). Silk degumming I. Degradation of silk sericin by alkalies. *Texture Research Journal, 5*, 92.
Wang, Q., Chen, Q., Yang, Y., & Shao, Z. (2013). Effect of various dissolution systems on the molecular weight of regenerated silk fibroin. *Biomacromolecules, 14*(1), 285–289. http://dx.doi.org/10.1021/bm301741q.
Wang, Q., Yang, Y., Chen, X., & Shao, Z. (2012). Investigation of rheological properties and conformation of silk fibroin in the solution of AmimCl. *Biomacromolecules, 13*(6), 1875–1881. http://dx.doi.org/10.1021/bm300387z.

Wang, Y. J., & Zhang, Y. Q. (2011). Three layered sericins coated around the silk fibroin fiber from *Bombyx mori* cocoon and their amino acid composition. *Advanced Materials Research, 175–176*, 158–163.

Wang, H. Y., & Zhang, Y. Q. (2013). Effect of regeneration of liquid silk fibroin on its structure and characterization. *Soft Matter, 9*, 138–145. http://dx.doi.org/10.1039/c2sm26945g.

Weber, C., Coester, C., Kreuter, J., & Langer, K. (2000). Desolvation process and surface characteristics of protein nanoparticles. *International Journal of Pharmaceutics, 194*, 91–102.

Wen, D. J., Wang, H., Zhu, X. S., & Sun, J. P. (2005). Conformation and crystallinity of silk fibroin. *Acta Sericology Sinica, 26*(1), 110–112.

Yamada, H., Nakao, H., & Takasu, Y. (2001). Preparation of undergraded native molecular fibroin solution from silkworm cocoons. *Materials Science and Engineering: C, 14*, 41–46.

Yamaguchi, K., Kikuchi, Y., Takagi, T., Kikuchi, A., Oyama, F., Shimura, K., et al. (1989). Primary structure of the silk fibroin light chain determined by cDNA sequencing and peptide analysis. *Journal of Molecular Biology, 210*, 127–139.

Yan, H. B., Zhang, Y. Q., Ma, Y. L., & Zhou, L. X. (2009). Biosynthesis of insulin-silk fibroin nanoparticles conjugates and in vitro evaluation of a drug delivery system. *Journal of Nanoparticle Research, 11*, 1937–1946. http://dx.doi.org/10.1007/s11051-008-9549-y.

Yang, T., Choi, M. K., Cui, F. D., Kim, J. S., Chung, S. J., Shim, C. K., et al. (2007). Preparation and evaluation of paclitaxel-loaded PEGylated immunoliposome. *Journal of Controlled Release, 120*, 167–177.

Yoshimizu, H., & Asakura, T. (1990). Preparation and characterization of silk fibroin powder and its application to enzyme immobilization. *Journal of Applied Polymer Science, 40*, 127–134.

Yukio, H., & Hiroshi, H. (1981). Production of immobilized protease. *Japan Patent JP56018590.*

Zhang, Y. Q. (1997). Comparison analysis of degumming procedures for silkworm silk. *Acta Sericology Sinica, 28*(1), 75–79.

Zhang, Y. Q. (1998). Natural silk fibroin as a support for enzyme immobilization. *Biotechnology Advances, 16*, 961–971.

Zhang, Y. Q., Shen, W. D., Gu, R. A., Zhu, J., & Xue, R. Y. (1998). Amperometric biosensor for uric acid based on uricase-immobilized silk fibroin membrane. *Analytica Chimica Acta, 369*, 123–128.

Zhang, Y. Q., Shen, W. D., Xiang, R. L., Zhuge, L. J., Gao, W. J., & Wang, W. B. (2007). Formation of silk fibroin nanoparticles in water-miscible organic solvent and their characterization. *Journal of Nanoparticle Research, 9*, 885–900. http://dx.doi.org/10.1007/s11051-006-9162-x.

Zhang, Y. Q., Tao, M. L., Shen, W. D., Zhou, Y. Z., & Ding, Y. (2004). Immobilization of L-asparaginase on the microparticles of the natural silk sericin protein and its characters. *Biomaterials, 25*(17), 3751–3759.

Zhang, Y. Q., Wang, Y. J., Wang, H. Y., Zhu, L., & Zhou, Z. Z. (2011). Highly efficient processing of silk fibroin nanoparticle-L-asparaginase bioconjugates and their characterization as a drug delivery system. *Soft Matter*, 7, 9728–9736. http://dx.doi.org/10.1039/c0sm01332c.

Zhang, Y. Q., & Zhou, Y. Z. (1995). Preparation of the peroxidase-immobilized fibroin membrane and its characteristics. *Progress in Biochemistry and Biophysics, 22*, 162–165.

Zhou, C. Z., Confalonieri, F., Medina, N., Zivanovic, Y., Esnault, C., Jacquet, T., et al. (2000). Fine organization of *Bombyx mori* fibroin heavy chain gene. *Nucleic Acids Research, 28*, 2413–2419.

Zhou, Z. Z., & Zhang, Y. Q. (2011). Biosynthesis of β-glucosidase-silk fibroin nanoparticles conjugates and enzymatic characteristics. *Advanced Materials Research*, *175–176*, 186–191. http://dx.doi.org/10.4028/www.scientific.net/AMR.175-176.186.
Zhu, L., Hu, R. P., Wang, H. Y., Wang, Y. J., & Zhang, Y. Q. (2011). Bioconjugation of neutral protease on silk fibroin nanoparticles and application in the controllable hydrolysis of sericin. *Journal of Agricultural and Food Chemistry*, *59*, 10298–10302. http://dx.doi.org/10.1021/jf202036v.

CHAPTER NINE

Food-Grade Protein-Based Nanoparticles and Microparticles for Bioactive Delivery: Fabrication, Characterization, and Utilization

Gabriel Davidov-Pardo*, Iris J. Joye*,†, David Julian McClements*,‡,1
*Department of Food Science, University of Massachusetts, Amherst, Massachusetts, USA
†Department of Microbial and Molecular Systems, Faculty of Bioscience Engineering, KU Leuven, Leuven, Belgium
‡Department of Biochemistry, Faculty of Science, King Abdulaziz University, Jeddah, Saudi Arabia
[1]Corresponding author: e-mail address: mcclements@foodsci.umass.edu

Contents

Advances in Protein Chemistry and Structural Biology, Volume 98
ISSN 1876-1623
http://dx.doi.org/10.1016/bs.apcsb.2014.11.004

Abstract

Proteins can be used to fabricate nanoparticles and microparticles suitable for use as delivery systems for bioactive compounds in pharmaceutical, food, cosmetic, and other products. Food proteins originate from various animal or vegetal sources and exhibit a wide diversity of molecular and physicochemical characteristics, e.g., molecular weight, conformation, flexibility, polarity, charge, isoelectric point, solubility, and interactions. As a result, protein particles can be assembled using numerous different preparation methods, from one or more types of protein or from a combination of a protein and another type of biopolymer (usually a polysaccharide). The final characteristics of the particles produced are determined by the proteins and/or polysaccharides used, as well as the fabrication techniques employed. This chapter provides an overview of the functional properties of food proteins that can be used to assemble nanoparticles and microparticles, the fabrication techniques available to create those particles, the factors that influence their stability, and their potential applications within the food industry.

1. INTRODUCTION

Delivery systems for bioactive molecules have a wide variety of different internal and external morphologies, which determine their stability and functional performance. The compound(s) to be entrapped is/are referred to as the "active" material, whereas the material that forms the surrounding particle matrix is usually referred to as the "encapsulant" or "carrier" material (Madene, Jacquot, Scher, & Desobry, 2006). The diameter of the particles determines if they can be considered to be microparticles (1–1000 μm) or nanoparticles (<1000 nm). Many of the beneficial functional attributes of these types of particles as delivery systems are associated with their small dimensions and high surface/volume ratios, e.g., better sedimentation stability, improved aggregation stability, higher solubility, higher penetration rates through biological barriers, reduced light scattering, and faster diffusion rates. On the other hand, the high surface/volume ratio may increase the sensitivity of any encapsulated substance to reactions that take place at the particle surface, such as oxidation or degradation during digestion (Joye & McClements, 2013). A number of factors have driven research into the development of encapsulation technologies for functional food ingredients (McClements, Decker, Park, & Weiss, 2009; Shahidi & Han, 1993):

- Protection of an active ingredient from degradation by reducing its exposure to harsh conditions in the food matrix such as salts, elevated temperatures, humidity, oxygen, light, extreme pH conditions, and enzymatic activity.

- Improvement of the water dispersibility of an active ingredient so that it can more easily be incorporated into aqueous-based food or beverage products.
- Improvement of the handling and utilization of the active ingredient, e.g., by converting it from a bulk solid form into a powdered or fluid form.
- Creation of a controlled release profile so that the active ingredient is released after a specific time or in response to a specific environmental stimulus within the food matrix or gastrointestinal tract (GIT).
- Improvement of the oral bioavailability of an active ingredient, so that a greater fraction is absorbed by the human body.
- Masking of any off-flavors (tastes or aromas) associated with an active ingredient, such as bitterness, astringency, or metallic off-taste.
- Separation of two or more active ingredients in a mixture that would otherwise chemically or physically react with each other in an adverse manner.

Ever since the development of the first encapsulation system for vitamins for application in food systems (Taylor, 1939), a wide variety of encapsulation technologies and carrier materials has been developed, evaluated, and utilized. The choice of a suitable matrix material depends on various physicochemical, regulatory, and economic considerations, such as polarity, solubility, physical state, transition temperature, loading capacity, generally recognized as safe status, cost, supply consistency, and ease-of-use. The selection of particle fabrication method is also affected by considerations of economic viability as large-scale commercial production should be the end goal of delivery system development. In addition, the delivery system should not adversely impact the organoleptic and physicochemical properties of the food product, and the bioactive compound should remain biologically active at the moment of consumption.

2. PROTEIN FUNCTIONALITY

Proteins are linear biopolymers composed of α-L amino acids connected by peptide bonds. In nature, 20 different amino acids are commonly found that can be categorized according to the properties of their side groups as aliphatic, aromatic, charged (positive or negative), or polar (Damodaran, Parkin, & Fennema, 2008). Differences in the number, type, and sequence of amino acids present lead to proteins with distinctly different molecular and physicochemical properties. Proteins interact with each other or with other molecules in their environment through various types of interaction, including covalent, van der Waals, steric, hydrogen, hydrophobic, and electrostatic interactions (Walstra, 2009). Adjacent amino acids on a protein chain can adopt locally organized structures (so-called secondary structures such as α-helices or β-sheets), which may play an important role in the formation, stability, and properties of protein particles. In addition, protein molecules tend to adopt an overall three-dimensional (3D) structure (tertiary structure) that depends on their molecular properties and the surrounding environment. In foods, the 3D structure of a protein may be very different from the structure found in nature (the native structure) because of changes in environment (such as pH, composition, and temperature). The conformation of the amino acid chain determines, which side groups are exposed to the environment, and, hence, overall protein stability, interactions, and reactivity.

Proteins may play an important role in determining the overall physicochemical and nutritional traits of food products. Some proteins (enzymes) catalyze specific biochemical reactions that are important in determining final product quality (Nagodawithana & Reed, 1993). Other proteins alter the physicochemical and sensory properties of foods through their ability to thicken solutions, form gels, hold water, or adsorb to interfaces (Damodaran et al., 2008; Walstra, 2009). The factors that determine these functionalities are the molecular characteristics of the protein itself (e.g., conformation, charge, and polarity) and the characteristics of the surrounding matrix (e.g., pH, ionic strength, and temperature). It is, therefore, important to understand the interplay between protein structure and matrix characteristics to understand and predict protein functionality in complex systems. For example, protein solubility depends on the relative strength of the various molecular interactions involved (e.g., protein–protein, protein–solvent, solvent–solvent), which depends on protein structure and solvent conditions. Knowledge of the solubility characteristics of proteins is critical when developing effective fabrication methods for protein particles (Joye & McClements, 2013).

The ability of proteins to form gels is often utilized during the fabrication of protein nanoparticles and microparticles. A gel is defined as a 3D network of molecules or particles that associate to each other through physical or covalent cross-links. A gel typically entraps a reasonably high amount of the surrounding solvent, which is water in the case of hydrogels. Important gel properties, such as appearance (e.g., clear, turbid, or opaque), texture (e.g., soft to hard), responsiveness (e.g., to salt, pH or temperature), and water-holding capacity (e.g., low to high), are determined by the number, strength, and morphology of the structural units and bonds (Fig. 1; Damodaran et al., 2008; Walstra, 2009).

The possible thickening effect of proteins is another factor that may be important in the fabrication and stability of protein particles. Biopolymers that trap appreciable amounts of water in their structure have an effective volume that is much larger than that of the polymer chain itself, which results in a large increase in solution viscosity. Thus, proteins with rod-like structure or random coil structures (e.g., gelatin) tend to have good thickening properties, while proteins with compact structures (e.g., globular proteins) do not have good thickening properties (Fig. 2; Damodaran et al., 2008; Walstra, 2009). Higher viscosity can stabilize particle dispersions by slowing down particle movement, thereby reducing the rate of particle aggregation and gravitational separation. Another advantage of using natural polymers (proteins) rather than synthetic polymers is that the removal of residual monomers after particle formation is not necessary. Furthermore, many proteins display antioxidant properties (Matalanis, Decker, & McClements, 2012) and are often easily digested within the human GIT.

The presence of numerous functional groups on protein surfaces is advantageous from a particle fabrication point of view, since it provides a wide range of opportunities to modify protein properties so as to promote

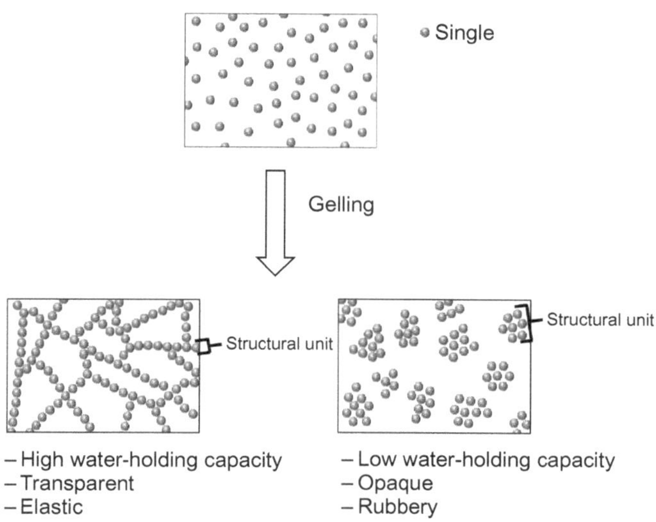

Figure 1 Schematic representation of different gel types (filamentous and particulate) and some of their properties.

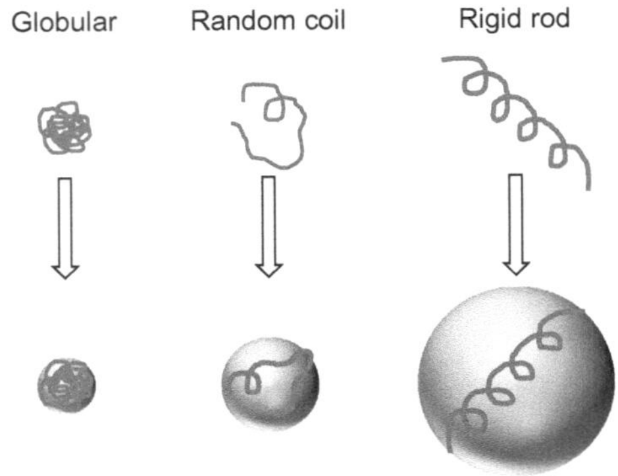

Figure 2 Effect of the biopolymer conformation on the volume ratio (volume of the sphere swept out by the biopolymer/volume of the biopolymer).

interactions with bioactive compounds, cross-linking agents, or coating substances. Thus, it is possible to custom-design protein particles with specific physicochemical and functional attributes for different applications (Jahanshahi & Babaei, 2008; Khan & Schneider, 2013).

3. PROTEIN SOURCES

As mentioned above, a high number of food-grade proteins exist with different molecular, physicochemical, and functional properties. The selection of the most suitable protein for a specific application depends on the desired particle properties (e.g., size, charge, surface characteristics, permeability, and degradability), on the properties of the bioactive to be encapsulated (e.g., polarity, solubility, and stability), and environmental conditions (e.g., pH, ionic strength, solvent quality, and temperature). After production, protein nanoparticles or microparticles often require a chemical or enzymatic hardening step to improve their stability. An enzymatic approach is usually preferred for food-grade systems, but there are often limitations associated with the sensitivity and cost of enzymes.

3.1. Animal proteins

Animal proteins are widely used for the formation of protein particles in the food industry, e.g., casein, whey protein, gelatin, egg proteins, and fibroin. Proteins isolated from milk are one of the most commonly used proteins from animal sources, e.g., caseins and whey proteins. If milk is acidified or treated with chymotrypsin, then the casein precipitates, while the whey protein remains soluble (Damodaran et al., 2008). Whey proteins are widely used in the food industry due to their high nutritional value and excellent functional attributes, e.g., emulsification, gelling, foaming, and thickening. Albumins are water-soluble globular animal proteins with a relatively low heat stability. These are attractive proteins for the production of microparticles or nanoparticles as they are widely available in a pure form and are biodegradable and nontoxic (Jahanshahi & Babaei, 2008). Gelatin is also readily available and is produced by controlled hydrolysis of collagen obtained from bone, cartilage, or skin tissue (Belitz, Grosch, & Schieberle, 2009). Gelatin displays interesting functional properties such as cold-set gelation and gel properties that are highly dependent on the gelatin concentration, pH, cooling rate, and holding temperature (Burey, Bhandari, Howes, & Gidley, 2008). Therefore, it is also a useful material to produce food-grade protein particles (Hedayati, Jahanshahi, & Attar, 2011; Khan & Schneider, 2013; Langer et al., 2003; Taheri, Jahanshahi, & Mosavian, 2012). Fibroin has been used in biomedical applications because of its biocompatibility, biodegradability, antimicrobial properties, and unusual high thermal stability (Subia & Kundu, 2013), but may also be suitable for some food applications.

3.2. Plant proteins

A number of different plant proteins are also available for fabricating protein particles, e.g., proteins isolated from wheat, corn, soy, and peas (Chen, Hebrard, Beyssac, Denis, & Subirade, 2010; Chen & Subirade, 2009; Pierucci, Andrade, Farina, Pedrosa, & Rocha-Leao, 2007). Soy protein is a reasonably hydrophilic plant protein that can be isolated from a widely available, cheap, and renewable source. Soy protein forms gels at high temperatures, but can also undergo cold-set gelation when it is predenatured and combined with Ca^{2+}. Zein and gliadin are of particular interest for the formation of food-grade protein particles because they are highly hydrophobic, unlike many other food proteins. This not only requires adaptations to the particle production process (such as the solvents used), but also enables one to encapsulate hydrophobic components without the need for oil phases. Indeed, hydrophobic plant proteins have already been widely studied for the production of particles to encapsulate bioactives (Chen & Subirade, 2009; Duclairoir, Orecchioni, Depraetere, & Nakache, 2002; Ezpeleta et al., 1996; Li et al., 2013; Patel, Hu, Tiwari, & Velikov, 2010; Podaralla & Perumal, 2012; Regier, Taylor, Borcyk, Yang, & Pannler, 2012). Furthermore, particles fabricated from these proteins usually do not need hardening to ensure their integrity in aqueous media, which substantially simplifies the production process. However, as these proteins generally have a high level of hydrophobic amino acids and a low level of charged amino acids, their aggregation stability is usually limited because of strong hydrophobic attraction and weak electrostatic repulsion. Consequently, stabilization of these hydrophobic protein particles by the addition of surfactants or copolymers is often required (Davidov-Pardo, Joye, & McClements, 2015; Patel, Bouwens, & Velikov, 2010; Podaralla & Perumal, 2012). Zein and gliadin particles have attracted attention for utilization in pharmaceutical and cosmetic products, because they can interact with epidermal keratin (Teglia & Secchi, 1994) and the mucus layer of the GIT (Arangoa et al., 2000; Patel, Hu, et al., 2010; Regier et al., 2012). The latter would substantially increase the residence time of the particles in the GIT and, hence, positively affect the absorption and bioavailability of any encapsulated bioactives.

Plant proteins offer some advantages over animal proteins for the formation of particle-based delivery systems as the risk of contamination and infection is reduced, they are not subjected to cultural dietary limitations, and they are often cheaper than their animal counterparts.

3.3. Protein combinations

Combinations of proteins from different sources have also been used to fabricate particles with tailor-made properties, e.g., zein–caseinate (Patel, Bouwens, et al., 2010), gliadin–caseinate (Davidov-Pardo et al., 2015), and zein–soy protein (Chen et al., 2010; Chen & Subirade, 2009). This approach can be used to fine-tune the hydrophobicity, density, and stability of the protein matrix. Furthermore, if the intended use of the particles is protection and delivery of bioactives through the food-processing chain and the GIT, altering the protein composition of the matrix will also affect the nature of the interaction/binding of the bioactive to the protein matrix. Research has shown that bioactives can interact with proteins through hydrophobic interactions, covalent bonds, hydrogen bonds, and/or van der Waals interactions (Acharya, Sanguansri, & Augustin, 2013; Hemar, Gerbeaud, Oliver, & Augustin, 2011; Joye, Davidov-Pardo, Ludescher, & McClements, in preparation; Ye, Thomas, Sanguansri, Liang, & Augustin, 2013). The nature of the interactions involved impacts the release characteristics and the behavior of the protein particles in the food matrix and inside the GIT.

4. FABRICATION METHODS

Numerous methods are available to produce food-grade delivery systems for bioactive compounds based on the nanoparticles or microparticles formed from proteins (Table 1). For the purpose of this review, the

Table 1 Summary of the protein nano- or microparticles' formation techniques discussed in this chapter

Fabrication technique	Materials	Driving forces	Improvement of stability
Coacervates	Proteins and polysaccharides	Electrostatic attraction	Gelling
W/W emulsions	Proteins and polysaccharides	Electrostatic repulsion	Gelling
Casein micelles	α- and β-caseins	Hydrophobic attraction Cross-linking	Cross-linking agents (phosphates) κ-Casein
Aggregation through denaturalization	Proteins	Denaturalization promotes hydrophobic attraction	Gelling Polymer coating
Antisolvent precipitation	Proteins	Loss of solvent quality causes nucleation	Gelling Polymer coating
Extrusion/emulsion templating	Proteins	Hardening by adding cross-linking agents or environment changes causes gelling	Gelling Polymer coating Cross-linking
Electrospinning	Proteins or proteins attached to a polysaccharide	High-voltage electric field causes a jet of protein solution or atomization	Complete evaporation of solvent
Spray drying	Proteins	Atomization and drying	Addition of surfactants, polymers or divalent ions Lowering the drying temperature

techniques will be separated into methods that can be carried out with only one type of protein and methods that need a combination of two or more different biopolymers for particle production.

4.1. Particles formed with one type of protein

4.1.1 Self-assembly of flexible proteins

The use of casein micelles as an encapsulation system for bioactive compounds can be considered to be an example of a biomimetic approach. In nature, casein micelles consist of submicellar structures that encapsulate calcium phosphate, which is essential for proper growth and functioning of infants. The core of the casein micelles is primarily formed by αs1-, αs2-, and β-caseins, while the outer layer consists of κ-casein, which helps stabilize them by generating a strong steric repulsion (Kaya-Celiker & Mallikarjunan, 2012; Fig. 3). Casein micelles can be produced *in vitro* by self-assembly of the casein units around a hydrophobic compound. Different salts (such as potassium phosphate, calcium chloride, and/or potassium citrate) are used to create the proper pH and ionic strength conditions to encapsulate the hydrophobic compound (Semo, Kesselman, Danino, & Livney, 2007). Attention has to be paid to the fact that a fraction of the bioactive compound can be attached to casein units that do not take part in micelle formation. Therefore, it is important to tightly control the conditions to maximize the self-assembly of the casein units into micelle structures. An alternative

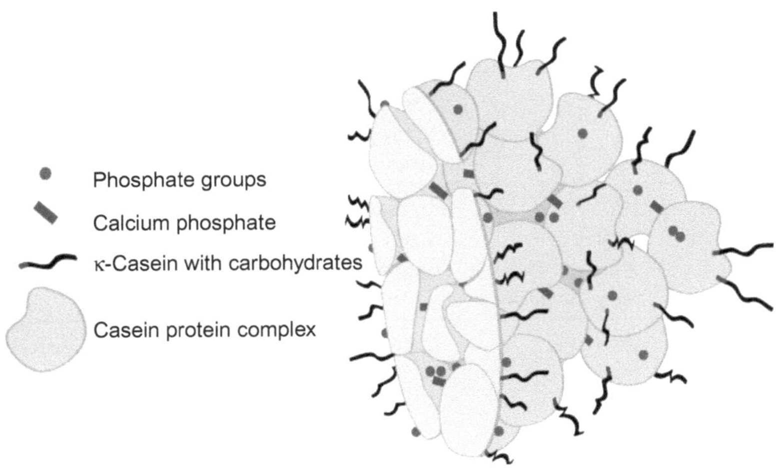

Figure 3 Schematic representation of a casein micelle. Source: *http://openwetware.org/wiki/User:Anthony_Salvagno/Notebook/Research/2009/11/11/Andy's_Poster. Last accessed 08/01/2014.*

method to produce casein micelles to encapsulate bioactive compounds is the disintegration/reassembly method. In this method, the casein units are disintegrated in alkaline conditions (pH ~ 12). After disintegration, the pH is readjusted to around 7.0 and the casein units reassemble themselves into micelles that embed the bioactive compound. Encapsulation of a hydrophobic bioactive compound (curcumin) using this strategy significantly improved its stability and bioavailability (Pan, Luo, Gan, Baek, & Zhong, 2014).

4.1.2 Thermal denaturation of globular proteins

Protein nanoparticles can also be formed by simply heating solutions of globular proteins. These proteins unfold above their thermal denaturation temperature, thereby exposing nonpolar groups normally buried within the protein interior. As a result, proteins start to aggregate through hydrophobic interactions and disulfide bonds (Jones, Decker, & McClements, 2009). The aggregation process is irreversible and can be stopped by rapidly reducing the temperature (Le Bon, Nicolai, & Durand, 1999). Ionic strength, pH, and temperature strongly influence the denaturation and aggregation of globular proteins and, therefore, the size of the protein particles produced. Particle size typically increases with increasing ionic strength, protein concentration, holding temperature, and holding time. The particles tend to be smaller when produced at a pH far from the pI of the proteins. At lower heating rates the particle size also increases (Bromley, Krebs, & Donald, 2006; Hoffmann, Roefs, Verheul, van Mil, & De Kruif, 1996). Polysaccharides can be added to protein solutions to form more stable biopolymer particles by creating electrostatic complexes, i.e., a protein core surrounded by a polysaccharide shell. Furthermore, it can be hypothesized that at ambient temperature the globular protein molecules are attached to the polysaccharide if the pH is below or above the protein pI. Once the proteins are heated close to their thermal denaturation temperature, they detach from the polysaccharide and aggregate with each other forming protein nanoparticles. After cooling, the particles are coated with the polysaccharides (Jones & McClements, 2010a; Fig. 4). A major drawback of this method is its unsuitability for the encapsulation of heat-sensitive bioactive molecules.

4.1.3 Antisolvent precipitation of proteins

Antisolvent precipitation (desolvation) methods are based on decreasing the solvent quality of protein solutions, thereby leading to protein self-assembly and particle formation. Nucleation and subsequent particle growth through

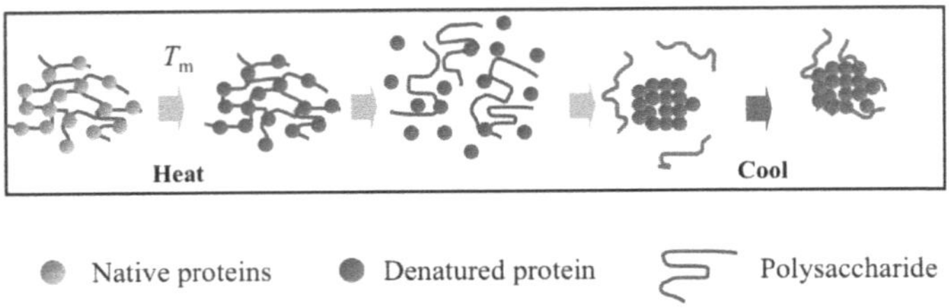

Figure 4 Proposed mechanism for biopolymer nanoparticle formation by thermal treatment of protein/polysaccharide complexes. *Reprinted with permission from Jones and McClements (2010a). Copyright (2014) John Wiley and Sons.*

condensation and coagulation are the driving forces that lead to the spontaneous formation of particles (Duclairoir, Nakache, Marchais, & Orecchioni, 1998; Oppenheim, Marty, & Speiser, 1978). The nucleation rate is particularly important because it determines final particle properties such as size, morphology, and composition. The key to producing ultrafine particles by antisolvent precipitation is to create conditions that favor rapid nucleation and slow particle growth.

The nucleation rate and therefore the final characteristics of the formed particles can be engineered by controlling several production parameters (Joye & McClements, 2013):

$$J = N_0 \nu \exp \frac{-16\pi\sigma^3 \nu_s^{\,2}}{3k^3 T^3 (\ln S)^2} \tag{1}$$

where J is the nucleation rate, N_0 is the number of solute molecules per volume, ν is the molecular transport frequency to the solid–liquid interface, σ is the interfacial tension at the solid–liquid interface, ν_s is the volume of one solute molecule, k is the Boltzmann constant, T is the temperature, S is the degree of super-saturation. From Eq. (1), it can be deduced that the main parameters affecting the particle characteristics are the ones that have an impact on the nucleation rate such as biopolymer concentration, antisolvent-to-solvent ratio, temperature, and solubility of the polymer in the antisolvent. To create the changes in solvent quality required to promote protein self-assembly, a protein is first dissolved in a solvent and then titrated into an antisolvent. Different types of proteins have already been used to produce nanoparticles or microparticles using this procedure (Ezpeleta et al., 1996; Gulseren, Fang, & Corredig, 2012; Joye, Nelis, & McClements, 2015; Teng, Luo, & Wang, 2012). Particle stability can be improved by cross-linking the proteins in the particles using enzymes (such as transglutaminase) or chemical agents (such as glutaraldehyde; Schmitt & Turgeon, 2011). It can also be improved by adsorbing other biopolymers or surfactants to their surface to increase electrostatic repulsion, increase steric repulsion, or reduce hydrophobic attraction. The driving force for surfactant or biopolymer adsorption is typically either hydrophobic or electrostatic attraction. These stabilizing agents can be added after particle fabrication (Joye et al., 2015) or can be dissolved in the antisolvent phase prior to particle production (Kim, Ng, Dong, Das, & Tan, 2012; Zhang, Le, Wang, Zhao, & Chen, 2013). Figure 5 shows gliadin particles produced with and without pectin in the antisolvent phase (Davidov-Pardo et al., 2015). It can be seen that the particles with pectin in the antisolvent (Fig. 5B) preserve their spherical shape after freeze-drying, while the particles without pectin tend to aggregate (Fig. 5A). For the antisolvent approach, the encapsulated bioactive molecules should usually have the same solubility behavior as the proteins used. Therefore, hydrophobic proteins (such as gliadin and zein) are suitable for the encapsulation of hydrophobic bioactive molecules (Duclairoir, Orecchioni, Depraetere, Osterstock, & Nakache, 2003; A. R. Patel, Hu, et al., 2010), while hydrophilic proteins (such as gelatin or whey protein isolate) are more suitable for the entrapment of hydrophilic

Figure 5 Scanning electron microscope pictures of gliadin particles produced by antisolvent precipitation (A) without and (B) with pectin in the antisolvent phase.

compounds (Gunasekaran, Ko, & Xiao, 2007; Huq, Khan, Khan, Riedl, & Lacroix, 2013). Nevertheless, particles produced with hydrophilic proteins typically need some additional stabilization mechanism to maintain their integrity when they are present in aqueous solutions.

4.1.4 Injection of proteins

This method is based on the injection of a protein solution into a hardening solution that promotes particle formation through protein cross-linking resulting in the formation of a microgel. The gelation induced in the hardening solution might be due to a temperature change (heating or cooling), cross-linking agents (e.g., multivalent salts, enzymes, or glutaraldehyde), or the presence of another polymer to form complexes (e.g., polysaccharides; Jones & McClements, 2010b; Murakami & Takashima, 2003). Filled hydrogel particles are useful to encapsulate hydrophobic bioactive compounds, for this purpose, an O/W emulsion is made and then added to the protein solution to be titrated in the hardening solution (Matalanis, Jones, & McClements, 2011). The dimensions of the hydrogel particles formed depend on the size of the titrated drops, which are usually relatively large (>100 μm).

4.1.5 Emulsion templating

Emulsion-templating methods can be used to reduce and better control the size of the hydrogel particles when compared to injection methods. Initially, a water-in-oil (W/O) emulsion is formed that contains the protein (gelling agent) and a hydrophilic bioactive compound. The protein within the water droplets of the W/O emulsion is then cross-linked using enzymes (Cho, Shim, & Park, 2003), thermal denaturation (Betz & Kulozik, 2011), UV-light induced cross-linking (van Rijn, Mougin, Franke, Park, & Boker, 2011), or addition of cross-linking ions (Egan, Jacquier, Rosenberg, & Rosenberg, 2013). After hardening, the micro- or nanoparticles formed are harvested from the continuous oil phase, e.g., by centrifugation, filtration, and washing with an organic solvent. To encapsulate hydrophobic compounds, a double emulsion could be used. In this case, an O/W emulsion is emulsified in the secondary oil phase to produce an O/W/O emulsion. Microfluidic or membrane emulsification can be used to improve particle homogeneity, i.e., form more uniformly sized particles (Xu, Li, Tan, & Luo, 2008). The stability of the emulsions during this process is important and depends on the type of emulsifier(s) used. As a general rule of thumb, W/O and O/W emulsions are stabilized by hydrophobic and hydrophilic surfactants, respectively (McClements, 2005). Additional strategies to inhibit creaming or sedimentation of emulsion droplets during particle fabrication is to reduce their size, reduce their density contrast, or increase the continuous phase viscosity (McClements, 2005).

4.1.6 Electrospinning of protein solutions

Protein particles (fibers or capsules) can be produced using electrospinning methods, which involve pulling a protein solution through a small nozzle using an electric field. The protein solution and bioactive component are dispersed in a suitable solvent and then placed within a capillary tube. When a high-voltage electric field is applied, the protein solution is pulled out of the capillary tube and forms a thin jet (Weiss, Kanjanapongkul, Wongsasulak, & Yoovidhya, 2012). Depending on the strength of the electrical field and the nature of the protein solution either fibers or capsules may be formed on the collection plate (López-Rubio & Lagaron, 2012). It is important that the time the jet of protein solution takes to travel from the capillary to the collector is long enough for the solvent to evaporate, otherwise the protein particles will fuse together (Weiss et al., 2012). One way to form capsules rather than fibers is to control the polymer concentration based on the following equation (Weiss et al., 2012):

$$(n_e)_{soln} = \frac{cM_w}{M_e} \tag{2}$$

where $(n_e)_{soln}$ is the degree of chain entanglement, c is the polymer concentration, M_w is the molecular weight of the polymer, and M_e is the average molecular weight between entanglement junctions. To form fibers, the entanglement number has to be >3.5. The viscosity of the sample also influences the formation of capsules or fibers, with capsules tending to be formed in less viscous solutions (Bohr, Boetker, Rades, Rantanen, & Yang, 2014). Either one of the resulting morphologies (capsules or fibers) of the electrospinning process can be used as delivery systems. Previous researchers produced zein nanofibers (Fernandez, Torres-Giner, & Lagaron, 2009) and whey protein nanocapsules (López-Rubio & Lagaron, 2012), which were successfully used as encapsulation vehicles for β-carotene. Certain food-grade proteins cannot be used alone to create electrospun fibers due to their complex macromolecular and 3D structures in combination with their strong inter- and intramolecular forces. To overcome this problem, surfactants, plasticizers, or reducing agents can be incorporated in the protein solutions, as well as other biopolymers more suitable to be electrospun. These polymers would be attached to the proteins by hydrophobic or electrostatic attractions, and/or hydrogen bonds (Aceituno-Medina, Mendoza, Lagaron, & López-Rubio, 2013).

4.1.7 Spray drying

Spray drying is a unit operation that involves atomizing a fluid product in a hot gas (100–200 °C) so as to rapidly convert it into a powder. The fluid that is fed into the spray dryer can be a protein solution, an emulsion, or a suspension. Typically, the hot gas is air, but nitrogen may also be used for oxygen sensitive or flammable materials (Zhong & Jin, 2009). Spray drying can be used to form solid particles from a variety of different hydrophilic and hydrophobic proteins, however, only those that are insoluble in water can be utilized in aqueous-based foods without dissolving, e.g., zein. Zein microcapsules can be formed by dissolving zein in a concentrated ethanol solution and then spraying it into a hot gas to evaporate the ethanol. The size of the protein particles produced depends on the characteristics of the feeding solution and the drying conditions. Spray drying is the most commonly used dehydration technique in the food industry, as well as the cheapest to produce microencapsulated food ingredients (Gharsallaoui, Roudaut, Chambin, Voilley, & Saurel, 2007). The particle formation process can be divided into three major steps: (1) the atomized fluid shrinks due to solvent evaporation and so the proportion of solute remaining in the droplet increases, (2) a surface shell is formed when the rate of solute diffusion to the interior of the droplets is less than the rate of solvent evaporation, and (3) heat starts to transfer from the shell to the interior of the droplet and the remainder of the solvent is evaporated. Depending on the permeability or strength of the shell, the solvent vapor is released by either inflation, cracking, or explosion of the particle (Bohr et al., 2014). During spray drying, the exposure of the product to heat is limited because of the latent heat of evaporation and the short drying time (Buffo & Reineccius, 2001), which helps prevent protein denaturation. Nevertheless, during spray drying, the hydrophobic regions of the proteins may be exposed causing aggregation. This can be reduced by (1) the addition of a surfactant to prevent the formation of insoluble aggregates, (2) the addition of divalent ions to prevent the formation of soluble aggregates, and (3) increasing the protein concentration in the liquid feed (Ameri & Yuh-Fun, 2006). To achieve a full coverage of the active material and a homogeneous product, a stable emulsion or suspension must be fed into the spray dryer. The advantages of using this technique are low cost of operation, availability of equipment, and reliable performance. The disadvantages are potential variability in capsule size and morphology and the requirement of low viscosity solutions to feed to the spray dryer.

4.2. Particles formed using a protein and other biopolymers

Two broad approaches can be distinguished to fabricate nanoparticles or microparticles from a protein and another biopolymer (such as different proteins or polysaccharides). One approach is primarily based on the utilization of attractive forces between a protein and another biopolymer ("aggregative" particle formation), while the other approach is mainly based on the utilization of repulsive forces ("segregative" particle formation).

4.2.1 *Aggregative particle formation*

Solution pH is of critical importance in the formation of protein particles based on the aggregative approach. The surface charges on the protein and polysaccharide molecules can be modulated by controlling pH (Damodaran, 1997). As an example, the formation of biopolymer particles based on the interactions of a protein with an anionic polysaccharide is summarized below (Jones & McClements, 2010b; Fig. 6). When the pH is well

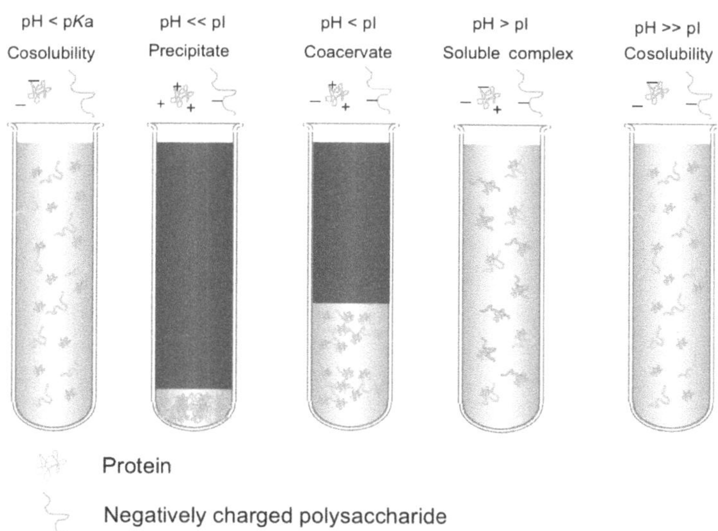

Figure 6 Schematic representation of the aggregative method to produce protein particles related to the pH of the system.

above the isoelectric point (pI) of the protein, the two biopolymers have negative charges and therefore strongly repel each, which prevents particle formation. When the pH is only slightly above the pI of the protein, the anionic polysaccharide binds to cationic patches on the protein surfaces and leads to the formation of soluble complexes. When the pH is reduced further, the soluble complexes may associate with each other and form electrostatic complexes known as "coacervates." Initially, coacervates are present as small biopolymer-rich droplets suspended in the aqueous phase (water-in-water, W/W, emulsions), but they are highly prone to coalescence due to their low net charge and tend to phase separate. This problem may be overcome by cross-linking the outer layer of the initial coacervate droplets using enzymes or chemicals (Schmitt & Turgeon, 2011), or by coating them with other biopolymers. Hydrophilic compounds can be directly encapsulated in coacervates (Bédié, Turgeon, & Makhlouf, 2008), while hydrophobic compounds first have to be encapsulated in oil-in-water (O/W) emulsions that are then mixed with the coacervates (Ru, Yu, & Huang, 2010). If the pH is reduced too far, the biopolymers become so closely packed together that they form precipitates that sediment. Finally if the pH is taken well below the pK_a value of anionic polysaccharides, no complex formation occurs due to the loss of charge of the polysaccharide.

4.2.2 Segregative particle formation approaches

In this case, thermodynamic incompatibility of two biopolymers mainly driven by repulsive forces is used to fabricate protein particles. Certain combinations of biopolymers spontaneously separate into two phases when they are added to aqueous solutions depending on the biopolymer types, biopolymer concentrations, and solution conditions (Tolstoguzov, 2003). One of the phases is rich in one biopolymer and depleted in the other and vice versa. The main driving force for phase separation is the so-called excluded volume effect (Jones & McClements, 2010b). The effective volume occupied by a biopolymer molecule in solution plays an important role in this effect, with biopolymers that have more open structures being more effective at inducing phase separation than those with more compact structures. Thus, high molecular weight biopolymers with open structures tend to induce phase separation when used at relatively low concentrations (such as xanthan), whereas low molecular weight biopolymers with compact structures only induce this effect when used at relatively high concentrations (such as globular proteins). In general, mixed biopolymer solutions have to be sufficiently concentrated to achieve phase separation. If the biopolymer concentration is not high enough, a single phase is formed consisting of an intimate mixture of the two repulsive biopolymers. However, if the biopolymer concentration exceeds a critical level, then two phases are formed. If this system is thoroughly mixed, a W/W emulsion can be formed, and after hardening of the droplets, particles are obtained (Zasypkin, Braudo, & Tolstoguzov, 1997). Droplets can be hardened into particles through enzymatic or chemical cross-linking agents, or by inducing environmental changes such as temperature or ionic strength. If the droplets of the dispersed phase are not hardened, they are prone to coalesce and eventually cream or sediment, if their density is lower or higher than the continuous phase, respectively. The type of W/W emulsion formed depends on the relative volume fraction of each biopolymer phase. The phase with the larger volume fraction will usually make up the continuous phase. When the two volume fractions are similar, a bicontinuous structure consisting of interconnecting regions of each biopolymer tends to be formed. A simple phase diagram can be used to determine which biopolymer solution will be the continuous phase and which will be the dispersed phase (Fig. 7). In this diagram, the relative volume fraction of the respective biopolymers is represented on the x- and y-axes. The binodal line represents the boundary between the monophasic and biphasic regions. This kind of delivery systems is useful for incorporating hydrophobic bioactive molecules. Initially, an oil phase is emulsified using an emulsifier that will enable the lipid droplets to preferentially accumulate into the dispersed aqueous phase, resulting in an O/W/W emulsion (Matalanis et al., 2011; Matalanis & McClements, 2013).

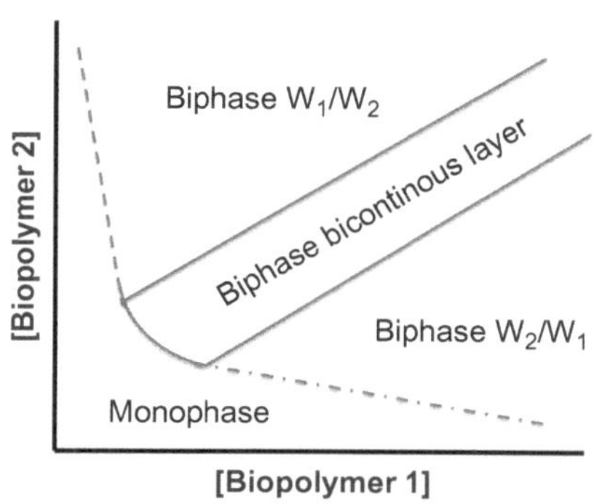

Figure 7 Theoretical phase diagram of mixed solutions between biopolymers 1 and 2 that experience incompatible phase separation. Hypothetical biphasic W/W emulsions.

5. STABILITY OF PROTEIN PARTICLES IN FOODS

5.1. Temperature

In the food industry, heating and/or cooling are common processing steps to create, modify, or preserve foods. Heating is also a common procedure in meal preparation by consumers at home. Protein particles may have to maintain their integrity during these processes so as to retain and protect any encapsulated compounds, or they may have to breakdown and release the encapsulated compounds in response to a specific thermal trigger. Globular proteins (such as those in whey, soy, and egg) unfold when heated above their thermal denaturation temperature thereby exposing nonpolar and sulfur-containing side groups, which promotes irreversible particle aggregation through hydrophobic interactions and/or disulfide bonds (Jones et al., 2009). Furthermore, heating typically speeds up the collision rate between particles which would result in an increased rate and degree of aggregation.

5.2. Ionic strength

Salt addition is often used to improve the flavor and extend the shelf life of food products, but it also simultaneously increases the ionic strength of the solution surrounding the protein particles, which can affect their stability. Assessing protein particle stability at elevated ionic strengths is therefore important for understanding how they behave when incorporated into actual food products. Usually the surface charge (ζ-potential) of proteins becomes smaller as the salt concentration increases due to the electrostatic screening and ion-binding effects, thereby resulting in the aggregation of protein particles (Salgin, Salgin, & Bahadir, 2012).

5.3. pH

Protein particles will be incorporated into a variety of food matrices with widely varying pH values, from acidic to alkaline. In addition, the pH of foods often changes during production, storage, or preparation, or after they have been ingested and pass through the GIT. Assessing the pH stability of protein particles is therefore important to predict their performance in commercial food products. One of the most important aspects of the pH for delivery system development is its influence on the electrical charge of protein particles. The pH at which positive and negative charge contributions on protein particles are equal (zero net charge) is usually referred to as the isoelectric point (pI). Below the pI, most of the carboxyl and amino side groups are protonated ($-CO_2H$ and $-NH_3^+$) and so the net charge is positive, while above the pI, most of these groups are non-protonated ($-NH_2$ and $-CO_2^-$) and so the net charge is negative (Damodaran, 1997). A minimal electrical charge is needed to ensure stability against particle aggregation based on the electrostatic repulsions (Jahanshahi & Babaei, 2008). A ζ-potential of ± 30 mV is considered as the limit above which a colloidal system is stable to this kind of aggregation, but this value depends on the nature of the other attractive and repulsive forces in the system (Uskokovic, 2012). The complexation of proteins in nanoparticles or microparticles will have an effect on the system pI, which will be different from individual proteins (Patel, Bouwens, et al., 2010).

5.4. Biopolymer coatings

As mentioned above, certain biopolymers can be used to stabilize protein particles by forming a protective coating around them that inhibits aggregation through electrostatic and/or steric repulsion. Nevertheless, if the biopolymer concentration is significantly lower or higher than the concentration required to completely coat the particles then bridging or depletion flocculation occurs (McClements, 2005). Bridging flocculation occurs when there is insufficient coating biopolymer to completely cover the protein particle surfaces. In this case, two or more particles are attached to one biopolymer molecule leading to the particle aggregation. Conversely, when there is an excess of free biopolymer molecules unattached to the protein particle surfaces they generate an osmotic attraction that drives the protein particles together. Consequently, the biopolymer concentration added must be carefully controlled.

6. PROTEIN PARTICLE CHARACTERIZATION

Particle characteristics (such as composition, size, shape, and charge) have a large impact on the functionality and stability of protein particles, as well as on final food quality. Hence, it is important to have suitable analytical techniques and protocols to characterize particle properties.

6.1. Particle size

The size of protein particles is usually determined by light-scattering or microscopy techniques. Two types of light-scattering instrument are commonly used to measure particle size: static light scattering (SLS) and dynamic light scattering (DLS). SLS analyzes the intensity of scattered light waves as a function of scattering angle and then finds the particle size distribution that gives the best fit to this scattering pattern using an appropriate mathematical model, such as the Mie theory. This approach assumes that the particles are non-interacting spheres with well-defined refractive indices, which is not always true in practice. In addition, artifacts can be introduced through sample preparation steps such as dilution or stirring. DLS monitors the direction and speed of particle movement in solution due to the Brownian motion, i.e., the zigzag motion caused by collision with solvent molecules. The movement of the particles causes the scattering intensity to fluctuate, which can be translated into the velocity of the Brownian motion. As described by the Stokes–Einstein relation, the Brownian motion is strongly dependent on particle size:

$$D = \frac{k_B \times T}{6 \times \pi \times \eta \times r} \tag{3}$$

With D, the diffusion constant; k_B, Boltzmann's constant; T, the temperature (in K); η, the viscosity of the solution; r, the hydrodynamic radius (Uskokovic, 2012). Analogous to what is the case for SLS, the results of DLS should be carefully interpreted. The model used to interpret the data assumes that the particles are spherical and move in a Newtonian fluid (Uskokovic, 2012). Typically, SLS is suitable for protein particles that are relatively large ($d > 100$ nm), whereas DLS is suitable for particles that are relatively small ($d < 10{,}000$ nm).

Reliable particle size measurements cannot be made using light-scattering techniques on samples containing complex particle morphologies, such as fibers. In this case, microscopy is typically used, such as optical, electron, or atomic microscopy. The dimensions of spherical protein particles can also be studied using microscopy. Conventional optical microscopy is suitable for the analysis of relatively large particles (>1000 nm), while electron and atomic force microscopy is more suitable for smaller particles (Pierucci et al., 2007). Two different modes of electron microscopy are generally used, i.e., scanning and transmission modes. Samples for scanning electron microscopy (SEM) are typically frozen, dried, or fractured and coated with a thin metal layer (Pierucci et al., 2007). Sample preparation has to be done carefully as this can introduce artifacts that are not present in the original sample. For transmission electron microscopy (TEM), sample preparation is generally gentler although biopolymer systems generally lack the density contrast required for high-quality imaging. Therefore, staining using heavy metals is often used to create enough contrast between the sample and the background (Kalab, Allan-Wojtas, & Miller, 1995). Atomic force microscopy generally requires mild sample preparation methods that substantially reduce the risk of damaging or altering the samples (Burey et al., 2008).

Another technique for particle sizing that recently experienced a revival is hydrodynamic chromatography. In this technique, the particle suspension is led through a packed bed of spherical particles and the rate of transport of the suspension particles through the bed is dependent, among other factors, on the size of the particles (Small, 1974; Striegel & Brewer, 2012).

6.2. Particle morphology

Microscopy is the most widely used technique to study particle morphology and aggregation state. As described earlier, conventional optical microscopy can be used to study microparticles ($d > 1000$ nm). Particle substructures can be visualized by dyes or fluorescent probes that selectively associate with one component in the particle suspension or that have been covalently linked to one of the components before particle production. Confocal fluorescence microscopy can also be used on fluorescent samples and generally provides a better quality image. The resolution of confocal scanning laser microscopy, however, is similar to that of conventional optical microscopy, which is, hence, inadequate to study the morphology of nanoparticles. In this case, atomic force or electron microscopy is better suited (Pierucci et al., 2007). SEM may provide information on surface characteristics, such as composition and relief (Jahanshahi & Babaei, 2008; Montes, Gordillo, Pereyra, & Martinez de la Ossa, 2011; Pierucci et al., 2007). The inner morphology of particles can be studied with TEM after fixation, dehydration, and slicing (Chen & Subirade, 2005). Atomic force microscopy, on the other hand, is useful for observing the surface topology of nanoparticles and microparticles.

6.3. Particle charge

Particle charge is often reported as the ζ-potential, which is a fairly easily measurable quantity (McClements, 2005). ζ-Potential is measured by applying an electric field and analyzing the resulting electrophoretic mobility of the charged particles. The mobility is typically determined by measuring fluctuations in the intensity of scattered light. ζ-Potential is then derived from this electrophoretic mobility using the Smoluchowski equation:

$$\zeta = \frac{\eta \nu}{\varepsilon E} \tag{4}$$

with η, the viscosity coefficient of the medium; ν, the electrophoretic velocity; ε, the dielectric constant of the medium; and E, the gradient of the applied electric field.

Surface charge has been used for centuries to engineer the stability of colloidal suspensions. The importance of the (surface) charge to particle stability is described by the Derjaguin, Landau, Verwey, and Overbeek theory, which was first written down in the 1940s. This theory takes into account repulsive electric forces and attractive (short-range) van der Waals forces (Uskokovic, 2012). As the pH and ionic strength of the system affect the ζ-potential of the particles, sample preparation steps before measurement, such as dilution, need to be carried out in a well-considered way. In addition, in systems consisting of a mixture of different biopolymer particles it may be difficult to interpret the data since they all contribute to the overall signal. Next to predicting the suspension stability, ζ-potential can also be used to study the location of bioactive components within the biopolymer particles, i.e., surface versus interior.

6.4. Particle composition and physical state

The small dimensions of protein particles make the determination of particle composition and component distribution challenging. If (fluorescent) dyes are available to specifically label particle components (confocal scanning laser) microscopy can be used (to some extent) to determine particle composition. Other techniques that have been used to study the chemical composition of particle surface and the overall particle are X-ray photoelectron spectroscopy and elemental analysis, respectively (Montes et al., 2011). The identification of certain components in the particles can also be performed with Raman spectroscopy (Beirao-da-Costa et al., 2013), which together with infrared spectroscopy is gaining importance for studying biopolymer interactions (Beirao-da-Costa et al., 2013; Bhattacharya et al., 2013). Fluorimetry assays combined with mathematical models such as the Stern–Volmer and the Van't Hoff equations (Keppler, Stuhldreier, Temps, & Schwarz, 2014) can be used to understand the nature and strength of the interaction to come to a more rational approach to the design of protein particles to encapsulate bioactive compounds.

7. UTILIZATION

Microparticles and nanoparticles have been investigated for various applications in food systems, such as delivery of bioactives, colors, flavors, antimicrobial agents, probiotica, and design of food structure and rheology. The latter application is especially important in the quest for the development and production of food products with a reduced fat or carbohydrate content (Duran & Marcato, 2013; Mihindukulasuriya & Lim, 2013; Woranuch & Yolsan, 2013).

7.1. Delivery systems

Encapsulation of bioactive compounds is not a new concept and the first report on protecting vitamins from oxidation by encapsulation stems from around 75 years ago (Taylor, 1939). Much more research has since then been devoted on the production of colloidal delivery systems for bioactive molecules, colors, flavors, microbial agents, and even living cells (de Vos, Faas, Spasojevic, & Sikkema, 2010; Huq et al., 2013; Sekhon, 2010).

When evaluating the suitability of protein particles as encapsulation system for a specific bioactive, the loading capacity, encapsulation efficiency, and retention efficiency are typically factors of interest.

The loading capacity can be defined as

$$\frac{\text{mass of the encapsulated molecule retained in the particles}}{\text{total mass of the particles}} \times 100$$

while the encapsulation efficiency can be defined as

$$\frac{\text{mass of the encapsulated molecule in the particle}}{\text{total mass of the encapsulated molecule in the system}} \times 100$$

Both parameters give an indication of the potential of a certain colloidal system to interact/bind/withhold a certain molecule of interest during production of the loaded colloidal system. However, the retention of this molecule under certain conditions encountered in food processing, storage, or ingestion is also of utmost importance. This can be described by the retention efficiency:

$$\frac{\text{mass of the encapsulated molecule in the particle after the treatment}}{\text{mass of the encapsulated molecule in the particle before the treatment}} \times 100$$

Delivery systems can be tested after heating, pH adjustments, alterations in ionic strength, or application of mechanical stresses (stirring, mixing, pumping) to determine their stability under conditions they may experience in food products. The nature of the interaction between the encapsulated molecules and the particle matrix determines the stability, retention capacity, and release behavior of the encapsulated molecules. Molecules can be physically entrapped in or chemically bound to the particle matrix. Physical entrapment generally requires the encapsulated molecule to be present in the ingredient mixture before the formation of the particles. If the encapsulated molecule is small (compared to the pores in the matrix of the particles) and/or when there is a strong interaction between the matrix and the encapsulated molecule, postproduction loading is also a possibility (Hejazi & Amiji, 2002; Jameela, Misra, & Jayakrishnan, 1994; Kumbar, Kulkarni, & Aminabhavi, 2002). For proteins, the loading is generally more efficient if it is carried out around the isoelectric point of the proteins. At this pH, the proteins are usually not very soluble and their absorption capacity is at its highest (Jahanshahi & Babaei, 2008).

For bioactive components, encapsulation no longer only focuses on the protection of the components in the complex food matrix or during processing. A lot of research has also been put into predicting and controlling the bioavailability of these molecules in the GIT. Bioavailability is the product of several factors (McClements & Xiao, 2012):

- the fraction of the nutrient that is still present in an active form in the food product at the time of consumption;
- the fraction of the nutrient that, after ingestion, is released from the food matrix and solubilized in the GIT fluids, i.e., the bioaccessibility;
- the fraction of the nutrient, released from the food matrix, that is transported across the mucus layer, i.e., the absorption;
- the fraction that is present in the GIT in a metabolically active form, i.e., metabolism.

A few of the factors that have been extensively studied are prolonging the gastric residence time and increasing the solubility and stability within the GIT. Proteins that display mucoadhesive properties, such as gluten, have been shown to increase the time a bioactive component spends in the stomach, thereby increasing the absorption of the component (Arangoa et al., 2000). Proteins are generally degraded in the upper parts of the GIT by peptidases and are hence more useful for target delivery and controlled release in the stomach or small intestine.

7.2. Food structure-building stones

Structuring of food ingredients on a nanoscale or microscale seems a promising strategy to improve taste, texture, and consistency of food products, e.g., for reduced fat content applications (Sekhon, 2010). The acceptability of particles in food products is largely dependent on their detectability in the mouth, which depends on particle size, hardness, morphology, and charge, as well as food matrix characteristics, such as viscosity (Burey et al., 2008). Depending on the food product, visual detection of particles is often undesirable. This is influenced by particle size and concentration and is especially important for transparent products such as many beverages.

Gelled protein particles are very attractive to food applications as they generally combine macroscopic structure formation with flowable properties and have a soft solid texture (Burey et al., 2008). For food products, which have a reduced fat level and, hence, also lack (part of) the functionality of fat, gelled particles seem a suitable alternative. Fat replacers should possess textural, sensory, and visual properties, which are similar to the fat droplets they replace. Oil-filled caseinate-rich hydrogel beads coated by a layer of anionic pectin were found to be suitable for the formulation of reduced calorie food products as they modulate the texture, appearance, flavor, and stability of emulsion-based products (Chung, Degner, Decker, & McClements, 2013). The trapping of fat droplets into these hydrogel particles further assists in balancing the flavor profile of products with a reduced fat content (Chung, Degner, & McClements, 2013).

Protein microparticles have also been used before to adjust the mechanical properties of products in which the protein concentration was changed and protein microstructuring seems to be an interesting tool to control the textural properties of high protein foods (Purwanti, Peters, & van der Goot, 2013).

ACKNOWLEDGMENTS

Dr. Gabriel Davidov-Pardo is a recipient of a postdoctoral fellowship by the Secretaria de Ciencia Tecnología e Innovación del Distrito Federal (SECITI, Mexico City). Dr. Iris Joye gratefully acknowledges financial support from the "Fonds voor Wetenschappelijk Onderzoek—Vlaanderen" (FWO, Brussels, Belgium) and from the European Commission 7th Framework Program (FP7-People-2011-IOF-300408). This chapter is based on the work supported by the Cooperative State Research, Extension, Education Service, U.S. Department of Agriculture, Massachusetts Agricultural Experiment Station (Project No. 831), and U.S. Department of Agriculture, NIFA, NRI Grant.

REFERENCES

Aceituno-Medina, M., Mendoza, S., Lagaron, J. M., & López-Rubio, A. (2013). Development and characterization of food-grade electrospun fibers from amaranth protein and pullulan blends. *Food Research International*, *54*(1), 667–674.http://dx.doi.org/10.1016/j.foodres.2013.07.055.

Acharya, D. P., Sanguansri, L., & Augustin, M. A. (2013). Binding of resveratrol with sodium caseinate in aqueous solutions. *Food Chemistry*, *141*(2), 1050–1054. http://dx.doi.org/10.1016/j.foodchem.2013.03.037.

Ameri, M., & Yuh-Fun, M. (2006). Spray drying of biopharmaceuticals: Stability and process considerations. *Drying Technology*, *24*(6), 763–768. http://dx.doi.org/10.1080/03602550600685275.

Arangoa, M. A., Ponchel, G., Orecchioni, A. M., Renedo, M. J., Duchene, D., & Irache, J. M. (2000). Bioadhesive potential of gliadin nanoparticulate systems. *European Journal of Pharmaceutical Sciences*, *11*, 333–341.

Bédié, G. K., Turgeon, S. L., & Makhlouf, J. (2008). Formation of native whey protein isolate–low methoxyl pectin complexes as a matrix for hydro-soluble food ingredient entrapment in acidic foods. *Food Hydrocolloids*, *22*(5), 836–844. http://dx.doi.org/10.1016/j.foodhyd.2007.03.010.

Beirao-da-Costa, S., Duarte, C., Bourbon, A. I., Pinheiro, A. C., Januario, M. I. N., Vicente, A. A., et al. (2013). Inulin potential for encapsulation and controlled delivery of Oregano essential oil. *Food Hydrocolloids*, *33*, 199–206.

Belitz, H. D., Grosch, W., & Schieberle, P. (2009). *Food chemistry*. Berlin, Germany: Springer-Verlag.

Betz, M., & Kulozik, U. (2011). Microencapsulation of bioactive bilberry anthocyanins by means of whey protein gels. *11th international congress on engineering and food (ICEF11) Procedia Food Science*, *1*, 2047–2056. http://dx.doi.org/10.1016/j.profoo.2011.10.006.

Bhattacharya, S. S., Banerjee, S., Chowdhury, P., Ghosh, A., Hegde, R. R., & Mondal, R. (2013). Tranexamic acid loaded gellan gum-based polymeric microbeads for controlled release: In vitro and in vivo assessment. *Colloids and Surfaces B: Biointerfaces*, *112*, 483–491.

Bohr, A., Boetker, J., Rades, T., Rantanen, J., & Yang, M. (2014). Application of spray-drying and electrospraying/electospinning for poorly watersoluble drugs: A particle engineering approach. *Current Pharmaceutical Design*, *20*(3), 325–348.

Bromley, E. H. C., Krebs, M. R. H., & Donald, A. M. (2006). Mechanisms of structure formation in particulate gels of β-lactoglobulin formed near the isoelectric point. *The European Physical Journal E*, *21*(2), 145–152. http://dx.doi.org/10.1140/epje/i2006-10055-7.

Buffo, R., & Reineccius, G. (2001). Comparison among assorted drying processes for the encapsulation of flavors. *Perfumer & Flavorist*, *26*(4), 58–67.

Burey, P., Bhandari, B. R., Howes, T., & Gidley, M. J. (2008). Hydrocolloid gel particles: Formation, characterization, and application. *Critical Reviews in Food Science and Nutrition*, *48*(5), 361–377.

Chen, L., Hebrard, G., Beyssac, E., Denis, S., & Subirade, M. (2010). In vitro study of the release properties of soy-zein protein microspheres with a dynamic artificial digestive system. *Journal of Agricultural and Food Chemistry*, *58*, 9861–9867.

Chen, L., & Subirade, M. (2005). Chitosan/beta-lactoglobulin core–shell nanoparticles as nutraceutical carriers. *Biomaterials*, *26*, 6041–6053.

Chen, L., & Subirade, M. (2009). Elaboration and characterization of soy/zein protein microspheres for controlled nutraceutical delivery. *Biomacromolecules*, *10*, 3327–3334.

Cho, Y. H., Shim, H. K., & Park, J. (2003). Encapsulation of fish oil by an enzymatic gelation process using transglutaminase cross-linked proteins. *Journal of Food Science*, *68*(9), 2717–2723.

Chung, C., Degner, B., Decker, E. A., & McClements, D. J. (2013). Oil-filled hydrogel particles for reduced-fat food applications: Fabrication, characterization and properties. *Innovative Food Science and Emerging Technologies*, *20*, 324–334.

Chung, C., Degner, B., & McClements, D. J. (2013). Designing reduced-fat food emulsions: Locust bean gum-fat droplet interactions. *Food Hydrocolloids*, *32*(2), 263–270.

Damodaran, S. (1997). *Food proteins and their applications*. Boca Raton, FL: Taylor & Francis.

Damodaran, S., Parkin, K. L., & Fennema, O. R. (2008). *Fennema's food chemistry*. New York, NY: Marcel Dekker, Inc.

Davidov-Pardo, G., Joye, I. J., & McClements, D. J. (2015). Encapsulation of resveratrol in biopolymer particles produced using liquid antisolvent precipitation. Part 1: Preparation and characterization. *Food Hydrocolloids*, (0). doi: http://dx.doi.org/10.1016/j.foodhyd.2014.11.023.

de Vos, P., Faas, M. M., Spasojevic, M., & Sikkema, J. (2010). Encapsulation for preservation of functionality and targeted delivery of bioactive food components. *International Dairy Journal*, *20*(4), 292–302.

Duclairoir, C., Nakache, E., Marchais, H., & Orecchioni, A. M. (1998). Formation of gliadin nanoparticles: Influence of the solubility parameter of the protein solvent. *Colloid and Polymer Science*, *276*, 321–327.

Duclairoir, C., Orecchioni, A. M., Depraetere, P., & Nakache, E. (2002). Alpha-tocoperol encapsulation and in vitro release from wheat gliadin nanoparticles. *Journal of Microencapsulation*, *19*, 53–60.

Duclairoir, C., Orecchioni, A. M., Depraetere, P., Osterstock, F., & Nakache, E. (2003). Evaluation of gliadins nanoparticles as drug delivery systems: A study of three different drugs. *International Journal of Pharmaceutics*, *253*, 133–144.

Duran, N., & Marcato, P. D. (2013). Nanobiotechnology perspectives. Role of nanotechnology in the food industry: A review. *International Journal of Food Science and Technology*, *48*, 1127–1134.

Egan, T., Jacquier, J.-C., Rosenberg, Y., & Rosenberg, M. (2013). Cold-set whey protein microgels for the stable immobilization of lipids. *Food Hydrocolloids*, *31*(2), 317–324. http://dx.doi.org/10.1016/j.foodhyd.2012.11.008.

Ezpeleta, I., Irache, J. M., Stainmesse, S., Chabenat, C., Gueguen, J., Popineau, Y., et al. (1996). Gliadin nanoparticles for the controlled release of all-trans-retinoic acid. *International Journal of Pharmaceutics*, *131*, 191–200.

Fernandez, A., Torres-Giner, S., & Lagaron, J. M. (2009). Novel route to stabilization of bioactive antioxidants by encapsulation in electrospun fibers of zein prolamine. *Food Hydrocolloids*, *23*(5), 1427–1432.http://dx.doi.org/10.1016/j.foodhyd.2008.10.011.

Gharsallaoui, A., Roudaut, G., Chambin, O., Voilley, A., & Saurel, R. (2007). Applications of spray-drying in microencapsulation of food ingredients: An overview. *Food Research International*, *40*(9), 1107–1121.

Gulseren, I., Fang, Y., & Corredig, M. (2012). Complexation of high methoxyl pectin with ethanol desolvated whey protein nanoparticles: Physico-chemical properties and encapsulation behaviour. *Food & Function*, *3*(8), 859–866. http://dx.doi.org/10.1039/c2fo10235h.

Gunasekaran, S., Ko, S., & Xiao, L. (2007). Use of whey proteins for encapsulation and controlled delivery applications. *Journal of Food Engineering*, *83*(1), 31–40. http://dx.doi.org/10.1016/j.jfoodeng.2006.11.001.

Hedayati, R., Jahanshahi, M., & Attar, H. (2011). Fabrication and characterization of albumin-acacia nanoparticles based on complex coacervation as potent nanocarrier. *Journal of Chemical Technology and Biotechnology*, *87*, 1401–1408.

Hejazi, R., & Amiji, M. (2002). Stomach-specific anti-H. pylori therapy. I: Preparation and characterization of tetracycline-loaded chitosan microspheres. *International Journal of Pharmaceutics*, *235*, 87–94.

Hemar, Y., Gerbeaud, M., Oliver, C. M., & Augustin, M. A. (2011). Investigation into the interaction between resveratrol and whey proteins using fluorescence spectroscopy. *International Journal of Food Science and Technology*, *46*(10), 2137–2144. http://dx.doi.org/10.1111/j.1365-2621.2011.02728.x.

Hoffmann, M. A. M., Roefs, S. P. F. M., Verheul, M., van Mil, P. J. J. M., & De Kruif, K. G. (1996). Aggregation of β-lactoglobulin studied by in situ light scattering. *Journal of Dairy Research*, *63*(03), 423–440.

Huq, T., Khan, A., Khan, R. A., Riedl, B., & Lacroix, M. (2013). Encapsulation of probiotic bacteria in biopolymeric system. *Critical Reviews in Food Science and Nutrition*, *53*(9), 909–916.

Jahanshahi, M., & Babaei, Z. (2008). Protein nanoparticle: A unique system as drug delivery vehicles. *African Journal of Biotechnology*, 7, 4926–4934.

Jameela, S. R., Misra, A., & Jayakrishnan, A. (1994). Cross-linked chitosan microspheres as carriers for prolonged delivery of macromolecular drugs. *Journal of Biomaterials Science, Polymer Edition*, *6*, 621–632.

Jones, O. G., Decker, E. A., & McClements, D. J. (2009). Formation of biopolymer particles by thermal treatment of β-lactoglobulin–pectin complexes. *Food Hydrocolloids*, *23*(5), 1312–1321.http://dx.doi.org/10.1016/j.foodhyd.2008.11.013.

Jones, O. G., & McClements, D. J. (2010a). Biopolymer nanoparticles from heat-treated electrostatic protein–polysaccharide complexes: Factors affecting particle characteristics. *Journal of Food Science*, *75*(2), N36–N43. http://dx.doi.org/10.1111/j.1750-3841.2009.01512.x.

Jones, O. G., & McClements, D. J. (2010b). Functional biopolymer particles: Design, fabrication, and applications. *Comprehensive Reviews in Food Science and Food Safety*, *9*(4), 374–397. http://dx.doi.org/10.1111/j.1541-4337.2010.00118.x.

Joye, I. J., & McClements, D. J. (2013). Production of nanoparticles by anti-solvent precipitation for use in food systems. *Trends in Food Science and Technology*, *34*(2), 109–123.

Joye, I. J., Davidov-Pardo, G., Ludescher, R. D., & McClements, D. J. (2014). Binding of resveratrol to zein and gliadin: A more rational approach of resveratrol encapsulation using water-insoluble proteins, submitted for publication.

Joye, I. J., Nelis, V. A., & McClements, D. J. (2015). Gliadin-based nanoparticles: Stabilization by post-production polysaccharide coating. *Food Hydrocolloids*, *43*, 236–242.

Kalab, M., Allan-Wojtas, P., & Miller, S. S. (1995). Microscopy and other imaging techniques in food structure analysis. *Trends in Food Science and Technology*, *6*, 177–186.

Kaya-Celiker, H., & Mallikarjunan, K. (2012). Better nutrients and therapeutics delivery in food through nanotechnology. *Food Engineering Reviews*, *4*(2), 114–123. http://dx.doi.org/10.1007/s12393-012-9050-3.

Keppler, J. K., Stuhldreier, M. C., Temps, F., & Schwarz, K. (2014). Influence of mathematical models and correction factors on binding results of polyphenols and retinol with β-lactoglobulin measured with fluorescence quenching. *Food Biophysics*, *9*(2), 158–168. http://dx.doi.org/10.1007/s11483-013-9328-x.

Khan, S. A., & Schneider, M. (2013). Improvement of nanoprecipitation technique for preparation of gelatin nanoparticles and potential macromolecular drug loading. *Macromolecular Bioscience*, *13*(4), 455–463. http://dx.doi.org/10.1002/Mabi.201200382.

Kim, S., Ng, W. K., Dong, Y., Das, S., & Tan, R. B. H. (2012). Preparation and physicochemical characterization of trans-resveratrol nanoparticles by temperature-controlled antisolvent precipitation. *Journal of Food Engineering*, *108*(1), 37–42. http://dx.doi.org/10.1016/j.jfoodeng.2011.07.034.

Kumbar, S. G., Kulkarni, A. R., & Aminabhavi, T. M. (2002). Cross-linked chitosan microspheres for encapsulation of diclofenac sodium: Effect of cross-linking agent. *Journal of Microencapsulation*, *19*, 173–180.

Langer, K., Balthasar, S., Vogel, V., Dinauer, N., von Briesen, H., & Schubert, D. (2003). Optimization of the preparation process for human serum albumin (HSA) nanoparticles. *International Journal of Pharmaceutics*, *257*(1–2), 169–180. http://dx.doi.org/10.1016/S0378-5173(03)00134-0.

Le Bon, C., Nicolai, T., & Durand, D. (1999). Kinetics of aggregation and gelation of globular proteins after heat-induced denaturation. *Macromolecules*, *32*(19), 6120–6127. http://dx.doi.org/10.1021/ma9905775.

Li, K.-K., Yin, S.-W., Yin, Y.-C., Tang, C.-H., Yang, X.-Q., & Wen, S.-H. (2013). Preparation of water-soluble antimicrobial zein nanoparticles by a modified antisolvent approach and their characterization. *Journal of Food Engineering*, *119*, 343–352.

López-Rubio, A., & Lagaron, J. M. (2012). Whey protein capsules obtained through electrospraying for the encapsulation of bioactives. *Innovative Food Science & Emerging Technologies*, *13*, 200–206.http://dx.doi.org/10.1016/j.ifset.2011.10.012.

Madene, A., Jacquot, M., Scher, J., & Desobry, S. (2006). Flavour encapsulation and controlled release—A review. *International Journal of Food Science & Technology*, *41*(1), 1–21.

Matalanis, A., Decker, E. A., & McClements, D. J. (2012). Inhibition of lipid oxidation by encapsulation of emulsion droplets within hydrogel microspheres. *Food Chemistry*, *132*(2), 766–772.

Matalanis, A., Jones, O. G., & McClements, D. J. (2011). Structured biopolymer-based delivery systems for encapsulation, protection, and release of lipophilic compounds. *Food Hydrocolloids*, *25*(8), 1865–1880. http://dx.doi.org/10.1016/j.foodhyd.2011.04.014.

Matalanis, A., & McClements, D. J. (2013). Hydrogel microspheres for encapsulation of lipophilic components: Optimization of fabrication & performance. *Food Hydrocolloids*, *31*(1), 15–25. http://dx.doi.org/10.1016/j.foodhyd.2012.09.012.

McClements, D. J. (2005). *Food emulsions: Principles, practice, and techniques*. London, UK: CRC Press Inc.

McClements, D. J., Decker, E. A., Park, Y., & Weiss, J. (2009). Structural design principles for delivery of bioactive components in nutraceuticals and functional foods. *Critical Reviews in Food Science and Nutrition*, *49*(6), 577–606. http://dx.doi.org/10.1080/10408390902841529.

McClements, D. J., & Xiao, H. (2012). Potential biological fate of ingested nanoemulsions: Influence of particle characteristics. *Food & Function*, *3*(3), 202–220. http://dx.doi.org/10.1039/c1fo10193e.

Mihindukulasuriya, S. D. F., & Lim, L. T. (2013). Oxygen detection using UV-activated electrospun poly(ethylene oxide) fibers encapsulated with TiO2 nanoparticles. *Journal of Materials Science*, *48*, 5489–5498.

Montes, A., Gordillo, M. D., Pereyra, C., & Martinez de la Ossa, E. J. (2011). Co-precipitation of amoxicillin and ethyl cellulose microparticles by supercritical antisolvent process. *Journal of Supercritical Fluids*, *60*, 75–80.

Murakami, R., & Takashima, R. (2003). Mechanical properties of the capsules of chitosan-soy globulin polyelectrolyte complex. *Food Hydrocolloids*, *17*(6), 885–888. http://dx.doi.org/10.1016/s0268-005x(03)00109-7.

Nagodawithana, R., & Reed, G. (1993). *Enzymes in food processing* (3rd ed.). San Diego, CA: Academic Press.

Oppenheim, R. C., Marty, J. J., & Speiser, P. (1978). Injectable compositions, nanoparticles useful therein, and process of manufacturing same. Google Patents.

Pan, K., Luo, Y. C., Gan, Y., Baek, S. J., & Zhong, Q. (2014). pH-driven encapsulation of curcumin in self-assembled casein nanoparticles for enhanced dispersibility and bioactivity. *Soft Matter*, *10*, 6820–6830. http://dx.doi.org/10.1039/C4SM00239C.

Patel, A. R., Bouwens, E. C. M., & Velikov, K. P. (2010). Sodium caseinate stabilized zein colloidal particles. *Journal of Agricultural and Food Chemistry*, *58*, 12497–12503.

Patel, A. R., Hu, Y. C., Tiwari, J. K., & Velikov, K. P. (2010). Synthesis and characterisation of zein–curcumin colloidal particles. *Soft Matter*, *6*, 6192–6199.
Pierucci, A. P., Andrade, L. R., Farina, M., Pedrosa, C., & Rocha-Leao, M. H. (2007). Comparison of alpha-tocopherol microparticles produced with different wall materials: Pea protein a new interesting alternative. *Journal of Microencapsulation*, *24*(3), 201–213.
Podaralla, S., & Perumal, O. (2012). Influence of formulation factors on the preparation of zein nanoparticles. *AAPS PharmSciTech*, *13*(3), 919–927.
Purwanti, N., Peters, J. P. C. M., & van der Goot, A. J. (2013). Protein micro-structuring as a tool to texturize protein foods. *Food and Function*, *4*, 277–282.
Regier, M. C., Taylor, J. D., Borcyk, T., Yang, Y., & Pannler, A. K. (2012). Fabrication and characterization of DNA-loaded zein nanospheres. *Journal of Nanobiotechnology*, *10*, 1–13.
Ru, Q. M., Yu, H. L., & Huang, Q. R. (2010). Encapsulation of epigallocatechin-3-gallate (EGCG) using oil-in-water (O/W) submicrometer emulsions stabilized by iota-carrageenan and beta-lactoglobulin. *Journal of Agricultural and Food Chemistry*, *58*(19), 10373–10381. http://dx.doi.org/10.1021/jf101798m.
Salgin, S., Salgin, U., & Bahadir, S. (2012). Zeta potentials and isoelectric points of biomolecules: The effects of ion types and ionic strengths. *International Journal of Electrochemical Science*, 7(12), 12404–12414.
Schmitt, C., & Turgeon, S. L. (2011). Protein/polysaccharide complexes and coacervates in food systems. *Advances in Colloid and Interface Science*, *167*(1–2), 63–70.http://dx.doi.org/10.1016/j.cis.2010.10.001.
Sekhon, B. S. (2010). Food nanotechnology—An overview. *Nanotechnology, Science and Applications*, *3*, 1–15.
Semo, E., Kesselman, E., Danino, D., & Livney, Y. D. (2007). Casein micelle as a natural nano-capsular vehicle for nutraceuticals. *Food Hydrocolloids*, *21*(5–6), 936–942. http://dx.doi.org/10.1016/j.foodhyd.2006.09.006.
Shahidi, F., & Han, X. Q. (1993). Encapsulation of food ingredients. *Critical Reviews in Food Science and Nutrition*, *33*(6), 501–547.
Small, H. (1974). Hydrodynamic chromatography, a technique for size analysis of colloidal particles. *Journal of Colloid and Interface Science*, *48*(1), 147–161.
Striegel, A. M., & Brewer, A. K. (2012). Hydrodynamic chromatography. In R. G. Cooks & E. S. Yeung (Eds.), *Annual review of analytical chemistry: Vol. 5* (pp. 15–34). Palo Alto, CA: Annual Reviews.
Subia, B., & Kundu, S. C. (2013). Drug loading and release on tumor cells using silk fibroin–albumin nanoparticles as carriers. *Nanotechnology*, *24*(3), 035103. http://dx.doi.org/10.1088/0957-4484/24/3/035103.
Taylor, H. F. (1939). *USPTO. Vitamin preparation and method of making the same.* Atlantic Coast Fisheries Co. US 2183053.
Taheri, E. S., Jahanshahi, M., & Mosavian, M. T. H. (2012). Preparation, characterization and optimization of egg albumin nanoparticles as low molecular weight drug delivery vehicle. *Particle & Particle Systems Characterization*, *29*, 211–222.
Teglia, A., & Secchi, G. (1994). New protein ingredients for skin detergency: Native wheat protein–surfactant complexes. *International Journal of Cosmetic Science*, *16*, 235–246.
Teng, Z., Luo, Y. C., & Wang, Q. (2012). Nanoparticles synthesized from soy protein: Preparation, characterization, and application for nutraceutical encapsulation. *Journal of Agricultural and Food Chemistry*, *60*(10), 2712–2720. http://dx.doi.org/10.1021/jf205238x.
Tolstoguzov, V. (2003). Thermodynamic considerations of starch functionality in foods. *Carbohydrate Polymers*, *51*(1), 99–111. http://dx.doi.org/10.1016/s0144-8617(02)00171-6.
Uskokovic, V. (2012). Dynamic light scattering based microelectrophoresis: Main prospects and limitations. *Journal of Dispersion Science and Technology*, *33*, 1762–1786.

van Rijn, P., Mougin, N. C., Franke, D., Park, H., & Boker, A. (2011). Pickering emulsion templated soft capsules by self-assembling cross-linkable ferritin–polymer conjugates. *Chemical Communications*, *47*(29), 8376–8378. http://dx.doi.org/10.1039/C1CC12005K.

Walstra, P. (2009). *Physical chemistry of foods*. New York, NY: CRC Press.

Weiss, J., Kanjanapongkul, K., Wongsasulak, S., & Yoovidhya, T. (2012). Electrospun fibers: Fabrication, functionalities and potential food industry applications. In Q. Huang (Ed.), *Nanotechnology in the food, beverage and nutraceutical industries* (pp. 362–393). Oxford, UK: Elsevier Science.

Woranuch, S., & Yolsan, R. (2013). Eugenol-loaded chitosan nanoparticles: I. Thermal stability improvement of eugenol through encapsulation. *Carbohydrate Polymers*, *96*, 578–585.

Xu, J. H., Li, S. W., Tan, J., & Luo, G. S. (2008). Controllable preparation of monodispersed calcium alginate microbeads in a novel microfluidic system. *Chemical Engineering & Technology*, *31*(8), 1223–1226. http://dx.doi.org/10.1002/ceat.200800027.

Ye, J.-H., Thomas, E., Sanguansri, L., Liang, Y.-R., & Augustin, M. A. (2013). Interaction between whole buttermilk and resveratrol. *Journal of Agricultural and Food Chemistry*, *61*, 7096–7101.

Zasypkin, D. V., Braudo, E. E., & Tolstoguzov, V. B. (1997). Multicomponent biopolymer gels. *Food Hydrocolloids*, *11*(2), 159–170.

Zhang, X. P., Le, Y., Wang, J. X., Zhao, H., & Chen, J. F. (2013). Resveratrol nanodispersion with high stability and dissolution rate. *LWT—Food Science and Technology*, *50*(2), 622–628. http://dx.doi.org/10.1016/j.lwt.2012.07.041.

Zhong, Q., & Jin, M. (2009). Nanoscalar structures of spray-dried zein microcapsules and in vitro release kinetics of the encapsulated lysozyme as affected by formulations. *Journal of Agricultural and Food Chemistry*, *57*(9), 3886–3894.

AUTHOR INDEX

Note: Page numbers followed by "*f*" indicate figures and "*t*" indicate tables.

C

D

E

G

H

I

K

Q

R

S

T

U

V

W

X

Y

Z

SUBJECT INDEX

Note: Page numbers followed by "*f*" indicate figures and "*t*" indicate tables.

B

C

O

P

R

S

T

V

W

Z

FSC
www.fsc.org
MIX
Papier | Fördert gute Waldnutzung
FSC® C083411

Kolibri 360 GmbH
Ferdinand-Jühlke-Straße 7
99095 Erfurt, Deutschland
produktsicherheit@kolibri360.de